Cambridge Studies in Biological and Evolutionary Anthropology 29

Primates Face to Face
Conservation implications of human–nonhuman primate
interconnections

Human and nonhuman primates share intertwined destinies. As our
closest evolutionary relatives, nonhuman primates are integral elements
in our mythologies, diets and scientific paradigms, yet most species now
face an uncertain future through exploitation for the pet and bushmeat
trades as well as progressive habitat loss. New information about disease
transmission, dietary and economic linkage, and the continuing international focus on conservation and primate research have created a surge
of interest in primates. Focus on the diverse interaction of human
and nonhuman primates has become an important component in
primatological and ethnographic studies. By examining the diverse and
fascinating range of relationships between humans and other primates,
and how this plays a critical role in conservation practice and programs,
Primates Face to Face disseminates the information gained from the
anthropological study of nonhuman primates to the wider academic and
non-academic world.

AGUSTÍN FUENTES is Associate Professor of Anthropology and
Director of the Primate Behavior and Ecology Program at Central
Washington University. His research interests include primate behavioral
ecology, the evolution of social organization, and conservation theory
and practice.

LINDA D. WOLFE is Professor and Chair of Anthropology at East
Carolina University. Her research focuses on primate sexual and so~
behaviors.

Cambridge Studies in Biological and Evolutionary Anthropology

Series Editors

HUMAN ECOLOGY
C. G. Nicholas Mascie-Taylor, University of Cambridge
Michael A. Little, State University of New York, Binghamton
GENETICS
Kenneth M. Weiss, Pennsylvania State University
HUMAN EVOLUTION
Robert A. Foley, University of Cambridge
Nina G. Jablonski, California Academy of Science
PRIMATOLOGY
Karen B. Strier, University of Wisconsin, Madison

Consulting Editors
Emeritus Professor Derek F. Roberts
Emeritus Professor Gabriel W. Lasker

Primates Face to Face

Conservation implications of
human–nonhuman primate interconnections

EDITED BY

AGUSTÍN FUENTES

and

LINDA D. WOLFE

CAMBRIDGE
UNIVERSITY PRESS

CAMBRIDGE UNIVERSITY PRESS
Cambridge, New York, Melbourne, Madrid, Cape Town, Singapore, São Paulo

Cambridge University Press
The Edinburgh Building, Cambridge CB2 2RU, UK

Published in the United States of America by Cambridge University Press, New York

www.cambridge.org
Information on this title: www.cambridge.org/9780521791090

© Cambridge University Press 2002

This publication is in copyright. Subject to statutory exception
and to the provisions of relevant collective licensing agreements,
no reproduction of any part may take place without
the written permission of Cambridge University Press.

First published 2002
This digitally printed first paperback version 2005

A catalogue record for this publication is available from the British Library

Library of Congress Cataloguing in Publication data

Primates face to face: conservation implications of human and nonhuman
primate interconnections / edited by Agustín Fuentes and Linda D. Wolfe.
   p.   cm. – (Cambridge studies in biological and evolutionary
anthropology; 29)
Includes bibliographical references and index.
ISBN 0 521 79109 X
1. Primates – Research.   2. Wildlife conservation.   3. Human–animal
relationships.   I. Fuentes, Agustín.   II. Wolfe, Linda D., 1942–
III. Series.
QL737.P9 P6775 2002
333.95′9816–dc21   2001035035

ISBN-13 978-0-521-79109-0 hardback
ISBN-10 0-521-79109-X hardback

ISBN-13 978-0-521-01927-9 paperback
ISBN-10 0-521-01927-3 paperback

Cambridge University Press has no responsibility for the persistence or
accuracy of URLs for external or third-party Internet websites referred to
in this publication, and does not guarantee that any content on such websites
is, or will remain, accurate or appropriate.

To all the primates (human and nonhuman) struggling to survive in a
rapidly changing world

Contents

Contributors

Frances Burton
Department of Anthropology, University of Toronto at Scarborough,
1265 Military Trail, Scarborough, Ontario, Canada M1C 1A4
Burton@scar.utoronto.ca

Loretta A. Cormier
Department of Anthropology, 338 Ullman Building, University of Alabama,
Birmingham, Birmingham, AL 35294-3350, USA
Lcormier@.uab.edu

Phyllis Dolhinow
Department of Anthropology, University of California, Berkeley, Berkeley,
CA 94720, USA
Dolhinow@qal.berkeley.edu

Ardith Eudey
Editor, *Asian Primates*; Vice-Chair for Asia, IUCN Specialist Group,
164 Dayton St, Upland, CA 91786-3120, USA
Eudey@aol.com

Deborah H. Fouts
Chimpanzee and Human Communication Institute, Central Washington
University, Ellensburg, WA 98926, USA
Foutsd@cwu.edu

Roger S. Fouts
Chimpanzee and Human Communication Institute, Central Washington
University, Ellensburg, WA 98926, USA
Foutsr@cwu.edu

Agustín Fuentes
Department of Anthropology and Primate Behavior and Ecology Program,
Central Washington University, Ellensburg, WA 98926-7544, USA
Afuentes@cwu.edu

Hiro Kurashina
Director, Richard Flores Taitano Micronesian Area Research Center,
University of Guam, USA

Manuel Lizarralde
Departments of Anthropology and Botany, Mail Box 5407, Connecticut College,
270 Mohegan Ave., New London, CT 06320-4196, USA
Mliz@conncoll.edu

Kelly G. Marsh-Kautz
Micronesian Area Studies, University of Guam, USA

Mary M. Pavelka
Department of Anthropology, University of Calgary, 2500 University Drive,
NW, Calgary, Alberta, Canada T2N 1N4
Pavelka@acs.ucalgary.ca

Poranee Natadecha-Sponsel
Chaminade University, USA

Anthony L. Rose
Wildlife Protectors Fund/The Gorilla Foundation/The Biosynergy
Institute/Bushmeat Project, P.O. Box 488, Hermosa Beach, CA 90254, USA
Bushmeat@biosynergy.org

Nukul Ruttanadakul
Department of Zoology, Prince of Songkla University, Thailand

Glenn H. Shepard Jr.
Instituto Nacional de Pesquisas da Amazônia (INPA), Department of Botany
(CPBO), C.P. 478, Manaus, Amazonas 69083–000, Brazil
ghs@inpa.gov.br / GShepardJr@aol.com

Pascale Sicotte
Department of Anthropology, University of Calgary, 2500 University Drive,
NW, Calgary, Canada T2N 1N4
Sicotte@acs.ucalgary.ca

Leslie E. Sponsel
Department of Anthropology, University of Hawaii, 2424 Maile Way- SSB 346,
Honolulu, HI 96822-2223, USA
Sponsel@hawaii.edu

David S. Sprague
Senior Researcher, Ecological Management Unit, Ecosystem Research Group,
National Institute for Agro-Environmental Sciences, Tsukuba, Ibaraki 305-8604,
Japan
Sprague@naies.affrc.go.jp

Rebecca Stephenson
Department of Anthropology, University of Guam, USA

Karen B. Strier
Department of Anthropology, University of Wisconsin-Madison, Madison,
WI 53706, USA

Prosper Uwengeli
Centre de Recherche à Karisoke, Rwanda

Gabriel S. Waters
Chimpanzee and Human Communication Institute, Central Washington
University, Ellensburg, WA 98926, USA

Bruce P. Wheatley
Department of Anthropology, University of Alabama, Birmingham,
AL 35294-3350, USA
Bwheat@uab.edu

Linda D. Wolfe
Department of Anthropology, Brewster A 215, East Carolina University,
Greenville, NC 27858, USA
Wolfel@mail.ecu.edu

Foreword

The intersection of anthropology and primatology is a complex one where our knowledge of human and nonhuman primates meet. Contributing to its complexity are the different sets of theoretical assumptions and methodological approaches that culturally-oriented anthropologists and biologically-oriented primatologists tend to bring to their studies. The different ways in which ethnographic and ethological findings are usually reported further increases the intellectual distance that anthropologists and primatologists alike must travel in their search for common ground.

There is also a curious asymmetry between anthropology and primatology that has developed along with the peculiar intellectual traditions of each. For example, although humans are primates, anthropologists who study humans rarely regard themselves as primatologists. Instead, primatologists, particularly in the social sciences, learn early on in their careers to respond to the persistent question of what primatology contributes to anthropology in terms of the comparative perspectives that nonhuman primates can provide. That this question is typically posed by scholars who focus on those aspects of cultural behavior that distinguish humans from other primates has often struck me as an odd detail because nearly all definitions of what makes humans human are implicitly or explicitly derived from comparisons.

Primatologists, for the most part, have been slow to turn the question of primatology's place in anthropology on its head. Yet, as this unique volume shows, the precarious status of so many primates makes the question, 'What can anthropology contribute to primatology?' a compelling one that primatologists concerned with conservation cannot afford not to ask.

For tens of thousands of years, ancestors of today's modern humans and primates have lived side-by-side throughout the tropics. The nature of their interactions has been mixed and subject to historical, demographic, and environmental dynamics. But whether primates are revered for religious purposes, or exploited for food and other products, or regarded as pests, there is no denying the fact that primates have long played a prominent role in the daily, if not spiritual, lives of many people. The diversity of past and present human attitudes toward primates affects how primates and their habitats are treated, and ultimately how effective any efforts to conserve them can be.

Unlike other primates, which are still largely restricted to the tropics and whose populations are dwindling at an alarming rate, human population expansion has led to our occupancy of much of the planet. In the past, the fates of primate populations were dependent on the behaviors, ecologies, economies, and belief systems of the local human communities that surrounded them. Now, in addition to these local influences, primates are also affected by much more far-reaching political agendas and economic forces that operate on a global scale.

As human pressures continue to grow and primate habitats to shrink, insights into the human side of human–nonhuman primate interactions have become even more critical to the development and implementation of informed conservation policies at all levels. Whether these insights come from ethnographers or primatologists is secondary to the far more urgent cause of securing a viable future for the world's endangered primates.

By assembling such an eclectic array of informative case studies about human–nonhuman primate interactions, Agustín Fuentes and Linda D. Wolfe identify a common ground where cultural anthropologists and primatologists can meet. In the process of exploring the interactions between human and nonhuman primates, they and their contributors demonstrate the implausibility of considering one kind of primate without the other, at least in those places where both can still be found.

Karen B. Strier

Acknowledgements

The editors thank Karen Strier and Tracey Sanderson for their interest and assistance in the publication of this volume.

Introduction

Human and nonhuman primates share intertwined destinies. Nonhuman primates are our closest evolutionary relatives and integral elements in our mythologies, diets, and scientific paradigms. The study of all primates (human and nonhuman) continues to be a rapidly expanding field. Recently, specific focus on the multifarious interaction of human and nonhuman primates, termed 'ethnoprimatology' or 'cultural primatology', is becoming a major component in primatological studies. We feel that it is possible to view human and nonhuman primates as co-participants in a rapidly escalating realm of ecological and cultural change. The fields of investigation into human ecology, nonhuman primate ecology and behavior, and conservation are traditionally considered distinct avenues of investigation. It is our contention, however, that conservation is most effective when human and nonhuman primate ecology and behavior are seen as interconnected and treated as a unified area of investigation. In this book we hope to illustrate that a constructive approach to assessing conservation realities can be obtained by including elements of human culture and ecology, nonhuman primate behavior and ecology, and creating a picture of a broad and dynamic interconnected cultural and biological ecosystem.

Whereas general ecology is seen as the set of interactions between organisms and their environment, cultural ecology can be envisioned as cultural models of the environment and the relation between a people and their ecological space. Although many human societies tend to envision the environment, or ecosystem, as external to themselves, there are also many that exist in a less dichotomous, more interconnected context. Humans are primates, and mammals, that interact with their environment via a wide set of biological and cultural adaptations and innovations. By examining instances of this interconnectivity, especially through our closest relatives, we are able to gain a broader conceptualization of the context of human ecology.

Currently, one primarily associates the term 'conservation' with the notion of biodiversity: conservation with a goal of maintaining and/or supporting the diverse and varied ecosystems on this planet and their inhabitants. Much of this work attempts to discover basal patterns of

1

behavior and ecological adaptation in 'pristine' or minimally impacted environments to better understand evolutionary systems prior to human impact. This is probably the best known and most practiced form of conservation, but cultural conservation is also significant, although less well defined. Cultural conservation ranges from the archiving of cultural heritage to the reintroduction of traditional or previous behavior patterns or adaptations of particular cultures. It is rare (in practice) for biodiversity and cultural conservation practices to significantly overlap in their application. In the chapters of this book we present a range of diverse instances where the interconnection between human and nonhuman primates would necessitate a merging of cultural and biodiversity conservation paradigms.

Primatological and cultural anthropologists bring a wealth of comparative analyses and a holistic approach to the ever-increasing complexity of studying nonhuman primates in the countries where they occur. It is in countries where social and economic realities clash with conservation and research priorities that anthropologists are especially suited to tackling the conflicts and problems that may arise. Nonhuman primates are, moreover, an important element in the overall anthropological paradigm. That is, the inclusion of nonhuman primates both as a comparative base for the study of human evolution and biology, and as important players in the folklore and practice of many human cultures, can enrich anthropology. We should note, however, that nonhuman primates and other animals have value in and of themselves and should be conserved for this intrinsic value. Because this book is focused on the interaction between human and nonhuman primates, we do not fully explore the issues of the potential conflict among members of national governments who covet financial gain from the products of their forests, global conservation organizations, the desires of local human inhabitants, and the rights of all organisms to co-exist. These are, however, important issues and we hope that the chapters in our book will contribute to a broader understanding of human–nonhuman primate interplay.

Ties between human and nonhuman primates have taken the spotlight recently. New information about disease transmission (the SIVcpz–HIV connection), dietary and economic linkage (i.e. the bushmeat trade and ecotourism), and the continuing international focus on conservation and primate research, have created a surge of interest in this topic. This book seeks to disseminate information gained from the anthropological study of the nonhuman primates to the wider academic and nonacademic world. Understanding the diverse and fascinating range of cultural, ecological, and evolutionary relationships between humans and other primates provides us with a greater basis to understand, appreciate and conserve the world around us. We hope that this book begins to fill the gap in our understanding of human and nonhuman primate interconnections and provides a broad base for understanding the critical role nonhuman primates play in cultural and ecological conservation practice and programs.

This book, however, is not only about conservation and ecological interconnections. We also address the overall multifaceted relationship between contemporary humans and nonhuman primates and the need to take the cultural practices of the local people, whether they are Guajá Indians or research scientists, into account when attempting to study the nonhuman primates. Because some studies of human–nonhuman primate interactions have been conducted longer than others have, the chapters are uneven in length and style, but not in quality. We did not, moreover, attempt to make the chapters all the same length nor do they all express the same viewpoints. We wanted to give each author(s) wide latitude to describe their research and express their views. We consider this feature of our edited volume a strength and hope that the reader finds this diverse compilation interesting and important.

The first part of this book provides a glimpse of the current central role that nonhuman primates play in our scientific culture. Two chapters discuss the relevance of nonhuman primates in anthropology and behavioral and evolutionary science in general and a third tackles the role of chimpanzees in the HIV–AIDS investigations. Part 2 delves into the important symbolism and cultural role monkeys and apes hold for some peoples in Amazonia, China, and central Africa. This section explores aspects of co-ecologies, or sympatric existence, between humans and the nonhuman primates that live around, or with, them. The third part presents a discussion of conservation in the modern world, examining aspects of economy, ecology, and politics in the context of endangered populations of primates in Africa and Asia. Finally, Part 4 focuses on specific socio-economic interrelationships between macaque monkeys and humans and a personal overview of a primatologist's role as political activist.

It is our hope that this volume will be of great benefit to those studying, or just interested in, conservation, biodiversity, cultural diversity and primate behavior. Anthropologists, biologists, social scientists, and resource managers should find significant overviews and descriptions of a wide array of primate-related ecological, cultural, evolutionary and political human practices. Hopefully, after finishing this volume the reader will have gained a better sense of the interconnections between primates across a range of ecological and cultural boundaries. In compiling this volume we realize that it is an attempt to embark on a new direction and that synthesizing such a broad array of information is a substantial undertaking. We hope to be able to look back in five or ten years and have the sense that this volume provided at worst interest in, or at best a call to action to embark upon, a relatively new and holistic way to talk and think about ourselves, the other primates and our communal place in the world around us.

Agustín Fuentes & Linda D. Wolfe

Part 1
Science and nonhuman primates

Because nonhuman primates are our closest living phylogenetic relatives, nonhuman primates have been put to use in many different aspects of scientific research, such as biomedical research, and as models for human evolution. In order for scientists to conduct biomedical research on non-human primates and extrapolate that research to humans, certain assumptions have to be made concerning the nature and evolution of prosimians, monkeys, and apes. For example, it is often assumed that the stress of caging a nonhuman primate has no effect on its physiology. Fouts, Fouts, and Waters argue in Chapter 3 that the stress of being caged coupled with isolation can affect an animal's physiology, thereby calling into question any data gathered on a stressed animal.

When using nonhuman primates for models of human evolution we run the risk of falling into the assumption that nonhuman primates have not substantially changed since our last common ancestor. However, most anthropologists would agree that adaptations have changed over the millennia and urge caution when extrapolating from the behaviors of extant nonhuman primates to past or present human behaviors.

The three chapters in this part of the book address the issue of how nonhuman primates are viewed by social scientists and biomedical researchers. Phyllis Dolhinow discusses the relationship between cultural anthropology and primatology. She advocates a cross-species comparative study of human existence, past and present. Dolhinow urges prudence, however, because of the unique human capacity for language. With words, she notes, humans create and construct realities. That is, with language, which no other primate has, humans can turn biological fiction into cultural realities such as is the case with human kinship systems. She also cautions that in the absence of data, one should not accept seemingly plausible explanations of primate, human or nonhuman, behavior. Here she is writing of those who have suggested that, among other things, human males are violent because of the evolutionary past that they share with modern chimpanzees. She finds that chimpanzees are rarely violent and the behaviors of extant chimpanzees may not have been present in the last common ancestor of chimpanzees and humans. Along these same lines, she reviews the assertions that gibbons live in monogamous groups because a male can only protect a limited number of infants from other infanticidal gibbon males. The problem with this argument is, of course, that male infanticide has not been observed among gibbons. Finally, Dolhinow argues that there must be cooperation between cultural anthropologists and primatologists if nonhuman primates are to survive.

Mary Pavelka (Chapter 2) discusses the problems and treatment of anthropological primatologists at the hands of cultural anthropologists. Pavelka is particularly concerned with two issues. First, she is perplexed by cultural anthropologists who question the presence of primatologists in departments of anthropology. As she relates, most of us now respond to such questions that primates are interesting in and of themselves without trying to explain further. Secondly, the resistance of cultural anthropologists to a cross-species, comparative viewpoint troubles her. Most of her chapter, in fact, is an exploration of the reasons behind the resistance of cultural anthropologists to cross-species considerations. She suggests that cultural anthropologists are dubious of evolutionary explanations of behavior and have the mistaken idea that primatologists want to simplify explanations of human and nonhuman primate behavior. Pavelka argues that cultural anthropologists are apprehensive of primatology because they fear that if some human behaviors have a biological base that removes the freedom we have to control our lives. She argues forcibly against the assumption that primatologists are biological determinists. Rather she argues that primatologists see behavior as the result of the evolutionary history, underlying biology, and current social conditions of the species. She concludes with the opinion that nonhuman primate conservation will benefit from a cross-species, holistic approach to anthropology and primatology.

Roger Fouts, Deborah Fouts, and Gabriel Waters (Chapter 3) write on the use of chimpanzees in HIV research and question the use of chimpanzees in biomedical research. According to these authors, ignorance of the nature of chimpanzees and of experimental design and methodology on the part of biomedical researchers render their research invalid with regard to its efficacy. The authors list several factors, such as the stress caused by the standard housing, that alter chimpanzee physiology and make the biomedical research conducted on chimpanzees questionable. They review the literature on the differences in the human and chimpanzee response to an HIV infection and conclude that data collected on HIV-infected chimpanzees cannot be extrapolated to humans. Finally, they question the ethics of using chimpanzees, whose behaviors are so similar to ours, in invasive research.

1 *Anthropology and primatology*

PHYLLIS DOLHINOW

Introduction

The relationship between cultural anthropology and primatology is more important today than ever in the past. It is critical. We are at a choice point in time and the future of many if not most of the living nonhuman primates hangs in the balance. The health, safety, and very survival of threatened, and soon to be threatened, species depend on our actions. If we do not act soon it will be too late to learn from them or to save what remains of natural nonhuman primate populations. Cultural anthropology's focus on the study of different cultures places it in a unique position to offer practical advice as to how best to effect strategies of conservation that will allow both humans and nonhuman primates to share the same land. It will be virtually impossible to assure the healthy survival of many populations of nonhuman primates without the understanding and cooperation of the people who live alongside them. It is obvious that there must be an interactive association between anthropology and primatology if we are to sustain the nonhuman primates. This cooperation requires an accurate and complete understanding both of the requirements of the nonhuman primates and of the human cultures that threaten the damage.

Anthropology encompasses comparative studies of human societies, past and present, as well as the behavior and biological study of living primates and their evolution. Human social systems have been studied in depth and over generations, including the countless aspects of our behavior that have their genesis in our unique, and from an evolutionary perspective of time, very recently evolved cognitive abilities. These abilities are expressed in the form of our language and the social systems and customs that only a human language could generate. Anthropology's holistic and comparative complex of investigative concerns represents a unique point of view on social systems and behavior. However, because our study is based on ourselves, it is very important to understand in every respect why it is necessary to remember how we are both similar to, and different from, the other primates. We must be aware of the extent to which, and how, our words create our realities. To complement what we know of ourselves, our study of the nonhuman primates offers anthropology comparative contrasts that allow us a perspective on human behavior.

7

The comparison becomes less and less a contrast as we move back in time towards our last common hominoid ancestor.

Those of us who focus our study on the nonhuman primates are collectively nested in many different departments. Reviewing the membership lists of various primatological societies, produces a list of home bases including anthropology, psychology, zoology, integrative biology, molecular and cell biology, animal behavior, comparative medicine, anatomy, psychiatry, environmental sciences, conservation, genetics, and pharmacology. Each discipline infuses its scholars with attributes, points of view, visions, and theories specific to the home discipline and its co-ordinates. That these differing contexts importantly affect out interpretive apparatus and our views is apparent from an inspection of scholars, their topics, and their conclusions. Not that anthropologists are a uniform group, far from it. Our departments vary tremendously in composition and with respect to the nature of each internal segment. However, we are unified in our goal of understanding the human primate, our present and our past as our ancestry has influenced the patterns of life we observe today.

Knowledge gained from studies of nonhuman primates assists anthropologists to reconstruct the most likely behaviors of our ancestors. There are many lines of evidence and modern techniques of investigation that provide us with information and insights to allow us to follow the major changes in the course of human evolution. Selection is a major operator in the evolution of all life forms, and it accomplishes its effects by acting on the behaviors of the populations in a species. Because our goal is to follow and to understand the path of survival of our ancestral species and to identify how changes over time produced new forms, we need to reconstruct the patterns of behavior upon which selection acts. One thing is certain, for the primates among the most important of selected patterns are social behaviors. Although the social behavior patterns of nonhuman primates are sometimes highly variable, even within a single species, at their most complex they are far simpler than the human situation. None the less, primate studies do provide valuable models of behavior and behavioral ecology and unequaled opportunity to develop and understand the genesis and nature of social systems. In addition, biological factors that are powerful influences on social behavior and ecological adaptations, are being identified in studies of the physiology of stress, affiliation, and reproduction. Because over recent decades the field of primate studies has become large and diverse, the last section of this chapter offers a few caveats or guidelines for the non-primatologist who wishes to use the primate literature wisely.

The authors of this volume share a common subject: our relations with the nonhuman primates. Our consensus is that, for better or for worse, human and nonhuman primates will continue to share living space and that we have a consequential responsibility to ensure their continued existence. They are part of our history, our cultures, and our world view (Carter and Carter, 1999). In this volume you will read about some of the ways humans study, hunt, eat, deify, work, and use them and how they occupy significant places in some cultures, from theological to culinary. In our own Western society they appear in captive contexts from zoological parks to laboratories. They are used to entertain, educate, investigate, and to help sustain species survival.

The anthropology connection

Anthropology is the study of ourselves. In addition to investigating prehistoric and modern humans, biological anthropology has investigated our past and traced our development from its roots in our ancestry 65 million years ago, to hominids during the past 5 million year span and in the present (Sussman, 1999). It is important to keep this time dimension clearly in mind because it was several million years after becoming bipedal that what we consider the most diagnostic features of being 'human' were evidenced in the fossil and archeological record. I refer to those changes in our central nervous system, as measured by increasing cranial capacity, that have occurred in the past two million years. Some have called this the encephalization of the genus *Homo*. Neither the great increase in brain size nor the cognitive powers it gave humans are shared by any of the nonhuman primates. Human cognition is unique, it is biologically based and it shapes every fiber of our lives, so much so that we are usually unaware of it. A result of our apparent obliviousness is how remarkably easy it is for us to project human experiences and behaviors onto a universe of nonhumans and even inanimate objects.

Our language is a unique system of communication, providing enormous complexity and power. Human language, and the biology that makes it possible, allows us to create and express all manner of meanings, motivations, plans, and much else that is unavailable to the nonhuman. We are uncommonly adaptive because of the cultures language makes possible. Culture, in turn, gives shape to words and the meanings they express and makes them amongst our most powerful tools. Meanings are the basis of our understanding and recognition of reality. Words direct our perception, focus our attention, and can enlarge or restrict our understanding of anything we consider. Culture can redefine reality or even recreate it, as, for example, when we can create kin relations instantly where no relation exists biologically. Culture also allows us to survive in any climate, to record events that took place billions of years in the past, and to travel in outer space.

Anthropology's traditional concern with the integration of complex systems of social constructions of behavior, and with dynamic mechanisms of change, provides primatologists with a formidable and effective set of tools for investigating and analyzing the characteristics of nonhuman social systems. Although the nonhuman primates cannot give us a spoken data base, their complex patterns of social behavior and resulting social systems, structured by subtle emergent and internal levels, often with hierarchies, coalitions, and alliances, can be accurately reflected in the interactions we record. It has been both appropriate and productive to apply many conceptual tools borrowed from the anthropological study of ourselves to the nonhuman primates. Of course a paradigm borrowed from another discipline does not come with whistles and bells to alert the user that it may be inappropriate to apply it to another species. Because that is so, it is very important for the primatologist to be constantly aware that a great proportion of human behavior has no analog in the behavior of nonhumans. Even though they are undeniably constrained by our basic biology, language-based complex and intertwined human systems, such as religion, economics, politics and kinship, are not biological in origin. These are all cultural constructs. Despite the fact that terms such as politics and economics have been applied to the behavior patterns of various nonhuman primates, it does not alter the fact that there is a world of difference between human constructed systems and behavior and those of any other primate.

The application of some social systems theory developed by studies of human behavior has been very useful in the analyses of the social system of another primate (Ray, 1999a, b). For example, hierarchy theory has been applied to demonstrate the emergence of both power systems and social systems and in so doing has identified the nature of the interconnection. It is becoming apparent that social systems constrain the expression of behavior and consequently power within the group. It remains for further studies of equal detail using the same methodology to be undertaken on other primates.

Kinship, a central part of the structure of all human cultures, is a social, not a biological, statement about culturally defined relationships. It is embedded in countless other aspects of culture and is defined and expressed by the terms or names human language assigns to each defined relationship. Humans are importantly concerned with how they are related to those around them, and for good reason: every culture defines a set of relationships that designate rights, duties, and obligations according to how people are said to be connected to others. Economics, religion, politics, and many other aspects of the social system are defined in large part by what you learn about how you are related to others. Many possible relationships such as clan, caste, or moiety are not based on a degree of genetic kinship. Any female may be designated and treated as a sister, and the same is true for the category of mother. Brothers may include men from other villages or on another continent, of any race or ethnic group. All this must be kept in mind when we consider the nonhuman primates who cannot think as we do because they lack the range of cognitive abilities necessary to comprehend relationships beyond those that are immediate and interactionally important.

Anthropology is vitally concerned with our evolutionary past, how we came to be what we are now and how what we were in earlier times carried over to affect what we are and do today (Washburn, 1961, 1972, 1978; Washburn and Dolhinow, 1972). Even though no modern counterpart exists today, various of the nonhuman primates are used as exemplars along with the information from the fossil record in reconstructing the probable attributes of stages in human evolution (Moore, 1996: Stanford, 1999). Primate field studies proliferated in the 1960s and anthropologists looked to the baboon for a model for early human behavior (Washburn and DeVore, 1961, 1963). Baboons were soon joined by the common chimpanzee (*Pan troglodytes*) as an even more appropriate model for our early behavior, especially with respect to cranial capacity and tool use. A third nonhuman primate, the pygmy chimpanzee (*Pan paniscus*), was considered by many to be an excellent if not the most informative model available for early hominid structure and behavior (Zihlman, 1996, 1999).

Washburn and others have considered how the behavior of our ancestors continues to influence modern human behavior in the form of propensities to respond or behave in certain ways in some situations (Hamburg, 1963; Washburn, 1966, 1971, 1972). Whereas many behaviors that proved adaptive for early humans continue to be adaptive even today in the face of drastically altered conditions of life, others such as conflict, although a perfectly normal component of our behavior, in excess or inappropriately expressed can be destructive. Our basic social nature has us not only spending our lives in groups but also depending upon the social context for learning all the skills, including language, that are necessary for us to survive and succeed.

Unfortunately, not everyone has used the primatological information wisely, as witnessed by some popular books. For example, one recent book ferrets out and elaborates creatively on extreme anecdotes from living apes, then transposes the resulting patchwork of far from normal or typical behavior via a purported continuous chain of DNA coding for these bad behaviors, directly to our early ape ancestors. The atrocious actions are followed forward once again to modern males, who are described as behaving very badly indeed: a modern saga of violence, rape, murder, and war (Wrangham and Peterson 1996). The fact that modern great apes rarely behave as the authors would have us think in no way impedes the creation of their myth of demonic males, chimpanzee or *sapiens*. Biology has long been offered as an excuse for the position that it is hopeless to change so-called immutable bad behaviors. Furthermore, a convenient failure to understand that DNA does not code for function further enables a grave misinterpretation of behavior.

Obviously, if they are to be at all useful, models to represent what was going on at various stages in our evolution must be built very carefully. Reconstructions are, at most, best guesses and when you rely on a living primate such as the chimpanzee for comparative insights, keep in mind that it has been approximately five million years since we and the chimpanzee had a common ancestor. The fact that chimps use and make surprisingly sophisticated tools in the present is no guarantee that their ancestors, who were also ours, did. If there is a caveat here it is borrower be aware.

And finally, and most timely, anthropology has the opportunity and obligation to play a crucial role in our efforts to conserve populations of threatened nonhuman primates. They live with us and the inevitably resulting competition for space and food generated by this proximity is rarely settled in favor of nonhumans. Although no known primate species went extinct in the twentieth century (Myers *et al.*, 2000) that is no guarantee and little comfort for the future. Together with the information primatologists provide about endangered species, cultural anthropology can provide the necessary understanding of the human side to create realistic solutions that will enable threatened primates to survive.

Conservationists are faced with the formidable task of designing strategies to aid the long-term viability of species and at the same time to be economically and politically compatible with local peoples and their governments. An example of such an organized effort is the integrated conservation and development project (ICDP). The ICDP

> attempts to link the conservation of biological diversity within a protected area to social and economic development outside the protected area. In ICDPs, incentives are typically provided to local communities in the form of shared decision-making authority, employment, revenue sharing, limited harvesting of plant and animal

species, or provision of community facilities, such as dispensaries, schools, bore holes, roads, and woodlots, in exchange for the community support for conservation.

(Newmark and Hough, 2000, p. 585).

ICDPs have, however, been only moderately successful in reaching their goals for many complex reasons. Sound advice based on knowledge of local peoples and governments is crucial to the success of any conservation project, and that advice can come from cultural anthropologists.

The primatology connection

The living primates are an impressive array of forms and sizes from the diminutive mouse lemur to the massive gorilla (Dolhinow and Fuentes, 1999). Some species live well in many different habitats whereas others are exceedingly limited in their ecological tolerances. We have both taxonomic and phylogenetic reasons for our interest in the nonhuman primates. We all belong to the Order Primates and in the panoply of living animals they are our significant others, our closest relatives. However, they are not quirky copies of ourselves, and, because of evolutionary history, certain of the living primates provide more appropriate models for helping us mirror ourselves and reconstruct our past than do others (Washburn, 1978). We are most closely related to the African apes, chimpanzee and gorilla. However, this degree of relationship should not be taken to mean that there is an uninterrupted continuum among the living primates in either biology or behavior, because there is not. The gap between humans and apes in cranial capacity and in cognitive abilities, including culture and language, regardless of how either is defined, is significant in both degree and kind.

The very impressive inter- and intraspecies variability in behavior and demographics among the nonhuman primates has been a focus of attention for decades (DeVore, 1965; Jay, 1968; Dolhinow, 1972). Relatively minor demographic differences between locations can be accompanied by a great range of behavior variability. The behavior of a particular species observed in one environment is one of a range of possible alternative behaviors available to that species, not an evolutionary adaptation to a particular environment. As an increasing number of species have been observed in a number of habitats living under different ecological conditions, our understanding of comparative primate behavioral ecology and demographics has encouraged us to generalize and identify useful comparative models based on this intra- and interspecies diversity. The comparisons show the importance of the biological nature of the actors in determining the social system, and the social system as the adaptive mechanism that permits the survival of individuals. The problem becomes biosocial, and this model can then be used in approaching human problems (Washburn, 1971). Primatologists have given a great deal of thought to creating models for the explanation and prediction of primate behavior, trying to identify, when possible, universals or general patterns. In the process we should not overlook the significance and meaning of exceedingly important avenues of variability in a species' behavior. In our search for unifying principles, diversity is untidy and can become an unwelcome guest, but one that is assuming increasingly impressive stature and is assisting us in the restructuring of many research questions.

Primatology has matured and gone in many directions over the past half century and it has been shaped by the training and interests of its practitioners. From early short studies of weeks' duration that produced fascinating but rather anecdotal information (Ribnick, 1982), we are now in a phase of development wherein long-term multi-year projects by teams of collaborators is more often than not our goal if not our practice. Studies are being undertaken with increasingly systematic methods of data collection, which in turn greatly facilitate comparative analysis. However, and fortunately, reducing behavior to units amenable to quantification generally has been accompanied by an appreciation for the range of adaptability and variability in behavior patterns.

Most primates live in social groups of varying degrees of duration and stability in structure and composition. Each group has its own special attributes based on many factors including its membership and ecological setting. Attempts to create general pictures of what social life is for any nonhuman primate runs the risk of overgeneralizing. Acknowledging this risk and urging the reader to seek information on the particular primate of interest, there are similarities in the social life of diverse primates. A group social system is constructed by the interactions of its members according to characteristics of the species (Ray, 1999a, 2000). Social systems are affected by many factors including resource availability, group demographics, composition, personalities of members, neighboring groups, and dispersal patterns. Habitat or ecological context directly affects group health and survival as well as determining the degree and kind of competition with conspecifics in nearby social units. Depending on a species' diet, climatic and phenological seasonality will structure the availability and quality of foods. Density of groups will affect the degree and kind of competition for food.

Maturation takes place within the social group. The length of infant dependency varies according to the maturation schedule of the species; depending on the species, care for a dependent immature may be given by many animals, including males, in addition to the mother. Few nonhuman primates match chimpanzees for the length of time mothers and offspring continue to associate preferentially. There has been some controversy as to the significance of genetic or kin relatedness for the nonhuman primate. Much evidence indicates that, rather than degree of genetic relatedness, it is familiarity based on proximity and interaction that is most significant in the continuing relations among individuals (Bernstein, 1999). Two hypotheses, that relatedness added to preferences based on familiarity, and that without being familiar monkeys would prefer another related monkey, did not survive testing (Fredrickson and Sackett, 1984; see also Chapais *et al.*, 1997 for relatedness thresholds and Chapais, 1988 for rank and relatedness).

Primate studies have been enriched by new techniques of investigation. Experimentation is standard in many field studies, as, for example, playback vocalizations are used to test for response and recognition of group members (Cheney and Seyfarth, 1990). DNA analysis has been applied to determine paternity and other genetic studies offer still more information. We can monitor reproductive and stress hormone responses noninvasively. Female ovulation cycles may be monitored by fecal steroid assay (Campbell, 2000; Wasser, Risler and Steines, 1988). Dietary analysis can be done from fecal materials and by monitoring stable isotope ratios (Schoeninger, Iwaniee and Glandes, 1997). Modern investigative techniques give us access to aspects of behavior previously unavailable except by speculation. Together these methods open windows into costs and impacts of social and reproductive events, and provide ways of measuring how successful individuals are over the life course. Being able to look inside the biological parameters of social life helps to answer questions about the internal states of the animals we are watching and lets us measure, at least in part, the impact of what happens to and near them.

An especially interesting example illustrates how understanding the biological basis of a nonhuman primate's behavior can be exceedingly useful for humans. Baboon stress was studied in an African game reserve and the results helped us to understand why human beings can differ in their vulnerability to stress-related disorders (Sapolsky, 1990). It had been known for a long time that, even in the same situations, some people were more vulnerable than others to diseases arising from stress. Sapolsky's observations of adult males living in a social group were supplemented by taking blood samples to track changes in testosterone levels under varying social circumstances. He also monitored cortisol levels among males of all social ranks. Comparing test results for dominant and subordinate males, he concluded that being able to predict and to control social interactions and having outlets for tension that builds from social stress all helped to limit the long-term negative effects of stress. Significantly, stress and the management of stress varied greatly among individual baboons. Every animal has his own 'style' of coping with stress and this manner of reaction served to increase or reduce the impact of tensions arising from conflicts. Sapolsky writes

> My ongoing research program has added strong support to a growing body of work suggesting that people's physiological and social characteristics (for example, their emotional makeup, personality and position in society) can profoundly influence their physiological response to stress.
>
> (1990, p. 116).

For both humans and the baboon, and most likely most of the primates, the great majority of sources of stress are social in origin. Knowing this is a giant step towards striving to solve the problem.

It would be awesome if our primate subjects could talk with us, but it is not possible. They use signs and symbols in the simplest manner and then only after much training in captivity. Because they are so limited in what they can communicate with us, at least for the time being, many if not most of our intriguing 'why' questions are going to remain unanswered. It seems that the more we watch them, the better we get to know them as individuals with distinct and ever-changing personalities, the longer grows our list of things we want to know about them. It is terribly frustrating to realize that after watching a male langur monkey for 10 years you still are more likely to be wrong than correct in trying to predict either what he will do and what will be the outcome when another adult male challenges his status or position in the group or how the overall hierarchy will shift and realign after a death or departure. Although the monkey's or ape's behavior is much simpler than ours, and their cognitive abilities far more limited, they still have a relatively rich and complex life and at every point have more options for action than we possibly can guess. Life experiences shape their decisions and actions; and since we can only glimpse a tiny portion of what they experience, and perhaps not very accurately at that in spite of our most determined efforts to be exact, we resort to guessing what their motivations are based on what we see happen. Although our efforts might provide us with a comforting feeling of having explained why something happened, this is not a scientific manner of proceeding and can be totally misleading. Best guesses remain guesses and by playing this game often we are only allowing what we consider to be an effect to identify or create its own cause.

Although nonhuman primates do not have language they communicate a great deal of information about themselves, others, and to a limited extent the environment. Vocalizations serve to identify the caller and inform the listeners much about the sender's emotional state, whereabouts, and in some instances about the presence and nature of potential danger (Cheney and Seyfarth, 1990). Gestures and body posture as well as quality and speed of motion can send very subtle messages about the individual doing the sending.

It is a long time since humans were defined as the tool-makers of the order Primates. Indeed, we have been joined by a number of other primates, including and most notably the chimpanzees with their array of different tools made and used in a number of contexts for specific purposes (McGrew, 1992) and in different ecological locations (Boesch-Achermann and Boesch, 1994). The New World capuchin monkey is also skilled in using objects to investigate and manipulate the environment in search of food (Panger, 1999). Object use and tool-making by nonhuman primates has given us valuable information to use in trying to reconstruct the period in our past when we were first experimenting with using objects in a number of contexts and then deliberately altering their shapes. Most likely early object use and even the shaping of instruments or tools for a purpose was done with perishable materials such as wood. If stones were used the wear was insufficient to mark or distinguish them as tools.

Caveats

The following caveats concern our study and understanding of the nonhuman primates for those who would undertake the transfer and application of the results of those studies to ourselves. You may wonder why this section is included. Part of the answer to this question lies in the fact that it has become virtually impossible for someone who is not a specialist in primatology to clamber through, let alone evaluate, the seemingly exponentially increasing volume of literature on primates. There really is no 'typical' primate (Strier, 1994). It is inadequate simply to suggest to the user to beware, because much of what is out there is compellingly written, and, since it is published, often repeatedly, has assumed the status of established wisdom. Although there is a great quantity of truly excellent information, some of the most popular and persuasively written is not reliable. This is not to suggest that you question observational reporting, but it is to encourage you strongly to consider closely the logic of arguments and interpretations concerning the observations so that you can identify conclusions that are without foundation or are illogical.

An anthropologist will tell you that human biology and culture conjointly act to configure our motivations, perceptions, logic, and all we experience. Our central nervous system structures all of the stimuli we perceive in the world that surrounds us, and our cognitive system how we structure our expressions of it. Language is our most powerful tool: it orders our thoughts and gives expression to all that we think. We are all too quick to transfer this ability of speech to other animals, creating talking birds, amphibians, reptiles, fish, and even insects. After all, who among us doesn't enjoy that brave dinosaur and root for the clever chicken? Needless to say it comes as no surprise that we create them to be able to express what we would in comparable circumstances. We make them also feel strongly all the emotions we have come to associate with our lives, loves, trials, and tribulations. No problem, just as long as we keep in mind what we are doing.

Words often have multiple meanings, both dictionary and popular uses. Words can arouse intense emotion, create illusions, immobilize or activate, and much more; and all this is true whether we like it or are even aware of it. We cannot afford to be as cavalier as Alice, who would have it that a word means what she wishes, no more, no less. It continues to be fashionable to analyze nonhuman primate behavior using terms more appropriate to our own, with results leading to profound misunderstandings of their behavior. Popularization of the nonhuman primates is understandable, and perhaps inevitable since it is so easy to project ourselves into them; but it has led to escalating anthropomorphism. There has been a deliberate blurring of the differences among primates in movements to 'humanize' the nonhumans, first psychologically and more recently legally (Marks, 2001).

Although we share much, we are sufficiently different that to translate directly, back and forth, human to nonhuman primates, is to invite some totally unnecessary and very messy muddles. Let me give you an example. Cultural anthropologists have always had a great deal to say about kinship. We can decree without respect to biological relatedness that you and I are brother and sister. We can then delegate our worldly goods, opportunities, rights and obligations accordingly. We can make a reality of a biological fiction. But what about kinship among the nonhuman primates? We cannot ask them anything but we can observe them closely. We see non-random patterns of interaction and much differential treatment of one another. We can measure these differences. If we stay long enough we can know which females give birth and follow the offspring through life, looking forward in time to the effects of early experiences. We can also apply modern techniques of paternity testing to create genealogies.

It has been demonstrated quantitatively that preferential relations and high familiarity do in fact correlate positively with some categories of kinship for at least periods of the life cycle, as for example mother and immature offspring. Mothers act to benefit their offspring and we can identify preferential relations among other sets of animals. Although these patterns of preferential behavior impact fitness, we are not justified in thinking, even for a moment, that it means the animals themselves have a clue as to whether or not an animal is related or whether they themselves or another animal is a parent. How would a nonhuman primate male know he is a father, let alone the father of any specific immature? The answer given to this question is usually couched as a metaphor or is expressed in 'he behaves *as if*' statements. This is a replay of the 'I wouldn't have seen it if I hadn't believed it' approach to animal behavior. Causes and effects are easily confused or incorrectly identified if we presume outcomes without a full analysis of the situation, including each and every assumption about what an animal needs to know for a specific scenario to be true.

As mentioned earlier, assignment of motivation to the actions of non-human primates is very risky and runs a high likelihood of being wrong. The literature abounds with assumptions, usually stated as fact, as to what an action or event means and therefore why it happened. For example, it is not infrequently asserted that animals are said to disperse from one group to another to avoid inbreeding (see Moore and Ali, 1984, for a critique). Imagine what a human would need to know to arrive at that choice of action for that reason. Another increasingly frequent assertion is that adult males deliberately target and kill the immature offspring of other males. Look more closely at this statement and consider what it means and what you must credit the killer male with knowing.

It is widely assumed and often stated unequivocally that adult male nonhuman primates are altogether concerned with knowing and controlling the paternity of individuals in the next generation. These statements are frequently expressed 'male X behaves as if he were or were not the father of animal Y'. Unsupported 'as if' interpretations mislead. At the very least they discourage investigation to determine the real influences on and development of behavior. One wonders why we attribute knowledge or understanding of the concept of paternity to an animal when many humans lack the knowledge. What mental mechanism must we assign to the nonhuman in order to use the notion of fatherhood: a concept encompassing events including insemination that produces a state (pregnancy) that culminates six or more months later in a birth event the male certainly does not witness, and a sure recognition that the generational product is his and none other's. A male langur monkey, no matter where he falls in the spectrum of langur intelligence, has no clue the act of copulation can result in a cascade of biological events culminating in the birth of an infant langur that is his offspring (Dolhinow, 1999).

It has been asserted that a male gibbon refrains from living with more than one adult female because he cannot possibly keep all the infants that would be produced safe from being killed by other adult males. This is said even though the notion is unsupported by any directly relevant data. Amazingly, gibbon infanticide has not been recorded. The fact that there was an absence of relevant data, not from a lack of excellent observation but because certain things simply were not seen, ought to suggest the scenario is wrong. Complexly detailed explanations are created to explain why primates structure their social systems and actions to avoid what is claimed to be an ubiquitous and terrible pitfall of social life, infanticide by adult males. The fact is, natural selection cannot operate in the absence of a behavior. What we have is a classic example of *a posteriori* reasoning. Presumed effects are identified and then the very existence of the effects are claimed to be proof of their ostensible causes. This is science in reverse, which, like an essay in reverse, doesn't make sense. Combine this with the assumption that current use leads us to discover what function something served in the past, that certain acts are adaptive strategies to accomplish specific ends, and you are napping on the freeway. The absence of data is not proof of absence of a trait, just as the absence of data is yet to prove the presence of a behavior.

In summary

Our close relatives, the nonhuman primates, are both similar to and importantly different from ourselves. We often rely on them for comparative information to reconstruct our past and to serve as models to investigate our present biology and behavior. However, we must be exceedingly cautious in our use of other primates as models because modern humans were produced by a set of biological changes that occurred late in hominid evolution, after we last shared a common ancestor with other primates, making us different from our monkey and ape relatives in many important respects. Human language, and the biology and cognitive mechanisms that make it possible, allows us to create and express elaborate time dimensions, meanings, motivation, plans, and much else in a manner unavailable to the nonhuman primates. Words have become our most powerful tools and their meanings are critical to understanding and dealing with our reality. But is our reality the same as that of the nonhuman, and will we ever know what they are thinking? Unless we can gain purchase on the array of choices available to a monkey or ape at moments of decision, it is unlikely we will be able to answer 'why' questions about their behavior any time soon. It has become fashionable to analyze nonhuman behavior using terms more appropriate to our own and the results can lead to misunderstandings of monumental proportion. The causes and effects of behaviors are often confused or incorrectly identified because we label outcomes without full analysis. Presumed effects may wrongly be used to confirm ostensible causes. The richness and diversity of primate behavior provides a formidable set of factors to weight and then to balance in the construction of any model or generalization for comparative studies. Formulating research questions and then creating models must be done carefully to guarantee appropriateness, and this is only possible when our differences are as clearly understood as our similarities.

Whatever the study location, New or Old World, an anthropological perspective will be a powerful force affecting the practice of primatology by providing additional theoretical foci and by sound advice concerning ways to effect conservation. We must conserve in concert with people who share their land with the nonhuman primates, and this includes understanding and respecting their vision of the world, the forest, and the animals. More often than not, the nonhuman primates lose in competition with us and become casualties of 'progress'. In a time of rapid culture change, with local and global economic and political pressures acting on all creatures, we must learn to co-exist and share. We control the present and future of all primates, an awesome responsibility. We must resolve to act to ensure an acceptable future for us all.

References

Bernstein, I.S. (1999). Kinship and the behavior of nonhuman primates. In *The Nonhuman Primates*, ed. P. Dolhinow and A. Fuentes, pp. 202–5. Mountain View, California: Mayfield Publishing Co.
Boesch-Achermann, H. and Boesch, C. (1994). Hominization in the rainforest: the chimpanzee's piece of the puzzle. *Evolutionary Anthropology* 3, 9–16.
Campbell, C. (2000). The reproductive biology of Black-Handed spider monkeys (*Ateles geoffroyi*): integrating behavior and endocrinology. Ph.D. dissertation, University of California, Berkeley.

Carter, A. and Carter, C.C. (1999). Cultural representations of nonhuman primates. In *The Nonhuman Primates*, ed. P. Dolhinow and A. Fuentes, pp. 270–6. Mountain View, California: Mayfield Publishing Co.

Chapais, B. (1988). Experimental matrilineal inheritance of rank in female Japanese macaques. *Animal Behavior* **36**, 1025–37.

Chapais, B., Gauthier, C., Prud'homme, J. and Vasey, P. (1997). Relatedness threshold for nepotism in Japanese macaques. *Animal Behavior* **53**, 1089–101.

Cheney, D.L. and Seyfarth, R.M. (1990). *How Monkeys See the World.* Chicago: University of Chicago Press.

DeVore, I. (ed.) (1965). *Primate Behavior: Field Studies of Monkeys and Apes.* New York: Holt, Rinehart and Winston.

Dolhinow, P. (ed.) (1972). *Primate Patterns.* New York: Holt, Rinehart and Winston.

Dolhinow, P. (1999). Understanding behavior: a langur monkey case study. In *The Nonhuman Primates*, ed. P. Dolhinow and A. Fuentes, pp. 189–95. Mountain View, California: Mayfield Publishing Co.

Dolhinow, P. and Fuentes, A. (eds.) (1999). *The Nonhuman Primates.* Mountain View, California: Mayfield Publishing Co.

Fredrickson, W. and Sackett, G.P. (1984). Kin preferences in Primates (*Macaca nemistrina*): relatedness or familiarity? *Journal of Comparative Psychology* **98**, 29–34.

Hamburg, D. (1963). Emotions in the perspective of human evolution. In *Expression of the Emotions in Man*, ed. P. Knapp, pp. 300–17. New York: International University Press.

Jay, P. (ed.) (1968). *Studies in Adaptability and Variation.* New York: Holt, Rinehart and Winston.

Marks, J. (2001). *What it Means to be 98% Chimpanzee.* California: University of California Press. (In press.)

McGrew, W.C. (1992). *Chimpanzee Material Culture: Implications for Human Evolution.* Cambridge University Press.

Moore, J. (1996). Savanna chimpanzees, referential models and the last common ancestor. In *Great Ape Societies*, ed. W.C. McGrew, L.F. Marchant and T. Nishida, pp. 275–92. Cambridge University Press.

Moore, J. and Ali, R. (1984). Are dispersal and inbreeding avoidance related? *Animal Behavior* **32**, 94–112.

Myers, N., Mittermeier, R.A., Mittermeier, C.G. Fonseca, G.A.B. and Kent, J. (2000). Biodiversity hotspots for conservation priorities. *Nature* **403**, 853–8.

Newmark, W.D. and Hough, J.L. (2000). Conserving wildlife in Africa: integrated conservation and development projects and beyond. *BioScience* **50**, 585–92.

Panger, M. (1999). Capuchin object manipulation. In *The Nonhuman Primates*, ed. P. Dolhinow and A. Fuentes, pp. 115–20. Mountain View, California: Mayfield Publishing Co.

Ray, E. (1999a). Social dominance in nonhuman primates. In *The Nonhuman Primates*, ed. P. Dolhinow and A. Fuentes, pp. 206–10. Mountain View, California: Mayfield Publishing Co.

Ray, E. (1999b). Hierarchy in primate social organization. In *The Nonhuman*

Primates, ed. P. Dolhinow and A. Fuentes, pp. 211–17. Mountain View, California: Mayfield Publishing Co.

Ray, E.M. (2000). Hierarchy theory and emergence of social organization: structure and power in an all-male group of Hanuman langurs (*Presbytis entellus*). PhD dissertation, University of California, Berkeley.

Ribnick, R. (1982). A short history of primate field studies: Old World monkeys and apes. In *A History of Physical Anthropology 1930–1980*, ed. F. Spencer, pp. 49–73. New York: Academic Press.

Sapolsky, R.M. (1990). Stress in the wild. *Scientific American* **262**, 116–23.

Schoeninger, M.J., Iwaniee, U.T. and Glander, K.E. (1997). Stable isotope ratios monitor diet and habitat use in new World monkeys. *American Journal of Physical Anthropology* **104**, 69–83.

Stanford, C. (1999). Great apes and early hominids: reconstructing ancestral behavior. In *The Nonhuman Primates*, ed. P. Dolhinow and A. Fuentes, pp. 196–200. Mountain View, California: Mayfield Publishing Co.

Strier, K.B. (1994). Myth of the typical primate. *Yearbook of Physical Anthropology* **37**, 233–71.

Sussman, R.W. (ed.) (1999). *The Biological Basis of Behavior: A Critical Review.* 2nd edn. Advances in Human Evolution Series. New Jersey: Prentice Hall.

Washburn, S.L. (ed.) (1961). *Social Life of Early Man.* Viking Fund Publication in Anthropology No. 31. New York: Wenner Gren Foundation for Anthropological Research; and Chicago: Aldine Publishing Co.

Washburn, S.L. (1966). Conflict in primate society. In *Conflict in Society*, pp. 3–15. London: Ciba Society.

Washburn, S.L. (1971). On the importance of the study of primate behavior for anthropologists. In *Anthropological Perspectives on Education*, ed. M. Wax, S. Diamond and F.O. Gearing, pp. 91–7. New York: Basic Books.

Washburn, S.L. (1972). Human evolution. In *Evolutionary Biology*, Vol. 6, ed. T. Dobzhansky, M. Hecht and W. Steere, pp. 349–60. New York: Appleton-Century-Crofts.

Washburn, S.L. (1978). The evolution of man. *Scientific American* **239**, 194–208.

Washburn, S.L. and DeVore, I. (1961). Social life of baboons. *Scientific American* **204**, 62–71.

Washburn, S.L. and DeVore, I. (1963). Baboon ecology and human evolution. In *African Ecology and Human Evolution*, ed. F.C. Howell and F. Bourliere, pp. 335–67. Viking Fund Publications in Anthropology, No. 36. New York: Wenner Gren Foundation for Anthropolgical Research.

Washburn, S.L. and Dolhinow, P. (1972). *Perspectives on Human Evolution*, Vol. 2. Holt, Rinehart and Winston, New York.

Wasser, S.K., Risler, L. and Steiner, R.A. (1988). Excreted steroids in primate feces over the menstrual cycle and pregnancy. *Biology of Reproduction* **39**, 862–72.

Wrangham, R. and Peterson, D. (1996). *Demonic Males: Apes and the Origins of Human Violence*. Boston: Houghton Mifflin Co.

Zihlman, A. (1996). Reconstruction reconsidered: chimpanzee models and human evolution. In *Great Ape Societies*, ed. W.C. McGrew, L.F. Marchant and T. Nishida, pp. 293–304. Cambridge University Press.

Zihlman, A. (1999). Fashions and models in human evolution: contributions of Sherwood Washburn. In *The New Physical Anthropology*, ed. S.C. Strum, D.G. Lindburg and D.A. Hamburg, pp. 157–61. Advances in Human Evolution Series. New Jersey: Prentice Hall.

2   *Resistance to the cross-species perspective in anthropology*

MARY M. PAVELKA

Introduction

Despite a fundamental commitment to the comparative cross-cultural approach, many anthropologists (and other social scientists) have long resisted, both implicitly and explicitly, the extension of a cross-cultural to a cross-species framework. No working anthropologist today, whether biological or cultural, will deny that this resistance exists, whether or not they themselves feel it. What does the study of nonhuman primates – prosimians, monkeys, and apes – have to do with anthropology anyway? This is a question that often lurks just beneath the surface in many departments where biological anthropologists (specifically primatologists) compete for resources with social and cultural anthropologists. Primatologists in anthropology departments grow accustomed, early in their careers, to being asked to relate their material to humans, and to explain how and why they belong in an anthropology department. Many become resentful of this sentiment and refuse to defend their right to membership in the discipline, taking the position that they are interested in nonhuman primates in their own right, as an end in itself. This is valid and understandable. Just as a Hadza researcher would tire of being asked to relate her findings to the !Kung, and a baboon researcher would tire of being asked to relate his findings to chimpanzees, those who study any of the 250 species of nonhuman primates tire of being asked to relate their findings to *one* other species: *Homo sapiens*.

Despite defensive proclamations to the contrary, however, the vast majority of primatologists employ a comparative perspective in which the ecology and behavior of other primate species, including humans, is directly and mutually relevant. Further, the inclusion of primatology within anthropology is not just an historical accident, and does imply that at least some aspects of the study of nonhuman primates will be relevant to humans. The study of nonhuman primates has typically been pursued by students of human nature (Richard, 1981). Over the past decade, the tension between biologically and culturally oriented anthropologists has been accompanied by growing concern over the potential disintegration of

25

the discipline and the subsequent loss of its distinctive holistic comparative approach to the study of human nature and society.

Since the late 1970s I have pursued the study of primate behavior within anthropology departments in which biological anthropology held a minority position, and primatology was often not represented at all. I became accustomed to the marginalized status of primate research, but remained committed to anthropology and to the cross-species perspective as an extension of the cross-cultural approach. Like many of my subdisciplinary colleagues I occasionally claimed my interests to be in monkeys as an end in itself, which was true, but I also knew that my primate behavior interests were truly anthropological. I have paid particular attention, over the years, to the difficulties that some of my colleagues have in accepting or recognising that primatology can really be anthropology. In a job interview one cultural anthropologist asked me 'why in the world an "anthropologist" would be interested in studying animals'. Another, well-meaning but obviously desperate to find something to talk about, asked, 'if you were a monkey, what would you study about people?'. I felt compelled to respond by asking them why in the world they were interviewing primatologists, and learned that the desire to have another on staff was almost entirely economic. The introductory primate behavior and human evolution classes drew large numbers of students and subsidized the smaller, more intimate cultural anthropology classes. The department was thus motivated to keep this teaching area healthy. The majority of the faculty may not have seen the value of primatology as a research area relevant to their own, but they had to accept that many students, unenlightened as to disciplinary boundaries and definitions, were often drawn to anthropology by popular course offerings in this 'marginalized' area. In this chapter I discuss the benefits of a cross-species perspective to the study of human behavior, and try to identify – based on my own experience as a primatological anthropologist and on many conversations with sociocultural colleagues – some of the reasons for resistance to this perspective in anthropology.

The value of the cross-species perspective

Primatology, as a discipline, had its origins in natural history, anatomy, and zoology in the nineteenth century, ecology and psychology in the first half of the twentieth century and anthropology in the past 50 years. Anthropologists are interested in understanding the range of human behavioral and biological expression both past and present, and from societies and cultures around the world. Where sociocultural anthropology is committed to holistic and cross-cultural comparison, primatology extends the comparative perspective to the level of species, putting *Homo sapiens* into the broader primate perspective. Anthropologists are interested in the study of nonhuman primates because humans are primates, and primatologists are interested in humans for the same reason. A full understanding of any given behavior pattern is enhanced by the broad and holistic comparative perspective offered by the subdisciplines of anthropology. Primatological inquiries attempt to gauge the plasticity of primate nature, and to understand the conditions under which it evolved and the conditions that explain variation.

Biomedical models, whether they involve *Drosophila*, mice, or monkeys, are obviously based on uncontested underlying commonality in the anatomy and physiology of humans and other animals. The study of animal behavior has made a number of contributions to other disciplines with a primarily human focus (Snowdon, 1999). For example, research on animal sensory systems has contributed to the development of sonar techniques with a range of human applications. Chimpanzee language studies have led to the use of computer keyboards being used to teach language to people with difficulties in verbal communication. Our understanding of the role of circadian and other endogenous rhythms in humans derives from previous animal research. In terms of resource management, the study of wild animals is critical to environmental monitoring, and the data collected by primate behavioral ecologists is the same data needed to successfully monitor and devise protection plans for their habitats and captive breeding programs for endangered species. Further, primate conservation requires that complex issues and conflicts of human and nonhuman primate land use be addressed, and for this, broad-based anthropological skills are a must.

My own research into the social and reproductive ramifications of aging in primates, and specifically the question of human menopause, is another example of the mutual benefits of the cross-species and cross-cultural approach. Research into the social manifestations of aging in Japanese monkeys lead to the realization that aging for monkeys is fundamentally different from aging for humans, owing in part to the apparent absence of an awareness of death, a division of labor, and menopause in nonhuman primates (Pavelka, 1991, 1999). This research highlighted some species-specific aspects of the human life, and spawned further enquiries into human menopause in a comparative life history perspective (Pavelka and Fedigan, 1991; Fedigan and Pavelka, 1994). Primate research identified the lengthy and universal post-reproductive lifespan as the derived or distinctive feature of the human lifespan, a point now central to understanding the evolution of menopause and the role of grandmothers in some traditional societies (Hill and Hurtado, 1991; Hawkes *et al.*, 1997, 1998; O'Connell, Hawkes and Blurton Jones, 1999). Increasing anthropological interest in menopause has led to a comparative investigation of reproductive termination in primate species (Caro *et al.*, 1995; Pavelka and Fedigan, 1999; Packer, Tatar and Collins, 1998; Johnson and Kapsalis, 1995, 1998; Takahata, Kayama and Suzuki, 1995), and the grandmother hypotheses for the evolution of human female menopause led us to consider the possible survivorship advantages accruing to individual monkeys who terminate reproduction prior to death (Fedigan and Pavelka, 2001). Over the past ten years, primatologists and social anthropologists have relied heavily on the mutual benefits of the cross-species and cross-cultural approach to the questions of aging and menopause.

But the menopause research is not primarily about behavior, and it is the cross-species approach to human behavior – one of the principal contributions to be made by primate behavior researchers to anthropology – that is the target of the greatest resistance from anthropologists. After all, people in general and sociocultural anthropologists in particular tend to define our species behaviorally. The value of the cross-species perspective for understanding shared *morphological* (i.e. physical) traits is not disputed. That our five fingers and opposable thumbs are shared with other primates and should be understood as a primate adaptation is not questioned. No one tries to explain the morphology of our hands exclusively in terms of the demands of human technology or of the industrial or information age, suggesting for example that we have the hands that we do because we need to be able to answer telephones and use keyboards. We clearly recognize that these are species-specific, indeed society-specific, applications of the general primate hand and take no offence at the suggestion that the primate hand originally evolved as a response to the environmental demands of the primate precursors. Likewise, there is no objection to the use of a cross-species evolutionary approach to *behavior* if humans are not involved. The nine species of lesser apes appear to share a monogamous social system (but see Fuentes, 2001), and although there is disagreement over the best explanation for this, no one is offended by the suggestion that this is an evolved structure. But when we include *humans* in our cross-species considerations of *behavior*, we seem to tread on cherished ground, and particularly if the behavior in question is one of political import.

The cross-species perspective on sexual aggression is a prime example of the value of primatological inquiry into questions of human behavioral expression. It is also strongly resisted by many anthropologists and other social scientists. Sex and aggression, like love and war, seem at one level to be distinct and incongruous behaviors. It is clear, however, from the widespread incidence of rape, spousal abuse, sadomasochistic sexual practices, and the use of degradation and violence in pornography, that there is a strong link between them in most human societies. This link is most commonly explained by feminist and social learning theories, which see sexual aggression as owing to male desire to control women and their sexuality in a male-dominated patriarchal society, and from the desensitising effects of pornography and objectification of women in the mass media, respectively. Feminist and social learning theories seek to explain the connections between sex and aggression in exclusively human terms, linking the behavior to unique aspects of human culture and society. Considerable support can be found for both (Ellis, 1989).

If the connection between sex and aggression is purely learned and grounded in patriarchal human society, a strong connection between the two would not be expected in nonhuman primate society. In fact, sex and aggression, particularly male aggression against females, is widespread among primates, and much of it occurs in the context of mating (Smuts and Smuts, 1993). This has long been recognised by individual researchers, and the primatological literature is filled with passing reference to male aggression against females and to the substantial increase in aggressive male attacks received by females in estrus (heat). What does it mean to our understanding of sexual aggression in humans to learn that sexual aggression is not limited to human society, but that it is phylogenetically widespread in primates? When a morphological feature such as grasping hands is found to occur in a whole group of closely related species, we assume it to be a trait inherited from a common ancestor. Likewise, the widespread occurrence of sexual aggression in primates – the recognition that sexual aggression is not unique to humans but is shared in close relative species – suggests that an explanation that relies entirely on human social and cultural institutions may be incomplete. A more complete understanding of the factors affecting the expression of behavior is always our goal as social scientists, but it is particularly important when the behavior in question is, like sexual aggression, one whose expression we may wish to influence.

Smuts (1992) analyzed the reported incidence of male aggression against females in primate societies and noted that, although widespread, sexual aggression was also highly variable in expression both within and between species. Situational factors predispose members of a particular society (human and nonhuman) toward or away from the use of sexual aggression. She used the primate data along with anecdotal ethnographic data to identify factors associated with varying rates of sexual aggression. These include factors common to human and nonhuman primates, such as the extent of the kinship bonds among the females and the degree of power inequality among males in a given society, as well as factors unique to a particular species. For example, gender ideology and other factors identified by the feminist and social learning theories clearly influence the expression of sexual aggression in human societies. A series of hypotheses about the incidence of sexual aggression in human societies were generated for future testing by sociocultural anthropologists. Although one of the most controversial applications of evolutionary theory to human behavior (discussed further below), this issue demonstrates clearly the potential for the cross-species perspective to inform the discussion and generate new insights into behavioral expression.

Why then the hesitation of some anthropologists to accept the value of primatological insights?

Resistance to the cross-species perspective

Biological determinism

If these suggestions about the sociological factors accompanying a high incidence of male sexual aggression had been derived solely from cross-cultural anthropology there would be no resistance to their further exploration. The hypothesis that societies in which there are strong power inequalities among men and weak kinship bonds among females will show higher incidence of sexual aggression, even if it was generated by the observation that this relationship exists in nonhuman primate societies, contains nothing to offend even the most biophobic of cultural anthropologists. It is the explicitly evolutionary nature of cross-species investigations that provokes negative reactions to this work, and subsequent failure to appreciate the value of the new questions and research directions for social anthropology that are generated by it. Objection to the evolutionary perspective on male sexual aggression is tied to the underlying adaptation arguments, which hypothesise that male sexual aggression can be a male reproductive strategy favored, under some circumstances, by natural selection. It is certainly consistent with the male desire to control women and their sexuality that is at the heart of the feminist perspective. But all of the insights into factors that explain variation in expression are overlooked owing to the fear of biological determinism that so often accompanies an evolutionary explanation for behavior.

Fear that the evolutionary approach to the study of behavior is inherently biologically deterministic is common in anthropology. It seems peculiar to me that completely ignoring the role of biological and evolutionary factors in favour of social determinism is so much less feared, as social factors must be responsible for some fairly undesirable and intractable behavior patterns. None the less, the suggestion that a behavior pattern has evolved due to selective advantages accruing to individuals who display it under certain circumstances – either now or in the species' past – is often assumed to mean that the behavior itself is inevitable and unavoidable. If the roots of a behavior are thought to be entirely social, the behavior pattern is assumed to be much more flexible and able to change, although I know of no good evidence that this is so. Social determinism should not, to my mind, offer much reassurance in terms of easily directed social change.

Evolutionary approaches face the greatest resistance when the area of behavior is one that seems to threaten our freedom: as a society we believe deeply, even desperately, in the freedom to design our societies and our lives. With the Enlightenment we were supposed to be released from gods and ancestors controlling our lives. Now, in a secular society, science itself threatens new restrictions on our freedom in the form of our inherited traits and evolutionary past.

Critics of the evolutionary perspective on male aggression against women have tended to depict it as an all or none proposition, one which ignores the learning, situational, and social structural factors which influence, for example, rape probabilities (Ellis, 1989). But nowhere in Smuts' work is there any suggestion that the behavior is inevitable or unavoidable: quite the contrary. Her review of male violence against women world-wide stressed that 'far from being an immutable feature of human nature, male aggression toward women varies dramatically depending on circumstances' (1992, p. 24). The whole point of the comparative evolutionary perspective is to offer insight into the variables that tend to be associated with behavior in diverse species and societies. Evolutionary theorists, far from ignoring social factors, have emphasized them. Although the potential to express the behavior is very likely rooted in the primate (and maybe mammalian) evolutionary past, its expression is not inevitable, and is not assumed to be so by an evolutionary perspective.

Primatology does not assume that the behavior of our primate relatives will always reveal to us the biological roots of our own behavior. Even if we find that we share a behavior pattern with other primates, it does not necessarily mean that the behavior is rooted in biology, separate from society and culture. If all of the species in question live in social groups, the behavior pattern might simply be co-existent with society or with social living (Fedigan, 1992). The behavior may be a by-product of social life. The nearly universal avoidance of mating between close relatives in both human and nonhuman primates may, at the proximate level, be rooted in early childhood familiarity that is known to erode the base of sexual attraction. In the absence of the social conditions that ensure childhood familiarity, the aversion might not be expressed. This social explanation for a cross-species behavior pattern is consistent with an adaptive perspective that would point to the selective advantages accruing to individuals and societies in which inbreeding is avoided.

Overly simplified explanations

Sociocultural anthropologists, along with many other social scientists, often make the related assumption that primatologists are attempting to *simplify* primate (especially human) behavior with a few simple explanatory devices that will account for all expressions of behavior. The comparative method in general runs the risk of oversimplification of the behavior in question; this is a problem faced by practitioners of both the cross-cultural and cross-species perspectives. But like the cross-cultural, the cross-species approach emphasises the contextual and species-specific factors involved in behavioral expression, whether it be in prosimians, monkeys, apes, or humans. Every species has unique characteristics that must be taken into account in understanding its behavior patterns. Primatology does not, therefore, discount the roles played by language, culture, ideology, and self-reflection in human behavioral expression, any more than we would ignore the roles played by social structure, group and individual history, and species-specific adaptations in explaining baboon behavior. As primatologists we start with the knowledge that we are primates. But as primatologists we are also trained biological and social scientists, not to mention humans ourselves. Thus we know that we are also what we believe, what we think, what we hope, and what we decide to be.

Naturalistic fallacy

The view that biological or evolutionary frameworks are automatically deterministic is related to another source of hesitation on the part of some anthropologists to accept these frameworks: a fear of the naturalistic fallacy. The naturalistic fallacy refers to an error in reasoning that involves moving illogically from describing what is to arguing for what should or ought to be. In biological terms, this takes the form of mistakenly assuming that any behavior pattern seen in 'nature', that is, in other species, is natural, and therefore good, and therefore the way things ought to be. To observe that a behavior pattern is common throughout the primates – whether that be sexual aggression or same-sex sexual interactions or mothers as exclusive caregivers of their own infants – does not tell us that these patterns are the way things should be in human society. It is an error in reasoning to confuse hypotheses about what is observed in nature with value judgements about what is normal and right in human life. By virtue of the naturalistic fallacy, evolutionary approaches, especially those that emphasize the adaptive nature of behavior, are seen to be offering *justification* for behavior patterns or inequalities that we would otherwise deem unacceptable. The naturalistic fallacy is both philosophically illogical and based on a misunderstanding of natural selection. Observed behaviors do not reveal the range of current or future expression. Unfortunately an underlying naturalistic fallacy is common in the popular literature, and even some evolutionary anthropologists, who should know better, are guilty of this error in reasoning.

Sociobiology

The fear that evolutionary explanations are biologically deterministic, justifying all observed behavior, corresponds historically to the adoption of sociobiology by some primatologists, and the common perception by sociocultural anthropologists that *all primatologists are sociobiobiologists*. Until the early 1970s, the main rationale for primate research in anthropology was the search for nonhuman primate models of early human evolution. Baboons were a favored genus in this regard because most lived in the savannah-like habitats thought be have been occupied by the earliest australopithecines. Although not that closely related to humans, as an analogous model they helped to reveal some of the adaptations made by a primate to these conditions. Chimpanzees, as our closest relative, were the favored homologous model, supposedly revealing best the behavior of our earliest ancestors prior to the evolution of culture. As far as I can tell, primatology as anthropology was less actively resisted at that time. The objectives of the research were much in line with those of archeologists and ethnographers trying to understand the basic man-the-hunter path of human evolution. However, the rise of sociobiology in the late 1970s, following the publication of *Sociobiology: A New Synthesis* (Wilson, 1975), gave primatology relevance to issues of contemporary human behavior, and it became much more pertinent (and threatening) to sociocultural anthropology.

I think it is fair to say that most sociocultural anthropologists hate sociobiology, and have done so since its introduction. For many it was seen as nothing more than a resurrection of the nature side of the old nature–nurture debate. Worse, existing patterns of behavior were now not only genetically based, but seen to have been favored by natural selection, thus giving them a kind of ultimate biological legitimacy. Further, with its apparently simplistic justifications for behavior based on difficult-to-demonstrate reproductive advantages, it further solidified the belief among some that those who think 'animal' behavior can shed any light on human behavior clearly have no appreciation for the richness, meaning, and diversity of human behavioral experience. The following quote from prominent cultural anthropologist Clifford Geertz (1980) captures the general sentiment in sociocultural anthropology (in a review of a book about the evolution of human sexuality):

> This is a book about the 'primary male-female differences in sexuality among humans,' in which the following things are not discussed: guilt, wonder, loss, self-regard, death, metaphor, justice, purity, intentionality, cowardice, hope, judgment, ideology, humor, obligation, despair, trust, malice, ritual, madness, forgiveness, sublimation, pity, ecstasy, obsession, discourse, and sentimentality. It could be only one thing and it is. Sociobiology.

The quote reveals the author's view that these are the important aspects of human sexuality and that an evolutionary approach automatically negates them all. But the evolutionary perspective does not negate the role of guilt, hope, ideology, ecstasy, or any of the others in understanding human sexuality. As noted above, it is critics like Geertz who inaccurately depict it as an all or none proposition, one that ignores the learning, situational, and social structural factors. It is true that this book emphasized reproductive aspects of male–female differences in sexuality, just as a book on human sexuality by Geertz would ignore them. A cross-species anthropologist might well review a typical cultural anthropology of sexuality and point out all kinds of relevant (not deterministic but relevant) factors that were not considered. Imagine this comment in a book review:

> This is a book about human sexuality in which the following things are not discussed: reproduction, life history, demography, sexual selection, competition, resources, mate choice, adaptation, anatomy, cognition, parental investment, hormones, pheromones, coercion, dimorphism,

meiosis, metabolism, energetics, morphology, endocrinology,
phylogeny, physiology, biology, neurology, ontogeny, genetics, and
limerance. It could be only one thing and it is. Cultural anthropology.

This imaginary comment is meant to illustrate that it is easy to criticize *any*
approach for not being *a different* approach.

Proximate mechanisms

The lack of plausible proximate motivators for evolutionary explanations
contributes to the difficulty that many have in accepting that natural
selection might underlie some behavior. To postulate that the immediate
motive for human behavior is a simple and uncontrollable drive to maximize reproductive success is unacceptable to most scientists, social and
biological. For one thing, it does negate the many ideological variables
that we all know play a role in our behavior. Second, it is easily contradic-
ted by apparent evidence that much of human sexuality is not about
reproduction. Evolutionary explanations for human and nonhuman primate behavior in the 1970s and 1980s did tend to downplay the role of
proximate motivation in favor of explanations that sought to understand
the selection pressures that favored the evolution of the behavior in the
past. After all, it is the effect of the behavior on reproductive success that
matters in evolutionary terms, not why the individual did it. A behavior
that increases the reproductive success of the individual who displays it *will*
appear at higher frequencies in subsequent generations (if there is even an
indirect genetic basis to the behavior), regardless of why the individual did
it. Or in the case of a human, why she thought she did it. Proximate
mechanisms, and the many proximate factors that explain variation in the
expression of behavior, did tend to be neglected in early applications of
sociobiology to human and nonhuman primate behavior. Further, it was
not uncommon for the hypothesised selective pressures of the evolutionary
past to be confused with the current motivations of the individual animals,
giving the impression that animals, maybe even humans, are consciously
motivated – here and now – to get their genes into the next generation.
Remote adaptiveness arguments can seem very implausible and fail at an
intuitive level to shed light on the motivation for human *and* nonhuman
primate behavior. Current evolutionary approaches are concerned much
less with exclusively evolutionary (also called ultimate) explanations and
post hoc adaptive explanations, as the emphasis has evolved into one of
understanding variation in the expression of behavior, and the proximate
social and ecological factors that accompany it.

Are we all sociobiologists?

Many primatologists in the 1970s and 1980s found that the unifying framework and hypothesis generating abilities of the theory gave new lifeblood to the discipline, enabling it to move beyond the description that had characterised the previous decades. Others objected to the zealous application of the adaptationist program to nonhuman primate behavior and the tyranny of this program over the discipline. In the early and mid-1980s, when the first wave of sociobiological applications were sweeping the field, I was a graduate student. I could not accept or adopt a view that the monkeys that I watched on a day-to-day basis, with their intricate social dynamics and often calm and relaxed social behavior, could really be hard-wired 'gene machines' constantly calculating to maximise their reproductive success. I had a hard time, intuitively, imagining that every behavior had a direct reproductive cost or benefit, and that the animals were motivated to maximize distant reproductive benefits with every behavioral action. I objected to the assumption of my sociocultural colleagues that as a primatologist I was necessarily a sociobiologist, and argued strenuously against it. I did not equate all cross-species (and hence evolutionary) research with the specific adaptationist thrust of early sociobiology, as my anthropology colleagues did. But for many of my sociocultural colleagues, my arguments against a narrow adaptationist interpretation of individual behaviors of individual animals were arguments against the evolutionary perspective altogether, and hence the very rationale for engaging in cross-species comparative work.

It is true that the most well-developed sets of hypotheses (kin selection, reciprocal altruism, parental investment) within the evolutionary framework are those that explain behavior in terms of individual striving for reproductive success. Whatever the impressions given by the early application of sociobiology, it was probably not that simplistically interpreted by those practising it, and is certainly not today. The sociobiological approach of the 1970s has evolved into one with a more explicit recognition of the great range of expression in behavior and the need to understand the immediate conditions under which behavior does or does not occur. We know that individual primates are behaving in accordance with a host of complex variables including demographic, ecological, and social factors, as well as individual personality and choices of the actor. In fact, sociobiology has merged with more ecological approaches in primatology to become behavioral ecology, which has multifactoral explanatory models for social structure and less emphasis on the costs or benefits of specific behaviors of individual animals. Primatology is not about finding adaptive explanations (or stories) to explain or justify every observed behavior pattern. If sociobiology is broadly defined as the use of evolutionary principles to gain further understanding of the expression of behavior in human and nonhuman primates, then maybe we are all sociobiologists today (although few use this label any more).

Genetic basis of behavior

The lack of direct evidence for a genetic basis for specific behaviors continues to be a stumbling block for those who reject evolutionary explanations of behavior, whether they be sociocultural anthropologists or primatologists. But the absence of evidence should not be interpreted as evidence of absence. Indeed, the more we learn about the complex pathways between DNA and the physiological and psychological processes that underlie behavioral impulses and decisions, the more certain it is that aspects of behavior are heritable. Most people know or believe, on a personal level, that children inherit more than just morphological features from their parents. But again, when a parent and child share similar mannerisms or personality traits, it is considered safer, or more prudent, to explain these similarities entirely on the basis of learning and the environment. But the possibility that these traits might be learned is not in itself evidence against the equally plausible possibility that inherited physiology and psychology underlie the shared behaviors. I think that the lack of evidence for a genetic basis for a behavior is often used as an argument against the evolutionary approach, when the resistance is actually based on some of the other fears and assumptions discussed in this chapter.

Rejection of evolution by natural selection

The resistance to the cross-species evolutionary perspective in anthropology may be tied to more than just opposition to the specific tenets of sociobiology. It may be tied to a failure to accept (or understand) evolution by natural selection. I have certainly encountered colleagues who have what can only be described as a pre-Linnean world view, so shocked are they at the suggestion that humans are actually animals, especially, as noted above, when it comes to behavior. But even those who do embrace the fact of our basic animal identity often do so with a distinctly progressive spin that places humans at the top of a hierarchy of beings, although this is almost certainly in conflict with natural selection.

A progressive view of nature, which sees humans as the pinnacle of all creation, has a history dating back to Aristotle at least. By the Middle Ages, it was widely believed that God had created all life forms in a chain or scale ranging from the simplest and least perfect to the most complex and most perfect. Humans represented the top of the chain on earth, surpassed in complexity and perfection only by heavenly creatures: angels and God himself. The chain or scale was seen to be entirely static, currently appearing exactly as it did at the time of creation.

Darwin's theory of evolution by natural selection, which eventually came to be widely accepted, should have entirely dispelled the notion of a great chain of being. Not just the static nature of the chain, but the chain altogether, in favor of, for example, a dense bush with millions of terminal branches representing the living descendants of species of the past. Natural selection leaves no room for a single or linear chain of being or ladder of progress, with evolution leading purposefully and progressively from early–lower–lesser forms to later–higher–better forms, like us. But many anthropologists – even biological anthropologists – apparently continue to see our place in nature in this framework. The persistent and continued use of terms such as 'higher primates' and references to species that are 'more highly evolved' clearly reveal our underlying allegiance to a medieval concept.

One might argue that we were helped on this path to misunderstanding natural selection early on by T.H. Huxley. In his effort to have natural selection accepted, Huxley worked as a kind of 'spin-doctor' and helped to soften the blow of our descent from apes rather than from heaven by interpreting evolution by natural selection as a progressive process leading inevitably and ultimately to us (see, for example, Huxley, 1906). According to S.J. Gould (1996), evolution by natural selection is widely misunderstood because the philosophical implications are so disturbing. Evolution by natural selection is purposeless and without direction, nonprogressive and heading nowhere in particular, and fundamentally materialistic, rejecting our longstanding commitment to dualism. Only by turning it into its near opposite – a progressive and purposeful process leading inevitably to us – can we bear it. But a realistic appreciation of evolution *by natural selection* is critical to an appreciation of the value of the cross-species perspective in anthropology. Rejection of its value may be tied to a conscious or unconscious rejection of evolution by natural selection.

Political misapplication

Ironically, the failure to accept an evolutionary perspective on human behavior (resistance to the primate perspective) is often based on the fear of political interpretations that support *laissez-faire* attitudes toward un-just social conditions, social Darwinism, and even eugenics. *In fact those interpretations only make sense in an incorrectly conceived progressive evolution.* If we truly accepted evolution by natural selection, and held an image of the evolution of life on earth like the dense bush described above, then we must regard ourselves as a tiny, late-occurring twig, rather than as the pinnacle of God's creation or natural selection's design. There is nothing inherent in natural selection to justify the political misapplication of it. It is the progressive spin that turns natural selection into its opposite that is most dangerous. The Great Chain of Being – which a full appreci-ation of evolution by natural selection would dispel – is at the heart of the thinking that some species are better, more worthy, more deserving of good or special treatment, than are others. Likewise the thinking that within a species some populations are more worthy than others, that some are closer to God while others are closer to animals, is at the heart of some of our most grievous social problems. A view of natural selection that supports social Darwinism, eugenics, racism, and sexism is a fundamental-ly incorrect view of natural selection.

Beyond biology

Whereas some object to evolutionary explanations for human behavior based on a rejection or misunderstanding of the process of evolution, others reject them for the almost opposite reason: they believe that humans have moved beyond the point where evolutionary forces affect us. A common sentiment in social and cultural anthropology is that primatology cannot begin to shine any light on human behavioral expression, because humans are *just too different*. By virtue of our language and culture we are just too far removed from our evolutionary past for the cross-species perspective to have any relevance to the understanding of human belief, behavior, and society. In other words, the objection to evolutionary explanations for human behavior may be tied to a belief that we have evolved *beyond biology*. Culture is so powerful and all-encompassing that our biological identity as primates is no longer relevant to understanding our behavior. Further, our cultural tradition of domination over our natural world leads us to see ourselves as emancipated from nature. Thus for some, a rejection of the value of cross-species insights into human behavior is justified within a biological and evolutionary framework, albeit a progressive one.

Post-modernism

Within the past decade, the rejection of primatology as anthropology has become quite explicit, as part of the wider postmodern rejection of the scientific and evolutionary tradition within anthropology (Morell, 1993; Cartmill, 1994). Anthropology has always been the only discipline explicitly and intentionally located at the difficult boundary of nature and culture: it has been the main unifying theme in an otherwise impossibly diverse discipline. Yet our two cultures – science and the humanities (Snow, 1959) – have been actively at war in anthropology throughout the 1990s. The war involves the attempted expulsion of biological and other mainstream anthropologists, who continue to value the scientific and evolutionary tradition, by those who now reject this tradition on both philosophical and moral grounds. In the middle of the last decade, it was a battle that many believed might lead to a disciplinary divorce on the grounds of irreconcilable differences. Although this tension has been part of anthropology since its inception, and will continue for as long as the discipline exists, it is my impression that the marriage will survive the crisis, although primatologists will continue to face resistance for the more traditional reasons outlined in this paper. Indeed, in some respects the threat of break-up seems to have made us stronger, since we had to go back and remind ourselves of why we were together in the first place. Most anthropologists recognize that valuable insights could be gained by the interpretative approach, but only if the 'discourse' itself didn't kill the discipline first. The American Anthropological Association has made a concerted effort to bring biological anthropologists back into the fold; indeed, a primatologist was until recently the editor of the *American Anthropologist*. However, the editor, Bob Sussman, is a well-known opponent of the narrow adaptationist program of some sociobiologists, and this may have eased the minds of those sociocultural anthropologists who were torn between a fear that the discipline was falling apart, and fear that primatologists are all biological determinists.

It is also my experience that those exploring a post-modern view of science are a much more intellectually diverse group than is often portrayed. In a luncheon address to the American Association of Physical Anthropology, Matt Cartmill (1994) made reference to the 'old mistake of seeing all culture as biology, and the new mistake of seeing all biology as culture'. But so few of us ever made the 'old' mistake of seeing all culture as biology that I am inclined to doubt the extent of participation in the 'new' mistake. Even the ultimate desconstruction that primatology received by Donna Haraway (1989) has not undermined primatology, but has enriched its discussions. A 1998 Wenner-Gren symposium brought together primatologists and science studies scholars to discuss the role of theory, methods, gender, and other factors to the changing view of primate society over the history of the discipline (Strum and Fedigan, 2000). The fact that science studies scholars have spent considerable time and energy deconstructing primatology has, in my experience, provided common ground for discussion with sociocultural colleagues. Since 1989 several have asked me my opinion, as a primatologist, of Harraway's book. Mainstream anthropology (including primatology) may have irreconcilable differences with some of the more extreme post-modern anthropologists, but this may be true for the extremes of any theory, approach, or subdiscipline.

Other methodological and theoretical differences contribute to the reluctance of sociocultural anthropologists to accept or relate to primatology. Methodologically, sociocultural anthropologists use largely qualitative ethnography while primatology is fundamentally committed to a modified scientific method of hypothesis generation, testing, and rejection using primarily quantitative research methods. Theoretically, exploration of the function of behavior is one of our central themes, and functionalism has been long rejected in sociocultural anthropology on teleological grounds. Further, a main thrust of sociocultural enquiry has been the demonstration of human social diversity, while primatologists, working with 250 highly diverse species, may focus more on common behavior features in any one species. So while the cross-cultural practitioners emphasize human diversity, the cross-species practitioners tended to emphasize unity.

Concluding comments

Nowhere are the mutual benefits of the cross-species and cross-cultural approaches of anthropology more complementary and more applied than in the area of ethnoprimatology and primate conservation. Primatologists studying nonhuman primates in the wild are required to deal extensively with the people of the tropics and subtropics, where most natural primate populations occur, and to deal with the complex issues of human–nonhuman primate interaction and competition for land and resources. Several of the papers in this volume demonstrate the benefits of the comparative analyses and a holistic approach of anthropologists in the study of the interactions of human and nonhuman primates in a cross-cultural setting and the implications for primate conservation.

In this chapter I have tried to identify, from my own experience, some of the reasons why the cross-species perspective of primatology is not always welcome in sociocultural anthropology. Before concluding, I should point out that for some, there is no resistance at all: some sociocultural anthropologists implicitly and explicitly accept the value of the primate perspective. For others, there is no active resistance, just lack of interest. Some areas of anthropological inquiry do seem rather remote from the life of nonhuman primates. Those investigators that focus on, for example, religious expression, ethnomusicology or the experiential component of any behavioral expression may be relatively uninterested in the work of colleagues whose research questions are so distantly related to their own. In resource-limited environments we expect competition, and thus in anthropology departments the sociocultural anthropologists might reasonably prefer resources to be directed toward teaching and research areas that more directly inform their own. For others, the rejection of evolutionary approaches in anthropology might be political at another level. Throughout its history, the political right has most often used a version of natural selection in support of political agendas. This continues today: in the social sciences (with the exception of primatology and psychology) sociobiology appeals most to right-wing social scientists. In my experience, most sociocultural anthropologists are left-wing in their politics. But tension between the subdisciplines of anthropology is widely recognized, and active resistance to primatology does exist. Some of the reasons for this resistance relate to a number of assumptions about primatology, including that we: are all sociobiologists and biological reductionists; do not appreciate the richness and distinctness of human experience; and are looking for single, simple explanatory devices to explain all expressions of behavior. For others, rejection of evolutionary approaches to human behavior may be due to rejection of the fact of human evolution for various reasons, including a lingering medieval view of our place in nature, and a postmodern rejection of the whole evolutionary and scientific tradition of anthropology. By arguing for the benefits of primatological anthropology, and bringing some of the sources of resistance to the surface, I hope to contribute to discussions about the value of the rich and diverse discipline of anthropology, and its importance to understanding human–nonhuman interactions.

References

Caro, T.M., Sellen, D.W., Parish, A., Frank, R., Brown, D.M., Voland, E. and Borgerhoff Mulder, M. (1995). Termination of reproduction in nonhuman and human female primates. *International Journal of Primatology* **16**, 205–20.

Cartmill, M. (1994). Reinventing anthropology: American Association for Physical Anthropologists Annual Luncheon Address, April 1, 1994. *Yearbook of Physical Anthropology* **37**, 1–9.

Ellis, L. (1989). *Theories of Rape: Inquiries into the Causes of Sexual Aggression.* New York: Hemisphere Publishing Corporation.

Fedigan, L. (1992). *Primate Paradigms: Sex Roles and Social Bonds.* Chicago: University of Chicago Press.

Fedigan, L.M. and Pavelka, M.S.M. (1994). The physical anthropology of menopause. In *Strength in Diversity: A Reader in Physical Anthropology*, ed. A. Herring and L. Chan, pp. 103–26. Toronto: Canadian Scholars Press.

Fedigan, L.M., and Pavelka, M.S.M. (2001). Reproductive termination in female Japanese monkeys: a test of the Grandmother Hypothesis. *International Journal of Primatology* **22**(2), 109–25.

Fuentes, A. (2001). Hylobatid communities: Changing views on pair bonding and social organization in hominoids. *Yearbook of Physical Anthropology* **43**, 33–60.

Geertz, C. (1980). Sociosexology. *The New York Review of Books* **26**(21/22), 3–4. [Review of *The Evolution of Human Sexuality*, D. Symons.]

Gould, S.J. (1996). *Full House: The Spread of Excellence from Plato to Darwin.* New York: Three Rivers Press.

Haraway, D. (1989). *Primate Visions: History, Race, and Gender in the World of Modern Science.* New York: Routledge.

Hawkes, K., O'Connell, J.F. and Blurton Jones, N.G. (1997). Hadza women's time allocation, offspring provisioning, and the evolution of long postmenopausal life spans. *Current Anthropology* **38**, 551–77.

Hawkes, K., O'Connell, J.F., Blurton Jones, N.G., Alvarez, H. and Charnov, E.L. (1998). Grandmothering, menopause, and the evolution of human life histories. *Proceedings of the National Academy of Science, USA* **95**, 1136–9.

Hill, K. and Hurtado, A.M. (1991). The evolution of premature reproductive senescence and menopause in human females. An evaluation of the grandmother hypothesis. *Human Nature* **2**, 313–50.

Huxley, T.H. (1825–95). *Man's Place in Nature and Other Essays.* London: Dent [1906].

Johnson, R.L., and Kapsalis, E. (1995). Ageing, infecundity and reproductive senescence in free-ranging female rhesus monkeys. *Journal of Reproductive Fertility* **105**, 271–8.

Johnson, R.L. and Kapsalis, E. (1998). Menopause in free-ranging rhesus macaques: estimated incidence, relation to body condition, and adaptive significance. *International Journal of Primatology* **19**, 751–65.

Morell, V. (1993). Anthropology: Nature-culture battleground. *Science* **261**, 1798–802.

O'Connell, J.F., Hawkes, K. and Blurton Jones, N.G. (1999). Grandmothering and the evolution of *Homo erectus*. *Journal of Human Evolution* **36**, 461–85.

Packer, C., Tatar, M. and Collins, A. (1998). Reproductive cessation in female mammals. *Nature* **392**, 807–11.

Pavelka, M.S.M. (1991). Sociability in old female Japanese Monkeys, Human versus nonhuman primate aging. *The American Anthropologist* **93**, 588–98.

Pavelka, M.S.M. (1999). Primate gerontology. In *The Nonhuman Primates*, ed. A. Fuentes and P. Dolhinow, pp. 220–4. Mountain View, California: Mayfield Publishing Company.

Pavelka, M.S.M. and Fedigan, L.M. (1991). Menopause: A comparative life history perspective. *Yearbook of Physical Anthropology* **34**, 13–38.

Pavelka, M.S.M. and Fedigan, L.M. (1999). Reproductive termination in female Japanese Monkeys: A comparative life history perspective. *American Journal of Physical Anthropology* Vol. 0, 1–10.

Richard, A. (1981). Changing assumptions in primate ecology. *American Anthropologist* **83**, 517–33.

Smuts, B. (1992). Male aggression against women: An evolutionary perspective. *Human Nature* **3**, 1–44.

Smuts, B. and Smuts, R. (1993). Male aggression and sexual coercion of females in nonhuman primates and other mammals: Evidence and theoretical implications. In *Advances in the Study of Behavior*, Vol. 22, ed. P.J.B. Slater, M. Milinski, J.S. Rosenblatt and C.T. Snowdon, pp. 1–61. New York: Academic Press.

Snow, C.P. (1959). *The Two Cultures and the Scientific Revolution*. Cambridge University Press.

Snowdon, C. (1999). http://www.animalbehavior.org/ABS/Education/valueofanimalbehavior

Strum, S. and Fedigan, L. (2000). *Science Encounters*.

Takahata, Y., Koyama, N. and Suzuki, S. (1995). Do the old aged females experience a long post-reproductive life span? The cases of Japanese macaques and chimpanzees. *Primates* **36**, 169–80.

Wilson, E.O. (1975). *Sociobiology: A New Synthesis*. Cambridge, Massachusetts: Harvard University Press.

3 The ethics and efficacy of biomedical research in chimpanzees with special regard to HIV research

ROGER S. FOUTS, DEBORAH H. FOUTS AND
GABRIEL S. WATERS

Among our fellow animals the chimpanzee is the species that most closely resembles ours biologically, behaviorally, and cognitively. At the biological level, the biochemical and genetic similarities are especially profound since chimpanzee beings are closer to human than to gorilla beings (King and Wilson, 1975; Lewin, 1984; Gibbons, 1990). King and Wilson (1975) were so struck by the similarity between chimpanzee and human that they referred to them as our 'sibling species'.

As the work of Jane Goodall (1986), and others like her, has shown, the behavior of the wild chimpanzee is just as strikingly similar to our own as are its biochemistry and genetics. Free-living chimpanzees in Africa as hominids are not so different from humans. They live in communities, they hunt, mothers care for their children and children care for their mothers, they use and make tools, and perhaps most important of all they can suffer from emotional as well as physical pain. Like human groups, there are also cultural differences between free-living chimpanzee communities; some medicate themselves with medicinal plants, others do not; some use hammers and anvils, others do not; in fact, some communities have never been observed to make or use any tools at all; and there are gestural dialects between different communities (McGrew, 1992). Recently, primatologists have collaborated to synthesize over 150 years of collective observations of chimpanzee behavior from seven locales (Whiten et al., 1999). From this synthesis, patterns of 39 different behaviors have emerged identifying chimpanzee cultures.

In addition to the striking similarities between their culture and ours, chimpanzees have also evinced striking cognitive similarities to our species. The research of R.A. and B.T. Gardner (1989) and our own (Fouts and Fouts, 1989; Fouts, Fouts and Van Cantfort, 1989) has demonstrated that chimpanzees can acquire the signs of American Sign Language, that they can use the signs to converse with humans, that they can use the signs to converse with each other with no humans present, that

45

they sign to themselves, and that they can pass this language on to their next generation without human tutelage, just to mention a few of their accomplishments.

As far as the chimpanzee is concerned, these findings have both a positive and negative side. The positive side is that the findings clearly demonstrate the Darwinian notion of continuity as opposed to the Cartesian notion of discontinuity. In other words, the difference between us humans and chimpanzees, as well as our other fellow animals, is one of degree, not of kind. The Cartesian gaps that have been assumed to exist have turned out to be gaps in human knowledge, not gaps in the phylogenetic scale. The implications are that the Cartesian model that held that our fellow animals were nothing more than unfeeling, unthinking machines is false. However, there are differences as well as similarities, to be sure, and one should not be considered without the other.

The negative side of being similar to our species is that the chimpanzee has become a prime candidate for invasive biomedical research. Biomedical researchers are not fiends who enjoy torturing powerless nonhuman animals, but they, like many others in this field, suffer from an ignorance. Too often this ignorance goes beyond that of the behavioral complexity and psychological needs of the chimpanzee, which can result in the chimpanzee being treated as nothing more than a 'hairy test-tube'. Unfortunately this type of ignorance can extend to an ignorance of experimental design and method. Whereas, the former can result in mistreatment of the chimpanzees in captivity, the latter can result in research that is invalid with regard to its efficacy. This chapter will focus on this latter point, namely those common confounds or overlooked basic rules of scientific method that can make a scientific study worthless. The presence of confounds in research or incorrect experimental assumptions can result in a waste of money and, more importantly, of the lives of chimpanzees as well as humans.

One reason for this ignorance of experimental design and methodology is that many scientists are trained in their particular specialty, but have little training with regard to experimental design and methodology; this is especially true of the medical profession. The questions that should be routinely addressed are: Were the proper sampling methods used? Did the particular design meet all the assumptions of the particular statistical test used? Were any confounding variables overlooked? Can the experimenter support the hypothesis studied with his or her design?

Common errors in research: a question of efficacy

When conducting research, there are basic rules and limitations that the investigation must follow for the research to have any value. If the researcher does not follow these rules, the results are useless. Because of the potential for outside or extraneous variables influencing the results, great care must be taken in the design of the experiment to make sure that those potential confounds are controlled for or eliminated. If the experiment has been designed properly and is free from confounds, and if the proper statistical test is used and all of its assumptions met, then the next concern is what the results mean and how they can be applied outside of the experimental situation (Kazdin, 1992).

The 'standard housing/rearing' confound

One of the most common errors that an experimenter can make is to ignore the history that the subject brings to the experiment. It often seems as if researchers assume that the subject has no history that might influence the results of the study. In some cases, for example when 90-day-old rats are obtained from an established breeder, the assumption that the animals' past experience is controlled for is correct. However, in chimpanzee research, ignoring the influence of early rearing and housing conditions can have a confounding effect on the results.

A great deal of biomedical research uses chimpanzee subjects who have been deprived at the social or environmental level. At best, they might have been taken from their mothers and reared in peer-group nurseries, or at worst held in isolation. The result is the same, some form of social deprivation, because even peer-group nurseries deprive the children of their mothers and their larger social community. Environmental deprivation is very common and is almost impossible to avoid. In the United States, standard housing for chimpanzees weighing at least 25 kg requires a cage 2.33 m × 2.33 m × 213.36 cm high. If they are younger and therefore smaller they can be kept in yet smaller cages. Frederick King (1986), the former director of one of the largest biomedical facilities to house chimpanzees, recognized this fact when he stated:

> There are many ways in which non-experimental stress can occur in the laboratory or in the animal housing situation and many ways in which stress can affect experimental results . . . environmental and social stress can influence neurological, endocrinological and immunological processes . . . Among the stresses which can produce abnormal

biological and behavioral changes are: ... immobilization, isolation, aggression, trauma and crowding. (p. 10)

The implication of using chimpanzees who have been housed in standard laboratory conditions as subjects of experimentation does not have a sound empirical basis because these standard conditions confound the results and negate the efficacy of the findings. Davenport (1979) also noted that laboratory chimpanzees are often housed in stressful, unnatural environments, which create abnormal behavior.

The social nature of chimpanzees as well as other primates is often not taken into consideration when standards for animal housing are drawn up. The isolation common in standard housing has a damaging effect on the primate and also serves as a potential confound for any research done using them. In this regard, Pereira and Altman (1983) stated that:

> ... laboratory-born primates usually begin life with their mother in a small isolation cage, get separated from her early, spend their juvenile years either alone or with a small group of peers in other barren cages, and then spend most of their adolescent and adult lives in a cage ... In such environments, we assert, nonhuman primates are not permitted to develop and behave normally. Indeed, we predict that under these conditions primates will exhibit aberrations not only in their behavior but also in all of their physiological systems. (p. 3)

Standard housing can prevent the chimpanzee from participating in play, a very common behavior in chimpanzees which is critical to normal neurophysiological and psychological development. Both young and adult chimpanzees have been reported to play. Mason (1965) found that young chimpanzees will spend 70% of their time in play and that they have as many as 16 times more social contacts than do adults, of which the majority are play interactions. In humans, play has been found to be critical in the cognitive development of creativity and divergent thinking (Lieberman, 1965) and to have a facilitory effect on problem-solving abilities in children (Rosen, 1974). Play has been found to be important in the cognitive, neurophysiological, psychological and social development of nonhuman primates as well. Caplan and Caplan (1973) compared monkeys exposed to an environment-rich setting with playthings and adult company to monkeys who had neither the rich environment nor the company. They found that the enriched monkeys had greater brain mass, were more alert, and were better socially adjusted compared with the deprived monkeys. Harlow and Harlow (1962) found that peer play was important in the development of young rhesus monkeys because it promoted social development and could help compensate for maternal deprivation. Poirier and Smith (1974) found that play serves to familiarize an individual with behaviors that are critical for his/her communication with conspecifics as well as his/her social group. Bateson (1955) states that it is during play that chimpanzees learn about nonverbal and metacommunication.

The above information with regard to play underscores the necessity of ensuring that chimpanzees have the opportunity to produce such species-typical behaviors. This opportunity is critical to their psychological well-being and to the development of their neurophysiology, social skills and cognitive abilities as well as their communication skills.

The extreme psychological and physiological damage of deprivation has been well documented in the literature. The damage to intellectual functions that results from being reared in a restrictive environment is persistent if not permanent. Finger and Stein (1982) have examined the types of neurophysiological damage that can occur in deprived conditions of standard housing. They state that research on the effect of enriched and deprived environments '... provide strong support for the contention that the central nervous system can be modified by the environment, and for the idea that subjects raised under different conditions differ both behaviorally and biologically' (p. 179). To support this position Finger and Stein (1982) cite research comparing 'enriched' and 'standard' environments in rats that indicate that deprivation conditions result in the following neurophysiological effects: (1) fewer dendritic branching in selected cortical and subcortical areas, (2) smaller synaptic buttons, (3) less cortical depth and mass, (4) differences in RNA–DNA ratios; and (5) a wide variety of brain enzymatic differences (acetylcholinesterase levels, etc.). They state that these differences between enriched animals and restricted animals are not just because the enriched animals had an opportunity for greater activity, but because the enriched animals had active interactions with their enriched environments.

There is evidence that restricted rearing conditions cause permanent neurophysiological damage to chimpanzees. Davenport (1979) found that although the chimpanzees were raised in a restrictive environment for only two to two-and-one-half years the deprivation caused long-term neurophysiological damage. For example, several years after the chimpanzees had experienced the deprivation conditions some of them were tested on intellectual problems. Davenport (1979) went on to state that:

> The persistence of cognitive deficits in the restricted-reared
> chimpanzees, even after 12 years of environmental enrichment,

> prolonged testing, and group maintenance, is interpreted to mean that
> deficits so acquired are not readily corrected. (p. 351)

With regard to the psychological and behavioral consequences of restrictive conditions, Davenport (1979) states that chimpanzees reared in extremely deprived environments under isolation conditions exhibited many behavioral abnormalities and problems not exhibited by chimpanzees who were reared by their mothers or by humans in a nursery. Davenport noted that female chimpanzees reared in the restricted environments did not show appropriate maternal behavior when they became adults. Out of the five animals studied, Davenport found that:

> At best, the infant was carried about in a hand or very low on the
> abdomen, and there was little indication of the mother's responsiveness
> to the distress cries and flailing of the off-spring. (p. 355)

In addition, Davenport found that the restrictive rearing conditions had a disruptive effect on later sexual behavior in the adult male chimpanzees. With regard to general social behavior, Davenport reported that the restricted chimpanzees either were entirely socially unresponsive or used socially inappropriate responses.

Any research that measures a response of the nervous system or of the immunological system must be concerned with these confounds. Therefore, any researchers who use chimpanzees who have been deprived in any fashion will not know whether the results they obtain are a function of the experimental manipulations or due to the confounding neurological or immunological dysfunction that the chimpanzee brought to the experimental situation. The efficacy of such research is therefore questionable.

The only chimpanzees who would not be subject to such confounds due to rearing and housing would be those in free-living communities in Africa. Researchers who have identified strains of SIV that are phylogenetically related to the HIV virus have suggested that screening from the free-living populations may contribute insights into the transmission of HIV to humans and the basis for the disease (Gao *et al.*, 1999). Although these suggestions may superficially circumvent housing confounds, they would decrease life-history knowledge. Such a procedure would unnecessarily disturb a free-living culture without addressing such confounds and likewise does not address issues which we will later address. We believe that intrusion into an active culture of individuals, chimpanzee or human, especially without addressing such confounds, is ethically reprehensible and scientifically invalid.

Same symptoms and different etiology

If some researchers proposed to study patriotism by examining some performing seals we would think that they were quite silly. But their reasoning might go something like this: patriotic humans are known to sing the *Star-Spangled Banner* quite regularly. Likewise, the seals at a local circus play the *Star-Spangled Banner* on their horns. These two behaviors being the same, it makes sense to study patriotism using seals who are known to play the *Star-Spangled Banner* with their horns on a repeated basis. We would all agree that the *Star-Spangled Banner* played by the seals has quite a different etiology than the *Star-Spangled Banner* sung by a human being.

A common mistake researchers make is that of assuming that similar surface characteristics denote a similar etiology. For example, if autistic human children display extreme stereotypies, then a researcher might apply for a grant to study autism in chimpanzees by raising them in isolation as Davenport (1979) did in anticipation that the chimpanzees would display extreme stereotypies. In other words, the researcher would attempt to justify the study based on the similarity of the overt behavior between the autistic child and the chimpanzee. This would be ignoring the fact that the two behaviors had entirely different etiologies.

All researchers must consider the etiology of a disease or behavior to be studied. If there are different etiologies, then the results from the study of one cannot be generalized to the other, no matter how similar it may appear on the surface. In regard to a chimpanzee model of AIDS one chimpanzee, Jerome, among the hundreds infected has developed AIDS-like symptoms 10 years after infection by a number of laboratory isolates of HIV (Novembre *et al.*, 1997). This single case has been used to validate the chimpanzee model; however, the virulence of the strain of HIV infecting Jerome was not representative of most primary infections in humans. Furthermore, Jerome's life history included previous biomedical procedures and a lifetime of captivity that was as likely to weaken his immune system as any of the HIV infections. These confounds are supported by the fact that although other chimpanzees were infected by the same virus and exhibited a decrease in their white blood cell counts not a single replication of the symptoms of AIDS was observed (Cohen, 1999).

Same stimulus, but a different response

Different species, and for that matter different individuals, may react quite differently to the same experimental manipulation. This is why Eibl-Eibesfeldt (1975) insists that a thorough knowledge of the biology is a necessary prerequisite before any research is started. When the diversity of response or the unique biologies are ignored, the efficacy of the research with regard to generalization beyond the species or individual is questionable. As LaFollette and Shanks (1996) state:

> Evolutionary theory tells us that animal models cannot be *strong* models of human disease; thus, we are theoretically unjustified in assuming that results in test animals can be extrapolated to humans. We have a theoretical expectation that there is an ontological problem of relevance. Although humans are not "essentially" different than rats, nor are we "higher" life-forms, we are differently complex. Species differences, even when small, often result in radically divergent responses to qualitatively identical stimuli. Evolved differences in biological systems between mice and men cascade into marked differences in biomedically important properties between the species (p. 118).

As previously mentioned, this particular problem is evident in the HIV research that uses chimpanzees. It is now a well-publicized fact that chimpanzees who were infected with the HIV did not develop the full suite of symptoms that characterize AIDS. To date only one chimpanzee has been reported to manifest symptoms that resemble AIDS. Yet many chimpanzees have been injected with HIV in several studies. (Alter *et al.*, 1984; Fultz *et al.*, 1986; Gajdusek *et al.*, 1985). Some of the chimpanzees did develop enlargement of their lymph nodes, but this disappeared. The conclusion is that whereas some of the chimpanzees do not keep the HIV alive at all, those that do have no observed response to the virus whatsoever.

The immunological systems of the chimpanzee and human are significantly different with regard to some of their characteristics, and how the two systems respond to the HIV. Eichberg, Lawlor and Alter (1984) reported that chimpanzees have fewer T4 lymphocyte cells than do humans. This is the primary target of the HIV. Fultz *et al.* (1986) noted that chimpanzees have a different T4:T8 ratio. In humans this ratio changes with HIV infection, but it has not been found to change in chimpanzees. In humans, HIV is carried in the plasma, but not in the chimpanzee, where it only has been found in the blood cells.

These differences are explained by an analysis of the cellular response of

HIV in humans and chimpanzees. Although the similarities between chimpanzees and humans are apparent on a macrobiological or organismal level, the 1.4% genetic difference translates to differences on a microbiological or cellular level that in turn greatly influence how the cells of the two species respond to the disease. HIV targets T4 cells in humans via two receptors, CCR5 and CD4 receptors. In order to penetrate a T4 cell both of these receptors, which can be thought of as two separate locks, must be unlocked. In contrast to this two-lock system in humans, SIV enters a chimpanzee white blood cell by binding to only the CCR5 receptor (Martin *et al.*, 1997). Owing to this difference, previous research into vaccines for HIV has concentrated on the CCR5 receptor with no success at the expense of research into the CD4 receptor. If in fact the key to a vaccine lies in denying access to the CD4 receptor then countless hours, resources, and lives have been wasted. To add insult to injury, a model for an HIV vaccine that effectively blocks entry into the white blood cells via the CD4 receptor is not even possible using a chimpanzee model, as the virus in nonhuman primate systems enters the white blood cell via only the CCR5 receptor, which is a co-receptor rather than a point of entry in human systems (Greek and Greek, 2000).

Revisiting the case study of Jerome highlights the importance of the response of the virus to cellular receptors. When HIV infects a human for the first time, it first enters the T4 cells via the CD4 and CCR5 receptors and then develops a preference for the CXCR-4 receptor. In Jerome, the virus bypassed both the CCR5 and CD4 receptors and gained access directly and exclusively through the CXCR-4 receptors (Cohen, 1999). In the one case presented to validate the chimpanzee model for HIV research, all three of the confounds already mentioned were not addressed.

Although it is well accepted that the chimpanzee is not a good model in the study of the pathogenesis of AIDS (Fauci *et al.*, 1989), some researchers continue to believe that the chimpanzee should be used to test vaccines. There is one important characteristic of the chimpanzee's reaction to the HIV that has apparently been overlooked by these researchers: the probability of obtaining a 'false positive' or a 'type I error' in such testing.

False positives, or type I errors

The Type I error is also known as a false-positive, or an 'alpha error'. A Type I error occurs when the sample result falls into the rejection region even though the null hypothesis should be accepted. Therefore, in a statistical test it is the 'alpha' probability that gives the risk taken of making a Type I error (Hays, 1963). The alpha level is set by the experimenter ahead of time. The convention for permissible size of the alpha probability comes from the particular sort of experimental setting. In certain types of experiments it is extremely important to avoid this kind of error. Hays (1963) states that '... an experimental setting where Type I error is clearly to be avoided ... [is one] ... with the goal of deciding if the medicine is safe for the normal adult population.' (p. 280). Hays goes on to state that:

> Such an error might be called "abhorrent" to the experimenter and the interests he represents. Therefore, the hypothesis "medicine unsafe" or its statistical equivalent is cast in the role of the null hypothesis, ... and the value of alpha chosen to be extremely small, so that the abhorrent Type I error is very unlikely to be committed. (p. 280)

Because the experimenter sets the alpha level he or she has complete control over Type I errors and over the chances of creating an 'abhorrent' situation for the human population.

Type I error must be considered when using chimpanzees to test vaccines. Because it was found that some of the chimpanzees do not keep the HIV alive and in others no response to it was observed whatsoever, there is an increased chance of committing a Type I error. When chimpanzees are used to test an AIDS vaccine they are injected with the vaccine and then challenged with the HIV. If the chimpanzee does not respond to or does not keep the HIV alive, then the experimenter will assume that the vaccine works when in fact it does not. In fact the chimpanzee's natural defenses may be responsible for the positive results. The ineffective vaccine would then be used on humans because of a Type I error.

The likelihood of a false positive is great when attempting to utilize the chimpanzee model for HIV vaccine research. SIV is believed to be the closest relative to HIV and has been hypothesized to be present in chimpanzees for several hundred thousands of years. SIV would then have been present in *Pan troglodytes troglodytes*, the hypothesized reservoir for the transmission of HIV to humans, prior to their divergence from *Pan troglodytes schweinfurthii* (Gao *et al.*, 1999). This makes the reliability of testing any vaccine at odds with the immune system of chimpanzees, which has had many generations of co-evolution with the virus.

Whereas the genetic similarities between chimpanzees and humans are acknowledged, the similarities between the viruses that infect the two species are not. HIV-1, the most common strain infecting humans, is only

40% homologous with SIV (Greek and Greek, 2000). This large variation between these viruses is likely due to the extremely high mutation rate of HIV, which plays an important part in its ability to escape destruction by the immune system (Nowak, 1990). This is complicated by the methodology of the research into HIV vaccines in chimpanzees. Although the virus that infects humans is cell-associated, the challenge virus used with chimpanzee models is cell-free to avoid the differences in the immunological responses of the two species (Letvin, 1998). Furthermore, the inoculum utilized with chimpanzees is considerably more than is responsible for transmission within humans, yet the one successful study of an HIV vaccine used an admittedly 'wimpy' virus as a challenge to the vaccine (Cohen, 1999).

One false positive has already had a significant effect on the history of treating HIV in human populations. Based on the slow reproduction of the virus in chimpanzees as well as differing ratios of helper T-cells to killer T-cells in chimpanzees, the prevailing wisdom of the time stated that there is a long latency period of HIV infection before AIDS develops. This led to a policy which, unlike that of today, did not include aggressive treatment from the onset. It was later discovered via clinical observation that in humans there is indeed a short latency and rapid reproduction of the virus. The fact that in the chimpanzee immune system killer T-cells do not rely as heavily on the helper T-cells that are affected by the virus was overlooked and the cost was the deaths of many humans. Such false positive results cast serious doubts on whether any generalizations should have been made at all (Greek and Greek, 2000).

False generalizations

One of the first things any psychology student learns is that the results of a research study can be generalized *only* to the population that was sampled. In other words, if one conducted a study on college freshmen at Central Washington University, you can only generalize those results to freshmen at CWU, and not even to the freshman one hundred miles away at the University of Washington in Seattle. Why? Because freshmen that choose a small rural campus to attend college may be qualitatively different from students that choose to attend a large university in a large urban area. Likewise, in psychology we have been careful to generalize our results only to genetically similar animals as well. Note that the limitation restricting the generalization to genetically similar animals is done out of a concern for strain differences, not to mention species differences, because there are striking differences between different strains of rats, dogs, and other animals. So experiential background and genetic similarity are two limitations that must be considered before any experimental results can be generalized beyond the immediate experimental situation.

Comparative studies, however, are quite different. In comparative studies we note both the similarities as well as the differences. With regard to the similarities within the field of psychology, we should not be concerned whether or not a behavior is homologous between species, because analogies work well for us. That is, typically we are concerned only with the effect of the laws of learning on a behavior and not its origin. For example, the airfoil principle in flight works equally as well for a bird and a bee, even though their wings are not homologous. However, in biologically based medical research the concern for homologies is very important. So whereas the psychologist can make general statements about the laws of learning, this may not be true for biological systems with regard to their specific reaction to an external agent.

When generalizations are made between vaccines using the chimpanzee model, it is important to realize that this is a generalization between a species whose immune system has co-evolved with SIV, chimpanzees, and a species, humans, who must be vaccinated against HIV. This is a generalization that includes more than two closely related species, but rather four separate species. This is especially precarious when one considers the high mutation rate of HIV: after ten years of infection the viral particles present in the host have undergone thousands of generations of change from the origin virus. This would be analogous to the genetic change in the human lineage after millions of years (Nowak and McMichael, 1995). This is a combination that is deadly to both humans and chimpanzees alike, especially when one adds the danger of vaccines that have been missed owing to a reliance on the chimpanzee model. This dangerous lesson should have already been learned from the history of the search for a vaccine for the polio virus. In fact, Albert Sabin (1984), the inventor of the polio vaccine, testified before the United States Congress that 'the work on prevention [of the polio vaccine] was long delayed by the erroneous conception of the nature of the human disease based on misleading experimental models of the disease in monkeys'.

Can the results from chimpanzee biomedical research be generalized to human populations? With regard to AIDS it is clearly not the case, as Fauci et al. (1989) admit. The species differences are too great with regard to the chimpanzee's immunological response. Then what population can the results be generalized to? It turns out that it cannot even be applied to chimpanzees in general. In fact the only population to which the results might apply is the one made up of chimpanzees who have been raised in the same captive situations as were the experimental subjects. The restrictive early rearing conditions and the housing under which captive chimpanzees were maintained restricts our ability to generalize the results beyond this very small population.

Conclusions

All research should meet three criteria for efficacy: first, the research must be empirically sound with proper design and free from all confounds; second, the research must be rational; and third, the research must meet society's ethical standards. This chapter has focused on the empirical criteria with regard to chimpanzee biomedical research. It has been noted that the rearing and housing conditions can have a confounding effect on any results that assume a normal nervous system or immunological system. It was also noted that just because the surface characteristics are the same, they may arise from very different etiologies and invalidate any research that assumes otherwise. Another experimental error is to assume that identical stimuli will produce identical responses in different organisms, the primary example being the HIV research using chimpanzees. Two other problems were more general ones that all experiments must deal with. One is the vulnerability to Type I errors in HIV studies with chimpanzees, and the other is the questionable efficacy of generalizing the results of any chimpanzee biomedical research beyond the captive cage-reared chimpanzee population in the United States.

The second criterion mentioned is that the research be rational. Given the confounds pointed out in this paper, the rationality of all biomedical research using chimpanzees must be carefully examined. But another parallel question is one with regard to the rationality of using an endangered species to help an overpopulated species become more overpopulated. If one was to expand this argument, given the evidence that chimpanzee communities possess unique cultures, it is even less rational to expose increasingly isolated communities to the risk of further decimation using a flawed experimental model as a rationale. The expansion of HIV research into wild populations of chimpanzees has been suggested, but only addresses confounds of housing while falling prey to each of the other confounds presented in this chapter. The risk of transmission of chimpanzee pathogens to human researchers and human pathogens to the chimpanzee communities is inherent in the increase of intimacy. The number of researchers necessary for such research would seriously threaten populations that are already in danger of disappearing. From a very objective and disinterested point of view, this alternative does not make sense.

It could be argued that discontinuing an avenue of research without a viable alternative would not be a rational choice either; however, the claim that the chimpanzee model is the only alternative for HIV research is a myth. The important advances in the knowledge of the virus as well as the most successful current treatments are the products of alternative models such as clinical studies, electron microscopy, and experiments *in vitro* (Greek and Greek, 2000). Another alternative that should be explored is the advancement of a chimpanzee genome project to serve as a counterpart to the human genome research. The resources necessary, such as blood and tissue samples, to begin work on mapping the chimpanzee genome are likely to be already present in laboratories. It is apparent after a review of the confounds inherent in a chimpanzee model of biomedical research that the rational alternative is to cease wasting resources on the needless torture and endangerment of chimpanzees and channel these resources into alternative methods and conservation efforts.

The third criterion has to do with the ethics involved in using chimpanzees in dangerous biomedical research just because they are so similar to our species. This similarity is a two-edged sword. Just as their biological similarity tempts us to use them, so too their extreme behavioral similarity brings into question the ethics of using a person such as a chimpanzee as a subject in invasive research.

References

Alter, H.J., Eichberg, J., Masur, H., Saxinger, W., Gallo, R., Macher, A., Lane, H. and Fauci, A. (1984). Transmission of HTLV-III infection from human plasma to chimpanzees: An animal model for AIDS. *Science* **226**, 549–52.

Bateson, G. (1955). A theory of play and fantasy. *Psychiatric Resident Representative* **2**, 39–51.

Caplan, F. and Caplan, T. (1973). *The Power of Play*. New York: Anchor Press/Double Day.

Cohen, J. (1999). Chimps and lethal strain a bad mix. *Science* **286**, 1454–5.

Davenport, R. (1979). Some behavioral disturbances of great apes in captivity. In *The Great Apes*, ed. D. Hamburg and E.R. McCown, pp. 341–56. Menlo Park, California: Benjamin/Cummings Publishing Co.

Eibl-Eibesfeldt, I. (1975). *Ethology: the Biology of Behavior*. New York: Holt, Rinehart & Winston.

Eichberg, J., Lawlor, D. and Alter, H. (1984). Cellular immune profile of the chimpanzee: Its potential usefulness in Acquired Immunodeficiency Syndrome (AIDS). *Federal Protocol* **43**, 1914.

Fauci, A., Gallo, R., Koenig, S., Salk, J. and Purcell, R. (1989). Development and evaluation of a vaccine for human immunodeficiency virus (HIV) infection. *Annals of Internal Medicine* **110**, 373–84.

Finger, S. and Stein, D.G. (1982). *Brain Damage and Recovery*. New York: Academic Press.

Fouts, R.S. and Fouts, D.H. (1989). Loulis in conversation with the cross-fostered chimpanzees. In *Teaching Sign Language to Chimpanzees*, ed. R.A. Gardner, B.T. Gardner and T. Van Cantfort, pp. 293–307. State University of New York Press.

Fouts, R.S., Fouts, D.H. and Van Cantfort, T.E. (1989). The infant Loulis learns signs from cross-fostered chimpanzees. In *Teaching Sign Language to Chimpanzees*, ed. R.A. Gardner, B.T. Gardner and T. Van Cantfort, pp. 280–92. Albany: State University of New York Press.

Francis, D., Feorino, P., Broderson, J., McClure, H., Getchell, J., McGrath, C., Swenson, B., McDougal, J., Palmer, E., Harrison, A., Barre-Sinoussi, F., Chermann, J.-C., Montagnier, L., Curran, J., Cabradilla, C. and Kalyanaraman, V. (1984). Infection in chimpanzees with lymphadenopathy-associated virus. *Lancet* **ii**, 1276–77.

Fultz, P., McClure, H., Swenson, R., McGrath, C., Brodie, A., Getchell, J., Jensen, F., Anderson, D., Broderson, J. and Francis, D. (1986). Persistent infection of chimpanzees with HTLV-III/LAV: A potential model for AIDS. *Journal of Virology* **58**, 116–24.

Gajdusek, D., Amyx, H., Gibbs, C., Asher, D., Gallo, R., Pren, S., Arthur, L., Montagnier, L. and Midvan, D. (1985). Infection of chimpanzees by human T-lymphotropic retroviruses in brain and other tissues from AIDS patients. *Lancet* **i**, 55.

Gao, F., Bailes, E., Robertson, D.L., Chen, Y., Rodenburg, C.M., Michale, S.F., Cummins, L.B., Arthur, L.O., Peeters, M., Shaw, G.M., Sharp, P.M. and Hahn, B.H. (1999). Origin of HIV-1 in the chimpanzee *Pan troglodytes troglodytes*. *Science* **397**, 436–41.

Gardner, R.A. and Gardner, B.T. (1989). A cross-fostering laboratory. In *Teaching Sign Language to Chimpanzees*, ed. R.A. Gardner, B.T. Gardner and T. Van Cantfort, pp. 1–28. Albany: State University of New York Press.

Gibbons, A. (1990). Our chimp cousins get that much closer. *Science* **250**, 376.

Goodall, J. (1986). *The Chimpanzees of Gombe: Patterns of Behavior*. Cambridge, Massachusetts: The Belknap Press of Harvard University Press.

Greek, C.R., and Greek, J.S. (2000). *Sacred Cows and Golden Geese*. New York: Continuum.

Harlow, H.F. and Harlow, M.K. (1962). Deprivation in monkeys. *Scientific American* **207**, 136–46.

Hays, W.L. (1963). *Statistics for Psychologists* New York: Holt, Rinehart & Winston.

Kazdin, A.E. (1992). *Research Design in Clinical Psychology*. Boston: Allyn and Bacon.

King, F.A. (1986). Keynote address: Animal research is here to stay. Presented to the National Capitol Area Branch of the American Association for Laboratory Animal Science (AALAS). Washington, D.C., September 10.

King, M.E., and Wilson, A.C. (1975). Evolution at two levels in humans and chimpanzees. *Science* **188**, 107–16.

LaFollette, H., and Shanks, N. (1996). *Brute Science: Dilemmas of Animal Experimentation*. New York: Rutledge.

Letvin, N.L. (1998). Progress in the development of an HIV-1 vaccine. *Science* **280**, 1875–80.

Lewin, R. (1984). DNA reveals surprises in human family tree. *Science* **226**, 1179–83.

Lieberman, J.N. (1965). Playfulness and divergent thinking: An investigation of their relationship at the kindergarten level. *The Journal of Genetic Psychology* **107**, 219–24.

Martin, K.A., Wyatt, R., Farzan, M., Choe, H., Marcon, L., Desjardins, E., Robinson, J., Sodroski, J., Gerard, C. and Gerard, N.P. (1997). CD4-independent binding of SIV gp 120 to rhesus CCR5. *Science* **278**, 1470–3.

Mason, W. (1965). Determinants of social behavior in young chimpanzees. In *Behavior of Nonhuman Primates*, Vol. 2, ed. A.M. Schrier, H.F. Harlow and F. Stollnitz, pp. 335–64. New York: Academic Press.

McGrew, W.C. (1992). *Chimpanzee Material Culture: Implications for Human Evolution*. Cambridge University Press.

Novembre, F.J., Saucier, M., Anderson, D.C., Klumpp, S.A., O'Neil, S.P., Brown, C.R., Hart, C.E., Guenthner, P.C., Swenson, R.B. and McClure, H.M. (1997). Development of AIDS in a chimpanzee infected with human immunodeficiency virus type 1. *Journal of Virology* **71**, 4086–91.

Nowak, M. (1990). HIV mutation rate. *Nature* **347**, 522.

Nowak, M.A. and McMichael, A.J. (1995). How HIV defeats the immune system. *Scientific American* (August), 58–65.

Pereira, M.E. and Altman, S.A. (1983). Statement for the public meeting held by the Institute of Laboratory Animal Resources Committee on the Care and Use of Laboratory Animals at the Holiday Inn, Rosemont, Illinois, July 12.

Poirier, F. and Smith, D. (1974). Socializing functions of primate play. *American Zoologist* **14**, 275–87.

Rosen, C. (1974). The effects of sociodramatic play on problem-solving behavior among culturally disadvantaged preschool children. *Child Development* **45**, 920–7.

Sabin, A. (1984). Statement before the subcommitte on Hospitals and Health Care, Committee on Veteran's Affairs, House of Representatives: Serial no. 98–48, April 24.

Whiten, A., Goodall, J., McGrew, W.C., Nishida, T., Reynolds, V., Sugiyama, Y., Tutin, C.E.G., Wrangham, R.W. and Boesch, C. (1999). Cultures in chimpanzees. *Nature* **399**, 681–4.

Part 2
Cultural views of nonhuman primates

In many areas of the world human and nonhuman primates are intertwined in important cultural and ecological contexts. While human and nonhuman organisms frequently share common environments, or space, some nonhuman organisms play an additional significant role in human 'place'. This place is the reality created by the human cultural incorporation, utilization and modification of the environment or habitat in which they live. In many parts of the circumequatorial world, especially Amazonia, nonhuman primates are central to human nutritional and social realities. In this section the authors seek to provide the reader access to the intricate, cultural interconnections between humans and a few species of monkey and ape.

Loretta Cormier's discussion of the Guajá of Eastern Amazonia (Chapter 4) provides a glimpse of a dramatic and complex series of social and mythological relationships involving humans and monkeys. Simultaneously playing central nutritional and cultural roles, the monkeys form an integral component of Guajá life. This example represents an extreme level of cross-primate connectivity and at the same time illustrates the amazing diversity and malleability of human cultural kinship systems.

In Chapter 5 Manuel Lizarralde contextualizes the discussion under the rubric of ethnoecology and presents a culturally based account of the role and importance of monkeys to the Bari of Venezuela. This chapter examines the multiple impacts of changing ecologies, economies, and political landscapes via the trends and patterns of monkey hunting and monkey taxonomy amongst the Bari. Unfortunately, as in many cases, external and internal changes threaten all of the primates discussed in this chapter and these changes can be directly tied to global economic and political issues.

Discussing aspects of hunting, ethnopharmacology, national parks, and sociopolitical realities, Glenn Shepard illustrates the complex nature of Matsigenka existence and the role that nonhuman primates play within it (Chapter 6). The Manu Biosphere Reserve of Peru and international conservation declarations have an impact on Matsigenka hunting and cultural ecology. Culture contact, changing resource use practices, and demography are affecting the way in which the Matsigenka subsist and this negatively impacts the monkeys of the region. This chapter emphasizes that although the monkeys and humans may be at odds (i.e. in a predator–prey relationship) their lifeways are truly intertwined.

Taking a different approach, Frances Burton (Chapter 7) reviews the centrality of the Monkey King mythology in China and argues that this cultural icon can be used to implement a more effective conservation policy. By examining the core concepts and characters in the Monkey King myth we see that many of the values and issues are well integrated into Chinese culture and may translate well into strategies for conservation of China's diverse and highly endangered primate population.

The final chapter in this section (Chapter 8) moves from Amazonia and China to the Central African nation of Rwanda. Pascale Sicotte and Prosper Uwengeli remind us that perceptions of 'nature' can vary dramatically across cultures and that these perceptions are critical to any conservation or management attempts. The peoples of Rwanda share some of their space (and place) with the conservation world's 'poster-primates', the mountain gorillas. Although these gorillas are amongst the most endangered primates in the world, the authors remind us that conservation action must be undertaken in the context of local perceptions of 'nature' and include the possibility (probability) that concepts may be rapidly changing and/or are changeable.

All of these chapters provide us with examples of a cross-cultural view of monkeys and apes, and also very real connections between the non-human and human primates.

4 Monkey as food, monkey as child: Guajá symbolic cannibalism

LORETTA ANN CORMIER

Human ecology and primate ecology share a concern with the relationship of primates to the environment, yet they have been segregated into treatment of human primates in cultural anthropology and nonhuman primates in biological anthropology. Although observational techniques from primate ethology have been adapted for use in human ethnography, and primate studies have treated interspecific primate relations, neither has attempted to integrate such research concerns. Sponsel (1997) was the first to systematically address the problem in his call for the development of a field of inquiry called 'ethnoprimatology', which would explore the interface of human ecology and primate ecology through comparative ecology, predation ecology, symbiotic ecology, cultural ecology, ethnoecology and conservation ecology.

The Guajá Indians provided an ideal opportunity to apply ethnoprimatology, as local primates are central to the Guajá way of life in material, social, and ideological aspects of their culture. The Guajá are a Tupi–Guarani-speaking group located in the high *terra firme* forest of western Maranhão, Brazil, on the eastern border of Amazonia proper. Although traditionally living in small foraging bands of five to fifteen people, most have been incorporated to varying degrees into one of three villages established after 1973 by the FUNAI (the Brazilian Indian agency) and are learning to cultivate domesticated plants (Balée, 1988; Gomes, 1996).

Few, if any, upland tropical forest peoples can be designated as unqualified hunter–gatherers, and pigeon-holing the Guajá economy is also difficult (see Bailey and Headland, 1991; Bailey et al., 1989; Headland, 1987; Headland and Bailey, 1991). Prior to their recent transition to agriculture, the Guajá have been most often referred to as hunter–gatherers. However, they have been known in the past to have acquired domesticated plants through crop-raiding neighboring tribes (Beghin, 1957; Nimuendajú, 1948). Further, Balée has provided convincing comparative ethnobotanical linguistic evidence that the Guajá practiced agriculture at an earlier point in their culture history (Balée, 1992, 1994a, 1999). At this time, the Guajá are learning (or perhaps relearning) swidden horticulture, but they remain difficult to classify because differences exist in the

63

extent to which some families are practicing agriculture. Some Guajá remain uncontacted or infrequently contacted and plant no domesticates at all.

With those caveats, the Guajá will be generalized as a group in transition to agriculture with hunting and gathering of nondomesticates (including fish) continuing to play an important role in their economy. Among those nondomesticates are seven species of primates that are used as food and kept as pets: the red-handed howler monkey (*Alouatta belzebul*), the black-bearded saki (*Chiropotes satanas*), the brown capuchin (*Cebus apella*), the recently identified Ka'apor capuchin[1] (*Cebus kaapori*), the owl monkey (*Aotus infulatus*), the squirrel monkey (*Saimiri sciureus*), and the golden-handed tamarin (*Saguinus midas*).

In this chapter, monkeys will be shown to be a key seasonal animal resource with their exploitation determining Guajá trekking behavior. Monkeys are also kept as pets and incorporated to some extent into the kinship system as children of women and serve to enhance the image of female fertility. The seeming contradiction found in monkeys as food and monkeys as kin is rendered logical in the endocannibalistic themes that pervade Guajá religious life and cosmology. The material, social, and symbolic roles of primates will be addressed in order to demonstrate that interacting with monkeys in multiple domains is a key to Guajá cultural identity. The research reported here was conducted between 1996 and 1997 on the Caru Reserve in eastern Amazonia, in the state of Maranhão, Brazil.

Material importance of monkeys

Generally, monkeys are a widely available source of food throughout lowland South America, but this fact alone is not always sufficient to predict the degree to which monkeys will be utilized as a food source by a given culture. Where quantitative measures have been recorded, the game mass of monkeys in the diet has ranged from 0.7% among the Machiguenga (Kaplan and Kopischke, 1992) to 36% among the Waorani (Yost and Kelley, 1983). Such differences cannot be accounted for entirely by environmental availability. The clearest example of this can be found with the Ka'apor Indians, who both share one of the Guajá reserves and are linguistically related to the Guajá. Although the Guajá and Ka'apor exploit essentially the same habitat, the Ka'apor do not use monkeys for food nearly to the extent of the Guajá, nor do monkeys have a prominent role in Ka'apor kinship and cosmology (Balée, 1984). In addition, various food taboos exist among lowland South American groups regarding the consumption of monkeys. For example, monkeys are taboo for pregnant Yanomamö women and their husbands, adolescent Jívaro girls for twelve months after their first menses, preadolescent Desana boys, and all Kayapó women (McDonald, 1977). Thus, the material importance of monkeys to a given lowland South American culture cannot be presumed to be simply a factor of the environmental availability of monkeys, but must be understood as an interplay of ecology, culture, and historical factors affecting both the environment and its inhabitants. The objective in this chapter was not to prove that the Guajá *could not* rely on other animal resources to meet their needs, but rather to demonstrate that among the Guajá, interacting with monkeys on several levels is key to what it means to be a member of their culture.

The role of monkeys in the Guajá diet

The original research plan was to evaluate the relative importance of monkeys to the diet through weighing of fish and game returned from a hunt[2]; however, random spot checks were found to be a more accurate measure. Random spot checks are a modification by cultural anthropologists of the method of random scan sampling used in primatology (see Altmann, 1974; Johnson, 1975). Among the Guajá, I made a daily round of the individuals in the group at a randomly selected time so that each individual would have one random spot check recorded each day. The original intent of the random spot checks was to determine the percentage of time spent in various activities. However, when the activity of eating was recorded, far more monkeys were observed to be eaten than were being brought in to be weighed. One reason for this was that when the Guajá returned from extended treks, they would often butcher, roast, and divide the monkeys while still in the forest, so there was not a whole animal carcass to weigh. In addition, because monkeys are a favored food among the Guajá, there may have been reluctance to advertise a big monkey catch owing to the high cultural premium on food sharing. Among the Guajá, one is never to refuse a request of food from another. Thus, the practice of roasting and dividing monkeys in the forest may not have been merely for practical purposes of reducing their weight, but also may have been a way to conceal this food source from others who could claim it if they wished.

Random spot checks were conducted from February through August of 1997 with 111 sampling days of a community of 110 people. Both the activity of 'eating' and the type of food being eaten were identified and

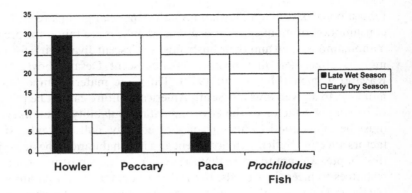

Figure 4.1. Late wet and early dry season random spot check frequencies for major species.

recorded when members of the group were in the village. A total of 248 episodes of animal eating were recorded. Species were identified through various zoological references of local animal life (e.g. dos Santos *et al.*, 1995; Emmons, 1990, 1997; Gery, 1977; Goulding, 1980; Mascarenhas, Lima and Overal, 1992; Queiroz, 1992). Significant differences were found between the types of game eaten during the late wet season (February through mid-May) and the early dry season (mid-May through August).

Two clear seasonal patterns emerged in animal consumption. Howler monkeys predominate as the species most utilized during the late wet season, and their exploitation is associated with increased trekking behavior (see Fig. 4.1). During the early dry season, increased fishing behavior is associated with increased sedentism and heavy exploitation of a detritus-feeding migratory characoid fish of the genus *Prochilodus*. This fish is the most heavily utilized individual species during the early dry season.

The patterns of fish and game use have both ecological and cultural correlates. The hunting of howler monkeys is associated with trekking behavior, whereas increased fishing is associated with increased sedentism. For the purposes of this study, trekking behavior is defined as hunting–fishing–collecting expeditions that involve sleeping away from the main village in the forest. Even the more settled families in the village still spend a substantial portion of their time trekking in the forest. Random spot checks from one such family indicated that its members spent almost five times more time trekking in the late wet season than in the early dry season. The family was composed of two adult males, two adult females, two adolescent females, and four children who reside in the same hut in the village, work cooperatively, and are related through the adult and adolescent females who are wives to the two adult males. A total of 1110 random spot checks were made of this family, with 610 occurring during the late wet season and 500 occurring during the early dry season. A significant difference existed in trekking between the two seasons ($\chi^2_{df=1} = 83.79$, $p < 0.001$). During the early dry season, members of this family camped away from the village only 5.60% of the time ($n = 28$ trekking, $n = 472$ in the village). During the late wet season, they spent 26.89% of the time camping in the forest ($n = 164$ trekking, $n = 446$ in the village). Further, all of the late wet season camping trips involved taking children, pets, and all possessions with them, whereas the early dry season trips involved either males alone or a man with his adolescent wife, leaving the children, possessions, and elder wives behind in the village.

The evidence that howler monkeys are determining trekking behavior comes not only from the quantity of howler monkeys that the Guajá are bringing back to the village for consumption, but also from the idiom of the howler monkey used when talking about hunting. Typically, when leaving the village to trek, the Guajá describe the activity as *wari-ika*, 'howler monkey hunting'. Further, their three terms for forest movements were described in terms of trekking for howler monkeys: *yačimə*, *watarəkwə*, and *iwə*. *Yačimə* refers to moving in any direction that leads away from the village and into the forest towards the howler monkeys, *watarəkwə* was described as tracking howler monkeys through the forest, and *iwə* refers to returning from the forest back to the village. Although the Guajá affirmed the directional terms would also be appropriate for hunting of other game, forest orientation was identified specifically in terms of the howler monkey. Another example of the howler monkey idiom comes from the only divorce during the study period where a woman left her husband, for another man. The story of their break-up was described as her leaving with only the following statement to her husband, which translates as, 'I'm leaving for the forest to eat howler monkeys'.

During the early dry season, fishing dramatically increases in importance. The characoid fish, *piračũ* (*Prochilodus* sp., *curimatá*[3] in Portuguese), is heavily exploited during its 'piracema', the migration from the receding waters in the tributaries to the Pindaré river at the start of the dry season (see Goulding, 1980, for a description of *Prochilodus* seasonal migration patterns). Diminishing rainfall also facilitates Guajá bow and arrow fishing since water turbulence decreases, and the *piračũ* are more readily visible. Later, as the waters recede further, the Guajá exploit a variety of fish trapped in pools, using fish poisons. Although the study period did not extend into the late dry season, fishing should certainly diminish again in importance as the waters recede further, with hunting perhaps again becoming more prevalent.

The early dry season is also an important ritual time for the Guajá, when the *karawara* ceremony is practiced. In brief, several times a week at night, the village gathers to sing in a ceremony involving contact with the divinities and dead ancestors who reside in the sky home, the *iwa*. While the women and children sing, the men dance as their spirits are transported to the *iwa* and their bodies are spirit-possessed by the divinities on the ground. The divinities have the power to heal the village and to bring game closer to the Guajá.

More importantly here, the divinities are consulted regarding marriage arrangements. The local, seasonal availability of fish during the early dry season allows the large group to congregate for a sufficient amount of time so that such arrangements can be made. Here, the idiom of fish becomes more important. Guajá girls marry at around six or seven years of age and men perform a type of brideservice, which involves bringing food to the family of the girl. Culturally this is not seen as a form of payment or exchange, but is explained as bringing fish to the girl so she will grow and not be afraid to live with her new husband. Although the cultural obligation exists to marry once an agreement has been made between a prospective husband and a girl's father, girls are not expected to leave the natal home until they are no longer afraid to do so. Stories of marriages tell of this period of waiting while the young girl eats fish. No special properties of fish were ascertained that would specifically help in this process, and informants agreed that the girl would be eating many other things as well, but 'fish-eating' is the way that this marital transition is described.

The monkey and Guajá ethnobotany

An additional measure of the importance of primates to Guajá material existence involves their role in Guajá botanical knowledge. During the research period, a plant collection was made and approximately 75% of the plant specimens identified through the Museu Goeldi and the Tulane University Herbarium. Of the plants identified, 275 are different, noncultivated, useful species with approximately one-fourth (23.27%) having multiple uses. The most prevalent Guajá knowledge of the botanical environment involved plants that were edible for game. Whereas 84.0% of the plants were eaten by animals, only 14.91% were eaten by human beings[4]. Clearly, knowledge of the plants eaten by animals is an important hunting strategy. Further, knowledge of what monkeys consume is extensive, with over half of the plants identified as being eaten by monkeys (51.94%). In terms of plants known by the Guajá to be eaten by mammals,

Other Mammals

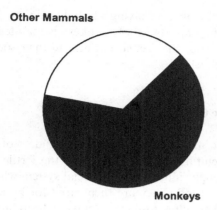

Monkeys

Figure 4.2. Knowledge of plants: monkeys versus nonprimate mammals.

65.21% are eaten by monkeys and 34.78% are eaten by other mammals ($n = 184$ total plants eaten by mammals) (see Fig. 4.2).

The botanical data provide much insight into how the Guajá perceive and utilize their environment. However, the botanical data are not meant to suggest that Guajá knowledge of the diversity of plant life eaten by prey corresponds directly to Guajá intensity of utilization of animals. One obvious reason is that not all animal species consumed by the Guajá are herbivores. The intensity of utilization of carnivorous species such as the caiman and the jaguar cannot be determined from the botanical data. In addition, diet breadth is variable among species. Some may consume a wide variety of plant species whereas others may rely on fewer reliable staples. This is likely true of the Guajá themselves who have historically relied heavily on the babassu palm as a staple (Balée, 1988). What the evidence suggests here is that, at least in terms of plant *diversity*, hunting and knowledge of plants eaten by animals is more important than gathering and knowledge of plants eaten by the Guajá. Moreover, the evidence clearly supports monkeys being important prey, as the Guajá devote much of their ethnobotanical range of knowledge to the diet of monkeys.

In sum, the seasonal exploitation of resources involves environmental availability and cultural behaviors. The neighboring Ka'apor Indians use different strategies in that they are agriculturists who rarely go on hunting expeditions that involve distances requiring them to sleep in the forest (Balée, 1985). The Guajá trekking–fishing pattern, and particularly their utilization of monkeys, cannot therefore be considered to be strictly dictated by the environment. Considering both the dietary and botanical evidence, the Guajá can be described as having a heavily monkey-oriented perception of the forest. The importance of monkeys is manifested not only in the material existence of the Guajá, but also in their social and religious life, to be discussed below.

The monkey and the Guajá social system

Young monkeys kept as pets are treated as dependent children of women and serve to enhance the culturally valued image of the fertile female. Orphaned monkeys are brought into the Guajá social system when their adult mothers are killed for food. They are then cared for by a female Guajá. The orphaned monkeys cling to their caretaker's head, are incorporated into her kinship system, and, unlike other pets, are often given personal names and a kinship term of address. All seven species of monkey are kept as pets and there were no households during the study period that did not include monkeys. Several households, in fact, contained more monkey than human inhabitants (for example, one household with eight monkeys and six people and another with twelve monkeys and ten people). A total of 90 monkeys were kept between February 1996 and August 1997 in a community of 110 Guajá.

Like infant Guajá children, infant monkeys stay in constant physical contact with the 'mother' and are cared for like an infant, including being played with, sung to, bathed, breast-fed, and generally nurtured. Pet monkeys are never eaten. Although a variety of other animals are kept as pets by the Guajá, none are nurtured as children to the degree of the monkeys. Dogs (including puppies) provide a striking contrast in that they are severely mistreated, with most suffering from malnutrition and frequent physical abuse.

The role of the monkey in enhancing female fertility involves monkeys serving as substitute children and relates to Guajá beliefs about paternity. Like many other lowland South American groups (e.g. the Suyá, Seeger 1981; the Mehinacu, Gregor, 1974; the Pakaa Nova, Vilaça, 1992; and the Tapirapé, Wagley, 1983), the Guajá believe that the fetus develops from the accumulation of semen. This has a number of consequences for the Guajá. First, Guajá men feel a tremendous obligation to provide semen to pregnant women so that the infant will grow and properly develop. A pregnant woman, with demonstrated fertility, is then an important sexual object for the men. Men describe the most attractive women to be those that are lactating and describe all the women in the sky-home as both pregnant and lactating. Second, owing to the belief in the requirement of semen for fetal growth, Guajá women seek out frequent sexual activity during their pregnancies and typically have more than one sexual partner, with all male sexual partners believed to be true, biological fathers. This results in an asymmetry in male and female perceived fertility with the average man having far more children than the average woman in a culture where children are highly valued. Here, a clear difference exists between the Guajá and many polygynous systems where some men have large numbers of children while others are excluded, because all of the Guajá men are believed to father children, and typically, many children. Women using monkeys as surrogate children mitigates the cultural perception of imbalance between male and female fertility to some extent. Further support for monkeys as substitute children comes from the practice of giving women who have miscarried an infant monkey to nurse, and that women beyond the child-bearing years tend to have greater numbers of monkeys than women with dependent children.

On one level, the monkey functions as a type of body art such as Burmese neck rings or Euro-American high-heeled shoes, creating a culturally desirable image of female attractiveness, even at the cost of physical impediment. But this can also be considered a type of performance art because it involves the interactive social behavior of Guajá women and the monkeys. The physical and behavioral similarity of another social primate is required for the interaction. When asked why other pets are not treated in the same way as monkeys, the Guajá explain this quite literally, saying, for example, that coatis fall off your head and fish cannot live out of water. However, they also believe it to involve a special, natural relationship between women and monkeys in that women are said to 'know' the monkeys in a way that the men do not and that the monkeys 'want' the women but they 'fear' the men. In this way, the social roles of female childcare provision and male hunting are underlined (both men and women are gatherers).

It should also be mentioned that the Guajá incur considerable cost in the practice of keeping monkeys in terms of their food burden on the group, the sheer weight of carrying them on the trek, and maturational changes, to be discussed later, that result in monkey destructiveness and physical aggression. The monkeys, and the other numerous pets of the Guajá, create a burden on the food supply of the Guajá, both directly through the need to procure a greater quantity of food and indirectly through the need of women who nurse monkeys to increase their caloric intake. Although one monkey may not create a tremendous burden, typically households have multiple monkeys that are being nursed and provided with food and water.

The weight of carrying the monkeys and other pets creates a consider-able impediment to a trekking lifestyle. One small monkey may have a negligible effect, but a single woman may try to carry three to four monkeys on her head as she leaves on a forest trek. The monkeys are difficult to manage not only due to their raw weight, but also because they are moving, living beings that are interacting with each other as well as attempting to interact with the woman as she is trying to move quickly through the forest, perhaps with an infant of her own in her carrying strap. In essence the practice can create virtually a small community living on the woman as she attempts to go about her tasks. Further, an early report of the Guajá prior to FUNAI contact described them as keeping large numbers of monkeys, particularly howler monkeys, as pets (Beghin, 1957) indicating that the Guajá were trekking with their monkeys prior to their recent transition to a more settled lifestyle.

Monkeys also play an important role in the socialization of children in learning childcare and hunting skills. Young girls typically have had experience in taking care of a number of infant monkeys long before they bear their first child. While girls, and boys to a lesser extent, learn childcare skills in caring for younger siblings, the monkey provides several advan-tages in the learning process. First, mechanically it is much easier for a child to care for a monkey baby than a human baby, not only because the monkey weighs less, but because the monkeys cling to the head of the girls instead of having to be carried. Related to this is the similar level of dependency among primates in comparison with other species. Whereas a girl can leave her baby turtle or agouti alone, the clinging of infant monkeys simulates the level of responsibility she will have in caring for a future infant. Further, girls can have direct experience in caring for mon-keys at a much earlier age than they can with human infants. Girls as young as two may begin carrying a baby monkey on their heads and girls as young as five may be the primary 'mother' of an infant monkey, meaning that the monkey clings to her throughout the day, leaving only to be breast-fed by the girl's mother or another lactating female.

For young boys, the monkey is most important in learning hunting skills. Keeping monkeys as pets in the household familiarizes the young boys with the calls of monkeys and aids them in recognizing and discrimi-nating these sounds in the forest. One young boy learned to imitate a howler monkey call before he spoke his first word. While female toddlers may begin to carry monkeys on their heads, male toddlers get toy bows and arrows as soon as they are able to walk. A boy's father will make him bows and arrows that gradually increase in size and quality as he grows older until he is proficient in the adult bow and arrow. On several occasions, young boys were observed taking practice shots at older pet monkeys that were past the clinging stage. Parents never reprimanded children for the behavior and on one occasion, parents were observed to actively encourage a young boy, and they even sang a song about monkey hunting while he practiced. This reflects both the changing role of the older monkey (to be discussed below) as well as the extreme importance of learning to hunt in that young boys' efforts are encouraged, even at the expense of pets. One final comment is that both young girls and boys learn to recognize monkey calls, and although women generally do not hunt with bow and arrow[5], their ability to recognize animal sounds on trekking expeditions would be an asset to their male companions. Young boys learn childcare skills through care of and play with baby monkeys, although they generally have less involvement with monkeys than do the girls.

Infant monkeys can perform and substitute for children through their natural behaviors until maturational changes occur in juvenile monkeys, causing them to become more independent and thus no longer able to fill the role of dependent child. Negative consequences result for both the Guajá and the monkeys as aggressive behavior begins to emerge among the monkeys. The level of aggression is species-dependent with capuchin monkeys creating the most serious problem, but aggressive behavior occurs in older monkeys of all species. The target of monkey aggression tends to be the Guajá 'siblings', the human children in the household. This occurs for a variety of interrelated reasons including the effects of food provisioning, natural maturational behavioral changes, and increasing social isolation.

Since the pioneering work of Jane Goodall (1971) with Gombe chimpanzees, primatologists have been aware that food provisioning increases levels of aggressive behavior in nonhuman primates[6]. The provisioned monkeys of the Guajá are not expending energy foraging in the forest. When they do 'forage', they tend to do so inside the Guajá huts on the Guajá foods. Juvenile monkeys that are allowed to move freely begin to create a nuisance to the Guajá as they race around the hut, get into the food supply, and pull apart the thatching on the walls and roof. This is particularly a problem with the capuchin monkeys, which engage in much manipulative activity, but occurs to some extent in juveniles of all species. Hormonal changes as the monkeys enter adolescence likely play a role in behavioral changes as well. The result is that most of the monkeys are tied up for increasing amounts of time after they are no longer clinging to women's heads. Tying up the monkeys compounds the social isolation that begins when they are separated from members of their own species as infants. Frequently, older monkeys are tied up alone with no physical contact with conspecifics or other monkey species. Their interactions with the Guajá are often reduced to being handed or tossed foods and occasionally grooming Guajá heads for lice. The monkeys that are tied up engage in much non-productive and often self-destructive behaviors such as pacing for hours on end, chewing their posts, pulling out their hair, and biting their fingers and toes. Essentially, these are the same abnormal behaviors often observed in zoo or laboratory monkeys in captivity (see, for example, De Waal 1989; Paterson, 1992).

Both monkeys and human children play a decreasing role in the household as they grow older. Girls marry and move out of the natal household in early childhood and young boys frequently stay on extended visits with various relatives outside the natal household. Older monkeys are no longer considered to be 'children' (*imimɨra*) and do not continue to have the same place of value that they had as infants and young juveniles.

Although the consequences for the monkeys in Guajá captivity are largely detrimental, it should be mentioned that for those whose mothers are killed in the hunt, they would likely die if not taken in by the Guajá. The monkeys have a high mortality rate in general in the community, so it would be most accurately described as a temporary extension of life. For example, the one-year survivorship rate for new infant howler monkeys brought into the Guajá community was 0% ($n = 6$) as compared with the 55% one-year survivorship rate found among *Alouatta palliata* in the wild (Clarke and Glander, 1984, p. 115). The capuchins tend to survive longer in captivity among the Guajá than other animals, and there are two known instances of capuchins reproducing among the Guajá. One capuchin birth occurred during the study period in 1997 and Balée (personal communication) noted the birth of another capuchin in 1995 in the village. In addition, pets in Guajá captivity may gain some protection from predators. However, it is difficult to describe this as a true benefit given their high mortality rate. In addition, predator protection is not absolute: one Guajá woman reported that her pet tamarin was snatched by a Harpy eagle.

In sum, the role of the monkey in the Guajá social system is predicated on the natural social behavior of monkeys. Infant monkey behavior is similar enough to human infant behavior that they can substitute for them and serve as a means to enhance the image of Guajá female fertility. However, the compounding effects of maturational changes, food provisioning, and social isolation make it impossible for monkeys to sustain this role resulting in older monkeys displaying aberrant and aggressive behaviors if they survive at all.

The monkey and Guajá cosmology

The seeming contradiction found in the monkey as food and monkey as child is rendered logical in the endocannibalistic themes that pervade Guajá religious life and cosmology. The Guajá practice a form of symbolic endocannibalism based on two related principles: the extension of kinship terms to nonhuman forest species, and a common theme of 'like eats like'. When nonhuman forms of life that are prey are defined as kin, then eating them becomes the eating of kin, or symbolic endocannibalism. Howler monkeys are central in the Guajá symbolic endocannibalistic complex. The Guajá consider howlers to be more closely related to them than any other forest species, and they are unique among the game animals in that howler monkeys are believed to have been created directly from the Guajá people.

Ka'arəpihárə: *the forest kin*

The validity of the nature–culture dichotomy is being increasingly challenged, particularly in Amazonian anthropology, not so much as a critique of structuralism, but owing to the failure of field data to support such a dualism in all indigenous perceptions (see, for example, Århem, 1996; Descola, 1992, 1996; Hviding, 1996; Rival, 1993, 1996). The relationship of nature to culture, or society to the environment, is being demonstrated to be one of juncture rather than opposition among many indigenous peoples. Pálsson (1996) has described three paradigms in human relationships to the environment: the orientalistic or Western paradigm where humans master and exploit nature, the paternalistic paradigm where humans master and protect nature, and the communalistic paradigm, which rejects the separation of nature and society and emphasizes reciprocity and continuity, often modeled on close personal relationships. The Guajá world view is consistent with the communalistic model in that their kinship relations do not define culture as opposed to nature, but rather serve as a means to explain nature in applying kinship relations to nonhuman forms of life.

Guajá life form classification and kinship classification are merged. Descola (1992) has called similar systems 'animic' and opposed them to totemic systems as described by Lévi-Strauss (1963) in that animic systems involve the social objectification of nature:

> ... animic systems are a *symmetrical inversion* of totemic classifications:
> they do not exploit the natural differential relations between natural
> species to confer a conceptual order on society but rather use the
> elementary categories structuring social life to organize, in conceptual
> terms the relations between human beings and natural species.
>
> (Descola, 1992, p. 114).

Fundamental to Guajá animism is social egalitarianism, which reflects the basically egalitarian nature of Guajá social life for same-sexed persons of the same age. Social equivalence in kinship relations to plants, animals, and environmental features are modeled on terms of reference for same-sexed siblings. The inhabitants of the forest as a whole are referred to as *ka'arəpihárə*, the forest siblings, and include three types: *harəpihárə* relations, *harəpiana* relations, and *harəpiana-te* relations. In the Guajá human kinship system, the term *harəpihárə* is used by both males and females to refer to their patrilateral same-sexed siblings; *harəpiana* refers to matrilateral same-sexed siblings. The Guajá consider patrilateral siblings to be their true brothers and sisters. However, it would be a distortion to call the system 'patrilineal' since they do not trace kinship through paternal 'lines' back in time, but rather reckon kinship horizontally through a network of fathers. Most plant and animal communities receive the matrilateral same-sexed sibling term, *harəpiana*.

In addition to the general *harəpiana* relationship that the Guajá community as a whole has with other forest biotic communities, individuals have special *harəpihárə* spirit siblings. Each Guajá derives his or her name from a plant, animal, or object with which the individual has a special spiritual connection and which is considered to be his or her *harəpihárə*, a true sibling. The divinities and pet monkeys also have such relationships.

While there is no cover term for non-kin, these forms fail to be recognized as *harəpihárə* or *harəpiana* and include domesticates recently introduced to the Guajá (such as manioc and dogs) as well as non-Indians, *karaí*, none of which are believed to be invested with a spiritual principle as are other forms of life.

The same-sexed sibling terms are used not only to describe the relationship of the Guajá to the natural world, but also describe the relationship among forms of life in the natural world. The Guajá emphasize the maternal same-sexed sibling term with the suffix *-te* as *harəpiana-te* (true or truly a maternal same-sexed sibling) to describe plant and animal communities that are considered more closely related to each other than just the general *harəpiana* relationship among forest communities.

Of particular significance is that the Guajá consider themselves to be *harəpiana-te* with only one other species, the howler monkey. Some

Table 4.1. *Harəpiana-te relationships among the primates*

Guajá name	Common name	*Harəpiana-te*
awa	Guajá	howler monkey
wari	howler monkey	Guajá, bearded saki
kwič^huə	bearded saki	howler monkey
ka'ihu	Ka'apor capuchin	brown capuchin
ka'i	brown capuchin	Ka'apor capuchin, squirrel monkey
yapayu	squirrel monkey	Ka'apor capuchin, owl monkey
aparikə	owl monkey	squirrel monkey, tamarin
etamari	tamarin	owl monkey

informants went so far as to classify the howler monkeys as *harəpihárə*, or consanguines, with the Guajá people. One similarity between them described by the Guajá is that they both 'sing', referring to the loud, extended territorial vocalizations of the howler monkey and the cultural importance of singing to the Guajá. In addition, as will be discussed shortly, the howler monkeys are believed to have been created directly from the Guajá people.

It should be noted that one type of organism may be *harəpiana-te* to two types of organisms that do not necessarily have this special kinship relationship to each other. This is logical to the Guajá since they believe that an individual can have more than one biological father. In the Guajá kinship system, *harəpihárə* and *harəpiana* are not classes of persons, but vary with each individual. In fact, none of the Guajá children in the community had exactly the same set of fathers, so each had a unique kin universe. The following diagram illustrates an example:

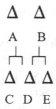

Here, A is the father of C and D, and B is the father of D and E, so C and D are *harəpihárə*, as are D and E, but C and E do not share the same father and would not be *harəpihárə*. In addition, D has two fathers. The same principle would apply to *harəpiana* relationships. The same logic follows with the Guajá community and other plant and animal communities. As seen in Table 4.1, although the Guajá are *harəpiana-te* to only the howler monkeys, howler monkeys are *harəpiana-te* to both the Guajá and the bearded sakis. The Guajá describe similar relationships among other species. For example, the white-lipped peccary and the collared peccary, the red-footed tortoise and the yellow-footed tortoise, and the inajá palm and the babassu palm are *harəpiana-te* pairs.

Guajá symbolic endocannibalism

The Guajá deny ever literally practicing endocannibalism or exocannibalism. However, of note is a description of a conflict given by two older men that occurred approximately thirty years ago. They reported that an enemy Indian group cannibalized the Guajá who were killed. The identity of these *kamara* (non-Guajá Indians) is unknown but possibilities include a number of Tupi and non-Tupi speakers or possibly even a distantly related group of Guajá. However, this description of cannibalism should not be accepted uncritically for two reasons. The first is that demonizing enemies and attributing atrocities to them is not an uncommon practice among human beings. Second, Guajá history is thoroughly mixed with mythology and cannot be relied on without reservation as actual fact (see Arens 1979).

Guajá symbolic endocannibalism is based on a pervasive theme of 'like eats like'. The principle does not employ neat cycles of energy exchange, but rather represents a logical association of like things eating like things and is seen in the Guajá creation myth, their belief in cannibal ghosts, and in the *harəpihárə* kinship association of the divinities with plant and animal spirit siblings. Among the Guajá, endocannibalism is conceptualized as characteristic of all forms of life, in that life forms eat other life forms to which they are related, and often specialize on the forms to which they are most closely related.

One example of this is found with the brown capuchin, which is believed to have been created from the *wã'i* palm (*Orbignya phalerata*), which the Guajá report the brown capuchins eat. The brown capuchins, thus, eat what they were made from. The howler monkeys are reported to eat the *yu* palm. A divinity named *Yu* is a spiritual sibling with the *yu* palm and is the controller of the howler monkeys, hunting and eating them in the *íwa*. Thus, the howler eats *yu* and *Yu* eats howlers. The divinity *Mariawa* is a spiritual sibling with a palm called *mariawa* (*Bactris setosa* Mart.). Squirrel monkeys are believed to have been created from the *mariawa* palm, a palm that the squirrel monkeys are also reported to eat. The *Mariawa* divinity is a spiritual sibling with the *mariawa* palm and the controller of the squirrel monkeys in the *íwa*. Thus, the squirrel monkey eats what it was made from and the divinity that consumes squirrel monkeys is a spiritual sibling with a plant the squirrel monkeys consume. In essence, the squirrel monkey is *mariawa*, it eats *mariawa*, and it is eaten by *mariawa*.

A somewhat similar construction exists with the Guajá, the howler monkeys, and the *aiyã* cannibal ghosts. In the creation myth, the creator divinity *Mai'íra* transformed a group of Guajá who were in a tree eating fruit into the howler monkeys. The Guajá believe that they themselves are preyed upon by cannibal ghosts called the *aiyã*. Upon death, the Guajá describe a bifurcation of the spirit. One image, or spirit, of an individual (*hatikwáyta*) goes to the sky home (*íwa*), which is described as a paradisiacal existence. Another image of themselves (*aiyã*) remains earthbound, near the corpse, stalking the Guajá at night in order to eat their souls. The Guajá believe that most illnesses are caused by the *aiyã*, particularly those that result in death. Thus, the howler monkeys are former Guajá and are eaten by the Guajá. Likewise, the *aiyã* cannibal ghosts eat the living Guajá, who represent former versions of the *aiyã*. The Guajá both eat and are eaten by transformed versions of themselves. It is tempting to describe the system as autocannibalism, but the Guajá use sibling terms to define these relationships of consumption, making endocannibalism more culturally cogent.

Guajá symbolic cannibalism is a process that links, integrates, and transforms the various forms of life in the Guajá cosmos. It is also a process that transforms earthly forms of life into sacred *íwa* forms. At death, the spirits or images (*hatikwáyta*) of plants and animals all go to the *íwa* (although only the Guajá generate the malevolent *aiyã* form at death). A somewhat similar process has been described for the exocannibalism of the extinct Tupinamba chiefdom of Brazil, where an enemy's cannibalization transformed him into a purely spirit being, giving him a heroic death, and was understood to be later reciprocated by one's own enemies (Viveiros de Castro 1992). Viveiros de Castro further states (p. 303), 'Cannibalism is an animal critique of society and the desire for divinization.' The assessment also rings true for Guajá endocannibalistic beliefs in that eating is not an act that merely satisfies hunger; it also has the transformational power to make another sacred. But rather than exocannibalistic reciprocity with the enemy, Guajá endocannibalism involves reciprocity through transformed versions of the self. Guajá transformations make them both predator and prey, and one version of the Guajá gives the gift of divinity to another version of the Guajá, knowing that the gift will at a later time be reciprocated.

Conclusion

Cultural anthropologists and primatologists can contribute to each other's disciplines, and this is perhaps nowhere more important than in the arena of conservation. In the tropical forests, cultural survival depends on preservation of the traditional habitats of indigenous peoples, and for the nonhuman primates, the literal survival of primate species depends on conserving natural habitats. Although it might seem counterintuitive that human predation on nonhuman primates can benefit conservation, the real threat to primate survival is development and habitat destruction. When indigenous hunting moves beyond subsistence activities to the commercial harvesting of primates, such as is seen with the African bushmeat crisis, the future survival of primates in Amazonia is in doubt.

Among the Guajá, if monkeys can still account for over 30% of the animal foods in the diet in the wet season, this suggests sustainability. In addition, the Ka'apor capuchin is only known to exist in the vicinity of the Guajá reserves (Queiroz, 1992). Although they are being hunted, the reserves provide a refuge for these monkeys because a habitat still exists for them there. Conservation efforts do need to take into account the beliefs and practices of the local peoples for them to be workable, practical, and hopefully beneficial to both the human and nonhuman primate communities sharing a habitat.

Identification, transformation, predation, and protection are recurrent themes in the Guajá material, social, and symbolic existence, with monkeys playing a central role in all domains. The predator and protector relationships that the Guajá have with forest life are mirrored supernaturally in their relations with the *aiyã* and *karawara*. The Guajá reenact the cannibalization of themselves by the *aiyã* transformed humans through heavy predation of other transformed humans, the howler monkeys. The *karawara* divinities have a protective relationship with the Guajá, feeding them by bringing game, healing illness, and sanctioning marriage alliances, and *karawara* protection is reenacted by the Guajá through caretaking of dependent young animals.

Guajá identification with the physical and behavioral similarities between themselves and the monkeys is prerequisite for infant monkeys to fill their social role as substitute human beings and in their symbolic role as a version of the self to be cannibalized. Howler monkeys are the preferred game because they are considered to be most like the Guajá. Ecological anthropology in the Neotropics has tended to treat monkeys as part of the natural landscape despite the importance of primate studies to anthropology as a whole. The Guajá are, in a sense, indigenous primatologists, looking to other primate species in order to understand themselves.

Notes

1. Emmons (1997) considers *Cebus kaapori* to be a subspecies of *C. olivaceus*; however, Queiroz (1992), who originally identified the species, considers *C. kaapori* to be a distinct species although with affinities to *C. olivaceus*. Kinzey (1997) also supports *Cebus kaapori* as separate from *Cebus olivaceus*.
2. See Cormier (2000) for a complete list of all fish and game masses.
3. The Portuguese term *curimatá* includes both prochilodins and curimatins. One of the key distinguishing features between these two similar genera is that the prochilodins have a sucking disk at their mouths, which the curimatins lack. The sucking disk is characteristic of the fish the Guajá call *piračũ* (see Gery, 1977; Smith, 1981).
4. Many plants the Guajá report as edible for animals have multiple uses, such as medicine, clothing, or other uses in their material culture.
5. Although women do not possess bows and arrows and were never observed to hunt game, they do borrow bows and arrows to fish. Occasionally women will use their husbands' bows and arrows and fish in open water, but more often they will borrow the toy bows and arrows of boys to spear fish that are trapped in pools after being stunned by fish poisons.
6. Although food provisioning increases aggression in nonhuman primates, Goodall's well known long-term studies among the chimpanzees have demonstrated that chimpanzee aggression is not merely the result of food provisioning.

References

Altmann, J. (1974). Observational study of behavior: Sampling methods. *Behaviour* **49**, 227–67.

Arens, W. (1979). *The Man-Eating Myth: Anthropology and Anthropophagy*. New York: Oxford University Press.

Århem, K. (1996). The cosmic food web. In *Nature and Society: Anthropological Perspectives*, ed. P. Descola and G. Pálsson, pp. 185–204. New York: Routledge.

Bailey, R.C. and Headland, T.H. (1991). The tropical rain forest: Is it a productive environment for human foragers? *Human Ecology* **19**, 261–85.

Bailey, R.C., Head, G., Jenike, M., Own, B., Rechtman, R. and Zechenter, E. (1989). Hunting and gathering in the tropical rain forest: Is it possible? *American Anthropologist* **91**, 59–82.

Balée, W. (1984). The persistence of Ka'apor culture. Ph.D. dissertation, Columbia University. [Microfilms International, Ann Arbor.]

82   L.A.Cormier

Balée, W. (1985). Ka'apor ritual hunting. *Human Ecology* **13**, 485–510.

Balée, W. (1988). Indigenous adaptation to Amazonian palm forests. *Principes* **32**, 47–54.

Balée, W. (1992). People of the fallow: a historical ecology of foraging in lowland South America. In *Conservation of Neotropical Forests: Building on Traditional Resource Use*, ed. K.H. Redform and C. Padoch, pp. 35–57. New York: Columbia University Press.

Balée, W. (1994). *Footprints of the Forest: Ka'apor Ethnobotany – The Historical Ecology of Plant Utilization by an Amazonian People.* New York: Columbia University Press.

Balée, W. (1999). Mode of production and ethnobotanical vocabulary: A controlled comparison of the Guajá and Ka'apor of Eastern Amazonian Brazil. In *Ethnoecology: Knowledge, Resources and Rights*, ed. T.L. Gragson and B. Blount, pp. 24–40. Athens: University of Georgia Press.

Beghin, F. (1957). Relation du premier contact avec les Indiens Guajá. *Journal de la Société des Américanistes, N.S.* **44**, 197–204.

Clarke, M.R. and Glander, K.E. (1984). Female reproductive success in a group of free-ranging Howling Monkeys (*Alouatta palliata*) in Costa Rica. In *Female Primates: Studies by Women Primatologists*, ed. M.F. Small, pp. 111–26. New York: Allan R. Liss.

Cormier, L.A. (2000). The ethnoprimatology of the Guajá Indians of Maranhão, Brazil. Ph.D. dissertation, Tulane University. [Microfilms International, Ann Arbor.]

Cristina dos Santos, M., Martins, M., Boechat, A.L., Pereira de Sá Neto, R. and Ermelinda de Oliveira, M. (1995). *Serpentes de Interesse Médico da Amazônia, Biologia, Venenos e Tratamento de Acidentes.* Manaus: Universidade de Amazonas.

Descola, P. (1992). Societies of nature and the nature of society. In *Conceptualizing Society*, ed. A. Kuper, pp. 107–26. New York: Routledge.

Descola, P. (1996). Constructing natures: Symbolic ecology and social practice. In *Nature and Society: Anthropological Perspectives*, ed. P. Descola and G. Pálsson, pp. 82–102. New York: Routledge.

De Waal, F. (1989). *Peacemaking Among the Primates* Cambridge: Harvard University Press.

Emmons, L.H. (1990). *Neotropical Rainforest Mammals, A Field Guide.* Chicago and London: University of Chicago Press.

Emmons, L.H. (1997). *Neotropical Rainforest Mammals, A Field Guide*, 2nd edn. Chicago: University of Chicago Press.

Gery, J. (1977). *Characoids of the World.* Neptune City, New Jersey: T.F.H. Publications.

Gomes, M.P. (1996). Os Índios Guajá: Demografia, Terras, Perspectivas de Futuro. Relatório de Pesquisas Realizadas em Fevereiro de 1996. [Unpublished manuscript.]

Goodall, J. (1971). *In the Shadow of Man.* London: Collins.

Goulding, M. (1980). *The Fishes and the Flooded Forest: Explorations in Amazonian Natural History.* Berkeley: University of California Press.

Gregor, T.A. (1974). Publicity, privacy, and marriage. *Ethnology* 13, 333–49.

Headland, T.N. (1987). The wild yam question: How well could independent hunter-gatherers live in a tropical rain forest ecosystem? *Human Ecology* 15, 463–91.

Headland, T.N. and Bailey, R.C. (1991). Introduction: Have hunter-gatherers ever lived in tropical rain forest independently of agriculture? *Human Ecology* 19, 115–22.

Hviding, E. (1996). Nature, culture, magic, and religion. In *Nature and Society: Anthropological Perspectives*, ed. P. Descola and G. Pálsson, pp. 165–84. New York: Routledge.

Kaplan, H. and Kopischke, K. (1992). Resource use, traditional technology, and change among native peoples of lowland South America. In *Conservation of Neotropical Forests: Working from Traditional Resource Use*, ed. K. Redford and C. Padoch, pp. 83–107. New York: Columbia University Press.

Johnson, A. (1975). Time allocation in a Machiguenga Community. *Ethnology* 14, 301–10.

Kinzey, W. (1997). Cebus. In *New World Primates: Ecology, Evolution, and Behavior*, ed. W.G. Kinzey, pp. 248–57. Chicago: Aldine.

Lévi-Strauss, C. (1963). *Totemism*. Translated by Rodney Needham. Boston: Beacon Press.

Mascarenhas, B.M., Cunha Lima, M. and Overal, W.L. (1992). *Animais da Amazônia: Guia Zoológico do Museu Paraense Emílio Goeldi*. Belém: Museu Paraense Emílio Goeldi.

McDonald, D.R. (1977). Food taboos: A primitive environmental protection agency (South America). *Anthropos* 72, 734–48.

Nimuendajú, C. (1948). The Guajá. In *The Handbook of South American Indians*, Vol. 3: *Tropical Forest Tribes. Bureau of American Ethnology Bulletin 143*, ed. J. Steward, pp. 135–6. Washington, D.C.: Smithsonian Institution.

Pálsson, G. (1996). Human-environmental relations. In *Nature and Society: Anthropological Perspectives*, ed. P. Descola and G. Pálsson, pp. 63–81. New York: Routledge.

Paterson, J.D. (1992). *Primate Behavior*. Prospect Height: Waveland Press.

Queiroz, H.L. (1992). A new species of capuchin monkey, genus *Cebus* Erxleben, 1777 (Cebidae: Primates) from Eastern Brazilian Amazonia. *Goeldiana (Zoologica)* 15, 1–13.

Rival, L. (1993). The growth of family trees: understanding the Huaorani perceptions of the forest. *Man* 28, 635–52.

Rival, L. (1996). Blowpipes and spears: the social significance of Huaorani technological choices. In *Nature and Society: Anthropological Perspectives*, ed. P. Descola and G. Pálsson, pp. 145–64. New York: Routledge.

Seeger, A. (1981). *Nature and Society in Central Brazil, the Suyá Indians of Mato Grosso*. Cambridge, Massachusetts: Harvard University Press.

Smith, N. (1981). *Man, Fishes and the Amazon*. New York: Columbia University Press.

Sponsel, L.E. (1997). The human niche in Amazonia: explorations in ethnoprimatology. In *New World Primates: Ecology, Evolution, and Behavior*, ed.

W.G. Kinzey, pp. 143–65. Chicago: Aldine.

Vilaça, A. (1992). *Comendo Como Gente, Formas do Canibalismo Wari'*. Universidade Federal do Rio Janeiro.

Viveiros de Castro, E. (1992). *From the Enemy's Point of View: Humanity and Divinity in an Amazonian Society*. (Translated by C.V. Howard.) The University of Chicago Press.

Wagley, C. (1983 [1977]). *Welcome of Tears, the Tapirapé Indians of Central Brazil*. Prospect Heights: Waveland Press.

Yost, J.A. and Kelley, P.M. (1983). Shotguns, blowguns, and spears: the analysis of technological efficiency. In *Adaptive Responses of Native Amazonians*, ed. R.B. Hames and W.T. Vickers, pp. 189–224. New York: Academic Press.

5 Ethnoecology of monkeys among the Barí of Venezuela: perception, use and conservation

MANUEL LIZARRALDE

Introduction

In this chapter, I describe the Barí, an indigenous people of Venezuela, their environment, and their perception, knowledge and use of monkeys. Most of the information provided in this chapter comes from extensive fieldwork and personal communications from my father, Roberto Lizarralde, and Dr. Stephen Beckerman, both of whom have participated in extensive anthropological fieldwork and have written extensively about the Barí ecology. In addition to several childhood visits with my father (starting in 1965), I have spent almost 30 months doing fieldwork with the Barí since 1988.

Until the present, the Barí have eaten monkeys, kept them as pets, and used their teeth for decoration. According to Beckerman (1980, p. 94) the monkeys in the area inhabited by the Barí accounted for 22% of the total mass of mammals and birds hunted with bow and arrows in the early 1970s. Over the past 30 years, changes in the environment and human–monkey relationships, and an increase in the Barí's population density, have been coupled with a drastic reduction in the size of the territory of the Barí. Owing to these changes, the Barí have been experiencing a depletion of major fauna, including all populations of monkeys.

The Barí are facing an unsure future as an ethnic group. They will not be able to practice the type of subsistence their ancestors had in the past. They continue to hunt many animals, including monkeys, but the numbers of many of these animals are rapidly decreasing (tapirs, peccaries, paca and monkeys). Living in a park (Parque Nacional Sierra de Perijá), where they are expected to set an example of the local flora and fauna conservation, raises many questions. In this chapter, I also describe the Barí's views on the conservation of their environment and their current ambiguous situation. There are no easy answers to the problems faced by the Barí and the plant and animals with whom they share this habitat. One of the main problems is the path to extinction of many species in their natural environment. These problems of habitat destruction, over-hunting and human

85

population increase need to be taken seriously before it is too late to save the plants and animals.

Habitat of the Barí and monkeys

The Barí live in the tropical rainforest of the southwestern part of the Lake Maracaibo basin in western Venezuela. They are swidden agriculturists growing sweet manioc and complementing their diet with fishing, hunting and gathering forest food. Agriculture supplies most of their food, with manioc providing 80–90% of the calories. Fishing provides 75% of the protein in their diet and hunting 25% (Beckerman, 1980, 1983, 1994; Beckerman and Sussenbach, 1983). Before the 1970s, 'Barí men spent about 1,500 person-hours/year in hunting and fishing combined' (Beckerman, 1994, p. 82). Today, some men spend no more than a fraction of the time they used to in hunting. Since the late 1960s the Barí have gradually entered the market economy by selling their labor and products to buy food and western goods such as tools, medicine and clothing.

R. Lizarralde conducted a census in both 1982 and 1992 as part of the Venezuelan Indian Census Program. He estimated that by 1997 there were approximately 1800 Barí in Venezuela spread among 37 villages and hamlets (Venezuela, 1985, 1993; Lizarralde, 1993; R. Lizarralde, personal communication). On the Colombian side of the border, another 500 Barí lived in two villages and a few hamlets (see Fig 5.1, R. Lizarralde, personal communication 1997; Beckerman, 1994, p. 83). The Colombian villages are all rural and most have rainforest near or around them (see Fig 5.2). These villages are located in lowland areas not higher than 250 m above sea level and rarely below 50 m above sea level.

In the sixteenth century, the original Barí territory was estimated to have 3 300 000 ha (R. Lizarralde, personal communication, 1997). Before 1960 and contact with Europeans, the Colombian and Venezuelan Barí territory was one extensive unbroken rainforest ecosystem of over 1 600 000 ha. Since this contact, this extensive rainforest covering has been rapidly destroyed and converted into cattle pasture lands, leaving only 31.9% of this territory (510 000 ha) with forest cover and in the hands of the Barí (Lizarralde and Beckerman, 1982, p. 49). At present, the Barí territory has four species of monkey. These are the white-bellied spider monkey (*Ateles belzebuth hybridus*), the red howler monkey (*Alouatta seniculus*), the white-fronted capuchin monkey (*Cebus albifrons*) and the night or owl monkey (*Aotus trivirgatus*). These four monkey species are found in the forest located in the present Barí territory,

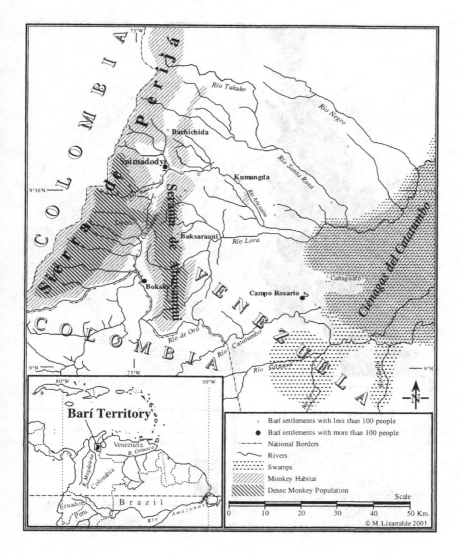

Figure 5.1. Barí Villages and areas inhabited by monkeys.

which measures about 190 000 ha (Lizarralde and Beckerman, 1982, p. 49).

Of the remaining land, which is where the Barí lived, more than 85% is still forested and is located inside and east of the mountain range called Sierra de Perijá (see Fig 5.2), most of it in the national park of Sierra de

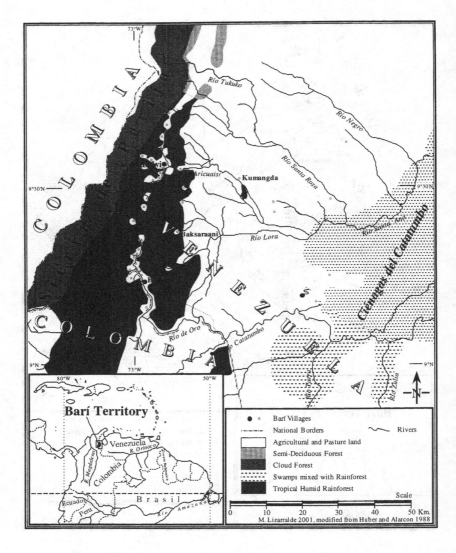

Figure 5.2. Barí territory and location of major types of forests.

Perijá. A small population of capuchin monkeys, night monkeys and red howler monkeys are also found in small isolated plots of forest of 400–500 ha on the large cattle ranches near and south of the village of Kumangda (see Fig 5.2, southeast of the Kumangda village, on the Aricuaisa river).

The region inhabited by the Barí is a humid tropical rainforest environment. The forest is basically an evergreen medium-tall rainforest with some trees observed in the field reaching 50–65 m high. A medium to high number of trees species, 105–150 per hectare, are 10 cm diameter or larger at breast height (Lizarralde, 1997). The rainfall is very high with an average of 3500 mm per year. In the south around the village of Bokshí, yearly rainfall can reach 4300 mm and in the extreme north of the present Barí territory, rainfall averages around 2500 mm per year (Gines and Jam, 1953; MARNR, 1997; MOP, 1973). Most of the traditional Barí territory is less than 100 m above sea level. Humidity averages around 90–95% and temperatures average 27.5 °C with an annual range of 18–38 °C.

Perception of monkey taxa and the ecosystem

The Barí view of the origin of monkeys is rooted in their mythology. When asked about the origin of monkeys, the Barí elders say that a long time ago there were only Barí and no monkeys on their land. *Sabasebaa*, their creator and the first Barí, and another Barí were looking for food in the forest and found two other Barí, the *Barísugshaa*, high in a tree eating fruits. *Sabasebaa* asked the *Barísugshaa* to toss them some fruit. However, they only tossed down the peels. He asked for fruits again but only got more peels. *Sabasebaa* got angry and shot both of them with his bow and arrows, knocking them down to the ground. Then *Sabasebaa* grabbed them and covered their bodies with tar and kapok (*asaa*; *Ceiba pentandra*), which resembles spider monkey fur. He broke his bow in half and placed one piece on the rear end of each, making monkey tails. He told these two *Barísugshaa* that they were now spider monkeys (*sugshaa*) and would live in trees and eat only fruits. *Sabasebaa* told the Barí they should now eat monkeys. Since then, the Barí have hunted spider monkeys for meat. Most monolingual Barí firmly believe in this myth and will retell similar stories when asked about the origin of monkeys.

At present, four species of monkey have been recorded in Barí territory (see Table 5.1). The Barí distinguish four species and give them generic taxon names: *sugshaa* (white-bellied spider monkey), *borou* or *kamaskougda* (red howler monkey), *barashii* (white-fronted capuchin monkey) and *kogchigbaa* or *baakbora* (night monkey). The two names for red howler monkey and night monkey are products of dialect variations in the Barí language. When asked about monkeys other than the four listed above, the ancestral Barí talk about a fifth generic taxon, *shaaroba*, described as a gigantic 500–800 kg red monkey which resembled a very robust spider monkey with powerful arms and large jaguar-like claws. The

Table 5.1. *List of monkeys and their Barí generic taxon names*

English name	Scientific name	Barí name	Preference	Mass[a] (kg)
White-bellied spider monkey	*Ateles belzebuth hybridus*	*sugshaa*	highest	5.9–10.4
White-fronted capuchin monkey	*Cebus albifrons*	*barashii*	high	1.2–3.6
Gray-necked night monkey	*Aotus trivirgatus*	*kogchigbaa* or *baakbora*	mid-high	0.8–1.2
Red howler monkey	*Alouatta seniculus*	*borou* or *kamaskougda*	medium	3.6–11.1

[a] *Source:* Emmons and Feer (1990), pp. 110–31.

Barí said that it had a long tail and small ears. Other Barí say that it resembles a cross between a cow and a jaguar. No living Barí have seen this monkey but they believe that it lived in the upper part of the high mountain range and was partially terrestrial. The Barí have asked me if I have seen this animal or if I have seen a photo of it in my books. As Shepard states in this volume (Chapter 6), this mythological being could possibly be an extinct animal that is retained in the folk memory of indigenous people like the Matsiguenga or the Barí (cf. Shepard, 1997). In fact, the Barí description is similar to that of an extinct ground-dwelling sloth (*Mylodon* or *Megatherium*) that used to exist in the region and became extinct some 10 000 years ago.

The Barí do not have a single term for 'monkey' as we do in English or Spanish, but they are aware that monkeys belong to a specific group of animals that live in trees, have prehensile tails in some cases and bear some resemblance to humans. In their myths described above, spider monkeys were once people. However, when asked about all the kinds or types of animals grouped under the Spanish gloss 'mono' (i.e. monkey) they will name the four species of monkeys plus the procyonids kinkajou (*bishwii*; *Potus flavus*) and the olingo (*bogsábuu*; *Bassaricyon gabbii*), which are considered to be related to night monkeys. When Barí informants are questioned as to why the kinkajou and olingo belong to the group of monkeys, they also note the differences in the face, fingers, ears, canines, and nose, thereby recognizing that they are not exactly 'monkeys.'

The geographical distribution of monkeys is basically throughout the Barí territory wherever there are large and continuous tracts of forest. This distribution has changed drastically in the past 37 years. Currently, owing to deforestation, primarily by cattle ranchers, these four species of monkey are largely restricted to three major areas and one very small pocket (see

Fig. 5.1 for distribution). To the east, the Ciénagas del Catatumbo park has nearly 300 000 ha of forest where mostly howler and capuchin monkeys live, according to the Barí. Once abundant in that area, spider monkeys now appear to be absent. The entire surrounding forest has been destroyed and spider monkeys were all hunted to extinction. However, in the west, on the top of the low mountain range of Abusanqui and Sierra de Perijá (see Fig. 5.1), there is extensive forest covering approximately 180 000 ha, consisting of both lowland tropical forest and cloud forest that have higher densities of all four species of monkeys. In the central part of the Barí territory, near the village of Kumangda, there are 400–600 ha of tall tree forest with a small population of capuchin and night monkeys. Spider monkeys are currently restricted to the central western part of the Barí territory where there are the large tracts of forest they need to survive and reproduce. According to both anecdotal and physical evidence collected in the field, the specific location with the highest density of monkeys, in particular many large numbers of spider monkeys, is in the area between the villages of Baksaraani and Okchídabu (see Fig. 5.1). Some of these spider monkey groups, according to the Barí, have more than fifty individuals and are quite easy to approach. In other areas within the Barí territory, spider monkeys travel in smaller groups of 8–12 individuals.

Barí uses of monkeys

The Barí people have three types of uses of monkeys: as a food source, as pets, and for their teeth, which are prized for Barí necklaces. The primary objective of most hunting trips is to get monkeys and the Barí exhibit a clear cultural preference for monkey meat over other kinds of animals in their forest. The Barí consume the meat of all four species of monkeys but they prefer spider monkeys. In fact, when the Barí are talking among themselves about food, they actually use the word '*sugshaa*' and '*mashu*' (meaning spider monkey and manioc, *Manihot esculenta*, respectively). The Barí word for spider monkey literally means 'meat'. They also enjoy eating capuchin and night monkeys. However, because the Barí are not fond of the meat of howler monkeys, they only hunt them in the absence of all others.

In addition to consuming their meat, the Barí also keep young (immature) monkeys as pets. Spider monkeys are the most common monkey pet species among the Barí (see Fig. 5.3). They are treated almost like a family member and given personal names, even though they are kept out of homes since they tend to be quite messy. Although capuchin monkeys are

Figure 5.3. A Barí child holding a young spider monkey pet in the village of Bokshi in June 1999. This pet had a personal name, 'Olga'.

frequently hunted, they are rarely kept as pets; I have only encountered one case. I have, however, observed night monkeys kept as pets on a few occasions. Because of their nocturnal habits, it is difficult to say how many Barí households do really have night monkeys as pets. It is possible these night monkeys might have been sleeping in the Barí houses I visited, and this might account for the low number of observations. In contrast, since they are more visible, diurnal and kept out of homes, spider monkey pets are more easily found in most villages in Sierra de Perijá.

When a Barí man goes hunting, his children will frequently ask him to bring back a monkey as a pet for them. It is not uncommon to see hunters return with baby spider monkeys after a successful hunt in which the babies' mothers were among those killed. I observed in the field that approximately 80–90% of spider monkeys killed were females. Some of these female spider monkeys appear unable to move quickly in the canopy owing to the weight of their infants and are therefore unable to escape the Barí hunters. In addition, the females tend to wait for their offspring before fleeing. Male spider monkeys are killed less often, possibly because they are free of the burden of carrying an offspring. On these hunts, approximately 75% of the infant spider monkeys die from falls or shotgun wounds. Therefore, when an infant spider monkey is found in a Barí village, it is not difficult to imagine that about eight to ten spider monkeys have been hunted.

The capture of infant capuchin monkeys is quite different from that of spider monkeys. In December 1991, after shooting at a troop of capuchin monkeys, I once observed a Barí man looking up in the trees where the capuchin monkeys were found. I asked him what he was doing and he replied that he was looking for young capuchin monkeys. According to him, capuchin mothers have the tendency to leave their young behind when they flee from Barí hunters. However, I never observed a single case of a Barí man returning from a hunt with a young capuchin monkey.

Howler monkeys are the only monkey species that I have never seen as pets in Barí villages. The Barí say that howler monkeys are hard to feed and rarely survive. Their teeth are not used in necklaces. According to informants, howler monkeys resemble the sloth, which the Barí perceive as being of low intelligence and slow speed. Moreover, the heads of howler monkeys are discarded and left in the forest where they were killed. The Barí do not eat sloth or the heads of howler monkeys or anteaters because they would not want to acquire these characteristics through the ingestion of their flesh. The Barí also believe that howler monkey heads contain potential disease. Some Barí believe that eating the heads of howler monkeys would make them *sanikani*, a Barí term that signifies madness. Almira L. Hoogesteyn (personal communication, 1998) has suggested that it is possible that the term *sanikani* is a metaphor for the health risk a person takes when consuming the brains and flesh of the heads of howler monkeys.

Because the teeth of monkeys and carnivores are believed to confer the characteristics of these animals, the use of monkey teeth and carnivore canines is important in Barí culture. The Barí prefer to use the canines of both carnivores and monkeys, although occasionally the incisors and to a lesser degree the molars are also sometimes used. The teeth of both the spider monkeys and capuchins are lumped together under the term *sugshaa aruru*, which roughly translates as 'spider monkey teeth'. The canines of spider monkeys and capuchins are similar and the differences are not always recognized by the Barí. The Barí use monkey teeth in their necklaces because of their desire to acquire monkey characteristics such as manual dexterity, speed, and prowess in climbing trees. The Barí frequently exchange monkey teeth amongst themselves for their necklaces.

In recent times the general use of monkeys has changed. The Barí who have been exposed to formal education have learned that 'white' people do not eat monkey meat. Some Barí have indicated that the similarity in appearance of monkeys to humans makes it difficult for them to eat monkeys. Traditional Barí, however, do not consider killing and eating monkeys any different from killing and consuming other animals, perhaps because they value eating them and wearing their teeth. However, owing to the obvious process of acculturation, westernized Barí no longer eat monkeys. Non-indigenous people tease the Barí by referring to them as bug and monkey eaters. These changes could ultimately benefit the future of monkeys.

The Barí are constantly testing non-Barí, especially those of European descent, to see if they would eat monkey meat. They see the tradition of eating monkey as an important element of their culture and perceive the eating of monkey meat as a boundary between them and non-indigenous people. If I eat a piece of monkey meat, they quickly say that I am like a Barí. Some Barí are proud to define themselves as 'ancient' Barí because they eat monkeys and use their teeth for necklaces as *Sabaaseba*, their creator, commanded. Moreover, the Barí seem to talk more about monkeys than other species of animals such as peccaries or agoutis that contribute similar amounts to their diet (cf. Beckerman and Sussenbach, 1983, p. 347). When Barí visitors come to a village, they will ask about the monkey hunting in the area. All these uses of monkey species are clear indications of how monkeys obviously play an important role in the lives of the Barí.

Barí hunting strategies

The knowledge of monkeys' favorite food trees obviously plays an important role in the success of hunting monkeys. The Barí men are familiar with the diet of all four species of monkey, especially spider monkeys. Hunters have the incredible ability to remember the location of an extensive number of spider monkey feeding trees. Lizarralde (1997) has demonstrated that the most successful hunters have a great knowledge of the forest trees. The other three species, however, are more likely to be found by chance or by their vocalizations rather than by searching their feeding trees. Occasionally, spider monkeys reveal their location by vocalizations too. In older times before the use of shotguns, I was told by the Barí that they would learn the spider monkeys' 'trails' in the canopy and ambush them with bows and arrows as the monkeys came to their favorite feeding trees, especially hog plum (*Spondias mombin*). The Barí hunters used to hide behind blinds set on the branches of those trees. The other strategy was to approach a group feeding on the fruits of a tree and shoot the monkeys as quickly as possible before they fled. In the past, the Barí found monkeys not far from their longhouses since they lived in a territory that was continuous tropical forest. Today the Barí have to walk long distances of 10–15 km from large villages like Saimadodyi and Bokshí until they meet a troop of spider or capuchin monkeys.

Hunting is generally divided into two main types: solitary hunting trips and group hunting expeditions. Many men go hunting regularly for a few hours before sunrise to catch both the nocturnal animals that are trying to find a sleeping place and the hungry diurnal animals that are trying to get their first meal of the day. Sometimes hunters will be gone alone all day to hunt and to check for other resources such as palm grubs, honey, fish, and game that requires the group to hunt. During solitary hunts, the Barí can more easily surprise monkeys and kill them. Hunting expeditions that require a group of Barí men include the hunting of peccaries (*Tayassu pecari* and *T. tajacu*), tapir (*Tapirus bairdii*), crocodiles and monkeys. During these group expeditions, the Barí will go far from the settlements and pursue spider and capuchin monkey groups in areas where they are still abundant.

Before 1960 many local Barí groups had longhouses for the dry season near good fishing areas. Similarly, wet-season longhouses were located where spider monkeys and oilbirds (*Steatornis caripensis*) were abundant and fat at the first peak of the rainy April–May season. These wet-season longhouses were located at an altitude of 700–1000 m above sea level on both sides of the mountain range of Perijá (R. Lizarralde, personal communication, 1997). The Barí continue to hunt in these locations and the monkeys are still abundant.

The contribution of hunting to Barí diet was 35.8 kg of meat per person per year in the 1970s (Beckerman, 1983, p. 98). The monkeys provided 8.9 kg of meat per person per year including 5.6% of all meat and 31.7% of all mammals (Beckerman, 1975). In four months of observation and 361 hours of hunting recorded by Beckerman (1975), the Barí captured 5 spider monkeys, 8 howler monkeys, 7 capuchin monkeys and 7 night monkeys. Extrapolating from Beckerman's data, a local group of fifty Barí could have hunted approximately 28 spider monkeys, 33 howler monkeys, 65 capuchin monkeys and 72 night monkeys in a year over a territory of 20–30 kha. Today many Barí, especially those who live near the Sierra de Perijá and Seranía de Abusanqui, continue to hunt monkeys. The Seranía de Abusanqui, which is located on the east side of the region between the Barí villages of Saimadodyi and Bokshí (see Map 1), has the highest densities of monkeys. Some of the Barí who hunt in this area claim to hunt between 50 and 100 per year. One Barí claimed to have killed 54 spider monkeys, 45 capuchin monkeys, 36 howler monkeys and 21 night monkeys, for a total of 156 monkeys, between the summer of 1997 and 1998. I believe the man's claims because he showed me a cup full of spider and capuchin canines. This belief was confirmed when this man's two sons and three other hunters brought back 16 spider monkeys (13 adults and 3 juveniles) from a day hunt a few hours away from the village of Baksaarani on June 26, 1999 (see Map 1 for location of villages). However, I would be surprised if an average Barí in the Sierra de Perijá gets more than a few monkeys per year. This exceptional man who hunted 156 monkeys had the obvious advantages of living in an isolated and extensively forested region, rich in fauna and he owned a shotgun. As was the case with this hunt, when such large quantity of meat is obtained, often the meat is smoked or salted and sent to relatives of the hunters.

In hunting expeditions, the Barí goal is to encounter spider or capuchin monkeys. Spider monkeys are more difficult to hunt because they are more aware of humans and will flee very fast into the forest canopy. In order to kill spider monkeys, the Barí have to run after them and try to shoot the monkeys as they quickly move from one tree to another. If the Barí do not find spider or capuchin monkeys and howler monkeys are heard, then they will search for them and hunt them. Howler monkeys tend to be quite slow, in contrast to spider monkeys, and the whole troop is hunted on most occasions. I had the unfortunate opportunity to see the sad group massacres of howler monkeys twice.

Night monkeys are fairly abundant on all parts of the Barí territory. They are nocturnal and their hooting always attracts the Barí hunters. It is fairly common to find the night monkeys approaching villages or forest camps during the evening. In fact, it is relatively easy to find them owing to their high density and ability to colonize new territories. They are commonly found as a pair of adults with sometimes one to three offspring. Because of their tendency to sleep in hollow trees, the Barí hunters also find them by banging on those trees during the day hunts.

Changes in Barí population, settlement and technology, and their impact on monkeys

The changes over the past 30 years in Barí settlement patterns, population densities and uses of technology have definitely affected the monkey population (Lizarralde, 1991). The Barí have lost a great proportion of their territory, and are now concentrated in a small portion of their former territory (12% of their original territory in 1900) (Lizarralde and Beckerman, 1982. p. 49). In 1950, the Barí population totalled 1300. By 1964, it had decreased by 65% to 850 individuals, primarily owing to the introduction of Western diseases (Lizarralde and Beckerman, 1982, p. 49). Since 1964, the Barí population has increased to nearly 2300.

Unfortunately, this population increase is also making their population density much higher than in the past, because they only have a fraction of their original territory. Therefore, the increase in population density creates higher demands on their local natural resources, which are unsustainable, with or without their traditional subsistence practices. Through the encouragement of missionaries and Venezuelan officials, and motivated by the medical and educational services offered to them, since the late 1960s they have been settling in permanent villages. By living in permanent places, they also degrade all forest resources within half a day's walking distance. This can be clearly observed in the village of Saimadodyi, where I have conducted most of my research.

The introduction of firearms has had a great impact on the faunal populations. First introduced in the late 1960s and early 1970s, and with the low cost of shells, firearms meant that the Barí hunted more efficiently and collected more meat than they would have done with bows and arrows. They have depleted the game around the villages, and have to walk 4–6 hours from the village to find animals to kill. I noticed that the teeth in the necklaces of people from the larger villages were mostly from juvenile individuals, since their roots are not entirely formed. In the past, older, yellowish, worn teeth would have been seen in necklaces, as is observed currently in remote isolated villages.

In addition to the decline in the animal population, deforestation is occurring in the Barí's current territory. Because the laws regarding holding land titles require deforestation, the Barí are deforesting plots of land to claim their own land titles. They are also deforesting large tracts of forest to produce cash crops in order to buy Western goods. The desire for Western goods has steadily increased, resulting in more production of cash crops to acquire cash. Two additional reasons the Barí need to earn money are their desire to send their children to high school in the missionaries' boarding schools and to purchase Western medicines, which are generally absent in the small state-run health centers in two villages. All these changes are contributing to the destruction of the habitat for the fauna and flora within Barí territory.

Conservation, biodiversity and the future of monkeys

A chain of events set in motion by the ranchers taking land from the Barí and the process of Westernization is changing the Barí way of life. Their perceptions and concerns have developed in the same directions as those of Western (developed) nations in recent years. They are aware that action is needed before their way of life is destroyed. In a meeting organized to bring together the headmen of all villages in Colombia and Venezuela in July 1993, three issues were raised: (1) the need to protect their current territory, (2) the need to seek strategies to recover lost territory, and (3) the need to protect the flora and fauna in their territory.

The Barí, like other indigenous peoples, are not the 'noble savages' that are idealized by national officials, park administrators and conservation organizations (cf. Lizarralde, 1992). They have acquired tastes for Western goods and abandoned their traditional technology. Their objective to conserve their flora and fauna will only be met if they (1) gradually stop hunting, (2) manage to recover part of their lost territory, and (3) convert recovered lands already deforested into plantations and cattle grasslands without sacrificing their own forested lands. This seems to be the only logical path that will guarantee the survival of a pristine environment including the monkeys.

It is important to point out that the future of the forest also depends on the survival of monkeys, since they are 'the most important seed dispersers for hundreds of plant species, especially canopy trees and lianas' (Emmons, 1990). The monkeys obviously play an important role in the ecosystem of the Barí territory. As Emmons states '[I]f they are eliminated from the forest by hunting, the plant species composition of the forest will eventually change' (1990, p. 110). Therefore, there is a clear interdependence between monkeys and people in the Sierra de Perijá. The Barí are aware of this reciprocity; their federation states that their forest and fauna need to be protected.

The Barí as a group are ready to make some sacrifices. However, they will need outside assistance. At the present, Venezuela is burdened with a heavy external debt (39 billion US dollars) and half of the population is living in extreme poverty. There are Venezuelan institutions and organizations that are aware of the ecological situation and the need to protect the habitat for endangered fauna, but the national financial situation limits their ability to assist the Barí and protect the fauna. Developed nations need to offer assistance, such as debt relief, to reach the root of the problems faced by the Barí. With this help, the Barí could concentrate their efforts on protecting their flora and fauna, continuing the symbiotic relationship with their monkeys.

Acknowledgments

Special thanks to Agustin Fuentes and Linda Wolfe for inviting me to write a chapter in this volume. I also thank all the Barí people for letting me conduct my research and reside in their cultural and natural environment. Funding for research to collect data in the field was provided by the Wenner-Gren Foundation (Grant No. 5519), PREBELAC funds from the Institute of Economic Botany of the New York Botanical Garden, and Shaman Pharmaceuticals, Inc. I am grateful to Anne-Marie Lott for her meticulous editorial assistance with this paper. I am further indebted to Agustin Fuentes, Linda Wolfe, Stephen Beckerman and Susanna Barsom for reviewing the manuscript and offering helpful suggestions.

References

Beckerman, S. (1975). The Cultural Energetics of the Barí (Motilones Bravos) of Northern Colombia. Ph.D. dissertation, Department of Anthropology, University of New Mexico.
Beckerman, S. (1980). Fishing and hunting by the Barí of Colombia. *Working Papers on South American Indians* **2**, 67–109.
Beckerman, S. (1983). Carpe diem: An optimal foraging approach to Barí fishing and hunting. In *Adaptive Responses of Native Amazonians*, ed. R.B. Hames and W. Vickers, pp. 269–99. New York: Academic Press.
Beckerman, S. (1994). Barí, in *South America*. In *Encyclopedia of World Cultures*, Vol. 7, ed. J. Wilbert, pp. 82–5. Boston: G.K. Hall and Co.
Beckerman, S. and Sussenbach, W. (1983). A quantitative assessment of the dietary contribution of game species to the subsistence of South American tropical forest peoples. In *Animals and Archaeology I: Hunters and their Prey*, ed. J. Clutton-Brock and C. Grigson, pp. 337–50. Oxford, England: British Archaeological Report.
Emmons, L. (1990). *Neotropical Rainforest Mammals: a Field Guide*. (Illustrated by François Feer.) Chicago: University of Chicago Press.
Gines, H. and Jam, P. (1953). El medio geografico. In *La Región de Perijá y sus*

Habitates, ed. H. Gines, pp. 15–22. Caracas: Sociedad de Ciencias Naturales La Salle.

Lizarralde, M. (1992). 500 Years of invasion: Eco-colonialism in indigenous Venezuela. *Kroeber Anthropological Society Papers* **75–76**, 62–79.

Lizarralde, M. (1997). Perception, knowledge and use of the rainforest: ethnobotany of the Barí of Venezuela. Ph.D. dissertation, Department of Anthropology, University of California at Berkeley.

Lizarralde, R. (1991). Barí settlement patterns. *Human Ecology* **19**, 428–52.

Lizarralde, R. (1993). *Población Indígena en el Parque Nacional "Perijá"*. Caracas: Oficina Central de Estadística e Informática.

Lizarralde, R. and Beckerman, B. (1982). Historia contemporánea de los Barí. *Antropológica* **58**, 3–51.

MARNR (Ministerio de Agricultura y de Recursos Naturales No Renovables) (1997). *Región Hidrologico No. 1: Datos Mensuales y Anuales de Precipitaciones de Perijá. Dirección de Hidrologia y Meteorologia*. Sistemas Nacional de Información Hidrologica y Meteorologica. Caracas: MARNR.

MOP (Ministerio de Obras Publicas) (1973). *Distrito Hidrologico 1: Isoyetas 1973*. Caracas: Division de Hidrologia, Ministerio de Obras Publicas, Republica de Venezuela.

Shepard, G. (1997). Monkey hunting with the Machiguenga: medicine, magic, ecology and mythology. Paper presented at the 1997 American Anthropological Association Meeting.

Venezuela, OCEI (Oficina Central de Estadística e Informática) (1985). *Censo Indígena de Venezuela 1982*. Caracas: Oficina Central de Estadística e Informática.

Venezuela, OCEI (1993). *Censo Indígena de Venezuela 1992*. Caracas: Oficina Central de Estadística e Informática.

6   *Primates in Matsigenka subsistence and world view*

GLENN H. SHEPARD

Introduction

Much ado has been made in recent years about whether indigenous people are conservationists, noble savages who act as stewards of nature, or whether they are efficient predators who maximize their exploitation of resources with the available technology, and thus represent a growing threat to biodiversity conservation in endangered ecosystems. A summary of arguments on both sides can be found in the *tête-à-tête* between Alcorn (1993) and Redford and Stearman (1993) in *Conservation Biology*. I will set aside a more thorough discussion of the complex debate for now, only to note that, in its essence, the polemic has to do with different notions of what constitutes Nature, and whether or not culture, or certain kinds of cultures, are part of the natural world.

The clash between alternative conceptions of nature was poignantly illustrated in 1993 during an interaction between a Matsigenka hunter and a Peruvian bureaucrat. The two met at a workshop[1] held in Manu National Park during which, for the first time since the creation of the park in 1973, representatives of the Peruvian government, the park administration, and the conservation community explained to Matsigenka inhabitants what a national park was, and why one had been created in Manu twenty years before. The bureaucrat in question was explaining the concept of threatened and endangered species, noting spider monkeys as an example, and went on to describe how national parks were created to protect such vulnerable species. Elias, one of the best young monkey hunters from the village of Yomybato, was listening attentively, but became perplexed. He raised his hand politely and asked the following question, which I translated from Matsigenka to Spanish, 'Let me get this straight. An endangered species is one that is almost extinct, right? And the spider monkey is an endangered species, right? Well here in my homeland, I and my father and my father's father, and all of my Matsigenka kinsmen as long as stories have been told since ancient times, have hunted and killed as many spider monkeys as we can every rainy season when the monkeys get fat. And there are still lots of spider monkeys. So that means they aren't

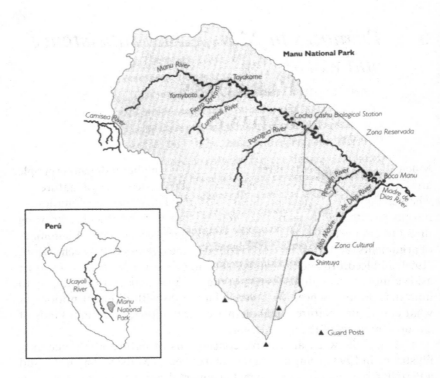

Figure 6.1. Map of study area.

endangered, right? At least not here. Maybe they're endangered where you live in Lima, but not here. Maybe you should have created the park in Lima rather than here in my homeland'. His tone was not sarcastic or hostile. Instead, he was trying to understand and assimilate novel terms and concepts, but had found a logical inconsistency in the argument. Later that evening, our Matsigenka hosts served spider monkey stew to all of the participants of the event. Elias asked the bureaucrat, sipping the stew, 'How does the endangered species taste?!'

The Matsigenka and the Manu Biosphere Reserve

Manu Biosphere Reserve is located in the department of Madre de Dios in southeastern Peru (Fig. 6.1). Despite persistent myths of the Manu as a natural paradise or 'living Eden', it was in fact the hub of brisk commercial activity controlled by the rubber baron Fitzcarraldo at the beginning of the twentieth century, and the scene of brutal massacres, enslavement, and other atrocities committed against indigenous populations, including the Matsigenka (Farabee, 1922; Reyna, 1941; Camino, 1977; Townsley, 1987; Shepard, in preparation). After the collapse of the rubber trade in 1917, most outsiders left the area. The scattered native populations who survived the rubber fever isolated themselves and assumed a hostile attitude towards outsiders. In the 1960s, outsiders returned to the Manu to harvest timber and hunt valuable animal pelts. The Manu was set aside as a national forest reserve in 1968, and was declared a national park in 1973. Manu National Park was elevated to the status of Biosphere Reserve by UNESCO in 1977, and in 1987 was declared a World Heritage Site by the International Union for the Conservation of Nature (IUCN).

Manu Biosphere Reserve, with an area of some 1.6 million hectares, embraces the entire watershed of the Manu River, and is divided into three nested zones with different levels of protection: the core area of Manu National Park, afforded the highest level of protection and considered 'untouchable' by outside human influences; the Manu Reserve Zone along the lower course of the Manu River, open to limited human activities such as scientific investigation and ecotourism; and the Cultural Zone, open to economic exploitation by legally recognized indigenous and colonist communities of the Alto Madre de Dios River. To the north of Manu National Park is an additional buffer zone, the 'Kogapakori–Nahua Reserve' set aside for isolated indigenous populations of the Upper Timpia, Camisea, and Mishagua Rivers. Within the Reserve Zone, the Cocha Cashu Biological Station has been the stomping ground for some of the most influential tropical ecologists of this generation (see Terborgh, 1983; Gentry, 1990), and has contributed to the Manu's almost legendary status within and beyond scientific circles.

The Manu Biosphere Reserve was created upon a fundamental contradiction (Shepard and Rummenhöller, 2000): the core area, considered untouchable and closed to human interference, is home to a substantial indigenous population. In addition to the two legally recognized Matsigenka communities of Tayakome and Yomybato, with a combined population of about 450, there are several remote Matsigenka settlements in the Manu River headwaters. The Yora (or 'Nahua'), Panoan speakers who were the Matsigenka's enemies for much of the past century, inhabited the Manu–Mishagua headwater region until they were decimated by Western diseases in the mid-1980s (Zarzar, 1987; Hill and Kaplan, 1990; Shepard, 1999a). They now reside in the upper Mishagua outside the limits of Manu Park, but return occasionally to hunt, gather turtle eggs, or visit and trade with other native communities. In addition to the Mat-sigenka and the Yora, there are unknown numbers of isolated indigenous populations of various cultural–linguistic groups living throughout the park and in nearby watersheds (d'Ans, 1972; Kaplan and Hill, 1984; Shepard, in preparation).

The Matsigenka belong to the Arawakan cultural–linguistic family, and currently have a population of about 13 000 people, distributed in ext-ended family settlements and small communities along various tributaries of the Urubamba, Madre de Dios and Manu Rivers. Historical records as well as folk tales indicate that the Matsigenka maintained trading relations with Andean populations since at least the time of the Inca Empire (Camino, 1977; Shepard and Chichon, 2001). The Matsigenka cultivate manioc, maize, plantains, sweet potatoes and other crops in small swid-dens that are abandoned to forest regeneration after a few years of active cultivation. The Matsigenka also hunt, fish, and gather a wide range of forest products. Near mission towns and other trading centers, some Matsigenka engage in small-scale commercial cultivation of coffee, cacao, and annatto, a natural red dye. In the past few years, ecotourism has emerged as an economic alternative in a few Matsigenka communities.

Research for this chapter was carried out in the Matsigenka communi-ties of Yomybato and Tayakome. The community of Tayakome was founded in the early 1960s by Protestant missionaries of the Summer Institute of Linguistics, who employed bilingual guides to contact Mat-sigenka from numerous small settlements throughout the Manu head-waters and attract them to a single large community with a small airstrip (d'Ans, 1981). The missionaries provided health care, literacy training, and Western goods and tools, including shotguns. After the creation of Manu National Park in 1973, the missionaries were expelled and the airstrip closed. About half of the native population of Tayakome accompanied the missionaries on their exodus from Manu to the adjacent Camisea River. Another segment of the population left Tayakome around 1978 to estab-lish gardens at Yomybato. Since the formation of the park, the Mat-sigenka of Manu have been prohibited from using firearms or engaging in commercial or extractive activities other than traditional hunting, gather-ing, and agriculture. Beginning in 1997, the Peruvian Institute of Natural Resources (INRENA) has received German foreign aid (GTZ) funding to implement a Matsigenka-managed ecotourism lodge at Cocha Salvador on the lower Manu River. Though still in its infancy, the project will certainly have a profound effect upon the economy, culture, and social organization of Matsigenka communities in Manu.

Table 6.1. *Nonhuman primate species of Manu*

Common name	Matsigenka name	Scientific name	Preference	Mass (kg)
spider monkey	*osheto*	*Ateles paniscus*	high	7.5–13.5
woolly monkey	*komaginaro*	*Lagothrix lagothricha*	high	3.6–10
red howler monkey	*yaniri*	*Alouatta seniculus*	medium	3.6–11.1
brown capuchin	*koshiri*	*Cebus apella*	medium	1.7–4.5
white-fronted capuchin	*koakoaniro*	*Cebus albifrons*	medium	1.2–3.6
squirrel monkey	*tsigeri*	*Saimiri sciureus*	medium	0.6–1.4
monk saki	*maramponi*	*Pithecia monachus*	low	2.2–2.5
dusky titi	*togari*	*Callicebus moloch*	low	0.9–1.4
night monkey	*pitoni*	*Aotus trivirgatus*	low	0.8–1.2
saddleback tamarin	*potsitari tsigeri*	*Saguinus fuscicollis*	low	0.3–0.4
emperor tamarin	*tsintsipoti, chovishishini*	*Saguinus imperator*	low	0.4
Goeldi's monkey	*potsitari tsigeri*	*Callimico goeldii*	low	0.5
pygmy marmoset	*tsigeriniro, tampianiro, tampiashitsa*	*Cebuella pygmaea*	low	0.1–0.2

Source: data from Emmons (1990) and Alvard (1993).

Primates as prey

The Manu hosts one of the world's most diverse primate communities, with a total of thirteen non-human primate species (Janson and Emmons, 1990; see Table 6.1). Of these, spider monkeys (*Ateles paniscus*) and woolly monkeys (*Lagothrix lagothricha*) are the preferred edible species of the Matsigenka. Howler monkeys (*Alouatta seniculus*) and two species of capuchin (*Cebus apella, C. albifrons*) are also hunted, though less frequently. In a collection of game animal skulls stored by a subset of Matsigenka hunters in Yomybato from December 1998 to December 1999, I and my biologist colleagues identified 17 woolly monkeys, 14 spider monkeys, three capuchins and one howler monkey (da Silva, Yu and Shepard, in preparation), a clear reflection of Matsigenka hunting preferences. Alvard and Kaplan (1991) found a similar pattern: 24 woolly monkeys, 17 spider monkeys, three capuchins, and two howlers were taken in the sample of hunts they observed during fieldwork conducted ten years earlier in Yomybato. Smaller monkeys such as squirrel monkeys (*Saimiri sciureus*) and tamarins (*Saguinus fuscicollis, S. imperator*), as well as less abundant or nocturnal species such as dusky titis (*Callicebus moloch*), monk sakis

(*Pithecia* cf. *monachus*), and night owl monkeys (*Aotus trivirgatus*), are not primary game species, but may be taken on occasion. Goeldi's monkey (*Callimico goeldii*), which is very rare, and the pygmy marmoset (*Cebuella pygmaea*), which weighs only 100 g, have never been observed to be hunted.

In the Manu, where hunters do not use shotguns, larger primate species such as spider and woolly monkeys are still quite plentiful, whereas medium-sized and small monkeys are seen as less desirable. In indigenous areas outside Manu park where shotgun hunting is common, larger monkeys become locally rare or extinct, and medium-sized monkeys become more important in the diet (Alvard and Kaplan, 1991; see also Peres, 1990). Relative body mass of different primate species is certainly important in understanding Matsigenka hunting preferences, but does not explain all of the data (see Table 6.1). Howler monkeys are equal in size to woolly monkeys and nearly as large as spider monkeys, aside from being the most abundant primate species and among the most abundant mammal species of Manu (Terborgh 1983, p. 26), and relatively easy to locate due to their loud vocalizations and slow movement. Yet both hunting surveys carried out in Yomybato (da Silva *et al.*, in preparation; Alvard and Kaplan, 1991), indicate that the Matsigenka take howlers approximately ten times less frequently than either of the two high-preference species. Furthermore, howlers are taken about half as often as capuchins, though they weigh twice as much. This glaring exception to an optimal foraging strategy of game choice is overlooked by Alvard and Kaplan (1991). Some Matsigenka claim that howler monkey meat is less tasty, which could reflect ecological considerations: the howler's leaf-eating habit, unlike that of most other fruit-eating monkeys, might make the meat distasteful to the Matsigenka palate. However avoidance of this monkey species also reflects cultural attitudes: because of their loud vocalizations, howler monkeys are considered in Matsigenka folklore to be *seripigari*, shamans, and thus pose spiritual hazards as well. Furthermore, howlers are considered slow and lazy (even Terborgh (1983, p. 32) concedes that howlers live in 'shameless lethargy'), a characteristic that is believed to pass on to children who eat the animal's meat.

The Matsigenka of Yomybato consider medium and small monkeys such as capuchins, squirrel monkeys, tamarins, and dusky titis to be mostly a form of target practice for adolescent boys. Adult men will generally only take these species if they happen to have an easy shot, or if nothing else has been killed that day (see also Alvard, 1993). Though small, squirrel monkeys are abundant and often accompany capuchin troops. For this reason, they are preferred over a number of similar-sized small monkey species.

Because capuchins are sometimes raised as house pets, the Matsigenka are well aware of their mischievous and sometimes destructive behavior. The Matsigenka name for capuchins, *koshiri*, means literally 'thief', and reflects the mythological origin of the genus in Matsigenka folklore (see below). Some Matsigenka claim that children should not eat capuchin meat because they might grow up to become dishonest.

The Matsigenka tend to avoid a number of other game species consumed safely by the neighboring Piro and Yora, for example capybara, deer, ocelot and caiman. Alvard and Kaplan (1991, p. 86) attribute such differences in prey choice exclusively to ecological considerations; however, I argue that culture also plays a role. The Matsigenka perceive capybara meat as having an unpleasant smell. The meat of caiman, jaguars, and other carnivorous terrestrial animals is avoided both because it is assumed to be unpleasant, and because the predatory spirits of the animals are feared. Deer are described in Matsigenka folk tales as demonic beings who rape human victims. Eating deer meat is believed to cause illness in some people. Some gastrointestinal and skin illnesses are attributed by the Matsigenka to the consumption or improper mixing of certain meats (Shepard, 1999a). Pregnant women, sick people, adolescent girls during puberty rites, and parents with newborn infants are considered to be spiritually vulnerable, and must avoid eating or coming into contact with certain game animals, carnivorous fish, toxic and urticating plants, and other foods or organisms having strong odors, noxious properties, or perceived dangerous spirits. Despite certain general patterns of dietary avoidance among the Matsigenka, there is also a great deal of individual variation. For example, some individuals or families avoid certain meats or other foods after allergic reactions, unpleasant associated experiences, or because their parents avoided that food.

Body mass and relative abundance of game species are not the sole factors shaping Matsigenka and other native Amazonians' hunting preferences. Ultimately, understanding dietary preference and avoidance in any society involves an appreciation of empirical and ecological as well as cultural and symbolic aspects of food consumption. Anthropologists have debated the predominance of cultural (ideological) versus ecological (material) factors in explaining variation in dietary choice among indigenous populations of Amazonia (see Ross, 1978; Milton, 1991). Whatever the theoretical explanation, the widespread occurrence and tremendous variability of fish, game, and other food taboos among traditional Amazonian populations reflects a certain degree of affluence: only those with plenty to eat can afford to be choosy.

Monkey hunting practices

Owing to Manu Park's prohibition of firearms, the Matsigenka of Tayakome and Yomybato hunt with palm wood bows and bamboo-tipped arrows. For most other mammal species besides primates, the Matsigenka use blinds and hunting dogs to get close to their prey, such that the bow shot is horizontal and at a relatively short range. When pursuing primates, Matsigenka hunters rely solely on their own tracking skills, stamina, and good aim. Large primate species have become locally rare in the immediate vicinity of the villages of Yomybato and Tayakome owing to hunting pressure and growing human populations over the past twenty or more years of steady occupation. However, small primates such as squirrel monkeys, tamarins, and even capuchins can often be seen within a few minutes' walk of village sites. Hunters must walk five kilometers or more (an hour or more walking) away from the central community area before encountering large monkey species. Many families maintain secondary gardens or hunting camps as far as 25 km from the community to serve as a base for hunting expeditions. None the less, as close as 8 km from Yomybato, it is possible to encounter spider and woolly monkey troops that appear fairly unafraid of human hunters. This supports the suggestion by Alvard et al. (1997) that primate populations in the vicinity of the two villages are sustained only by continuous in-migration from troops living in adjacent, non-hunted areas.

The Matsigenka use visual and auditory cues to locate monkey troops. Hunters of the same extended family unit exchange accounts of where monkey troops have been seen in recent days to improve their chances of encounter. Matsigenka hunters imitate woolly and spider monkey calls well enough to elicit responses or even attract naive troops. Upon encountering a monkey troop, Matsigenka hunters try to position themselves for a nearly vertical shot, straight up as far as thirty meters into the forest canopy. Hunters try to pick out the large adult males or kurakas (a Quechua loan word meaning 'leader') as targets for their first arrows. If the first arrow does not hit the animal in the chest, or if the hunter manages to scare the troop before getting a clean shot, he must pursue the fleeing animals, often targeting the slower-moving females burdened by young. Sprinting in the dense forest understory, the hunter must keep a constant eye on the monkeys while also navigating through tangled vegetation and uneven terrain. Spider monkeys are especially swift, acrobatic, and intelligent in their escape, and are said to sometimes remove arrows from their bodies and throw them, sharp end first, towards hunters below[2]. Many monkeys and other animals wounded by arrows are not recovered, though

most are unlikely to survive. Even fatally wounded monkeys are often able
to climb into a tall tree and get a firm grip on a branch. Matsigenka hunters
frequently recover their prey by climbing high into the forest canopy.
Genealogical interviews with Matsigenka families reveal significant mor-
bidity and mortality in adult men caused by accidents that occur while
hunting, for example falling from a tree, being struck by a stray arrow, and
snakebite (Shepard, 1999a). Bow hunting for monkeys is a highly strenu-
ous and potentially dangerous activity, to say the least.

Females with young represent a disproportionate amount of spider
monkeys taken by Matsigenka hunters. In interviews conducted with three
Matsigenka hunters from Yomybato based on game animal skulls stored
for a one-year period, 93% ($n = 14$) of spider monkey skulls in the collection were from females, most of which hunters remembered as having been
burdened with young. In a more detailed study based on direct observation, Alvard and Kaplan (1991) describe a similar pattern, where females
represented 88% of spider monkeys taken by Matsigenka bow hunters in
Yomybato. Natural spider monkey populations studied at nearby Cocha
Cashu Station were found to have a skewed sex ratio of 35 males per 100
females (McFarland-Symington, 1987), which means that about 74% of
spider monkeys available to Matsigenka hunters are females. Thus, Matsigenka bow hunters take significantly more females (93% and 88% in two
independent studies) than would be expected from the natural sex ratio of
spider monkey populations (da Silva *et al.*, in prepration). The bias is likely
not by choice, since Matsigenka hunters claim to prefer large, adult males.
Instead, female spider monkeys burdened by young are more vulnerable to
bow hunters, especially during pursuit. It must be noted that the reported
sex ratios represent only successful kills followed by retrieval. Considering
that large male monkeys are in fact preferred targets, and that wounded
animals may escape retrieval only to die later, then the female skewing in
the skull data at least partially represents a post-shot retrieval bias. The sex
ratio of all killed animals could be closer to parity with the natural ratio.
For woolly monkeys, the sex ratio of kills in our data was close to even:
47% were female and 53% were male ($n = 17$). Alvard and Kaplan (1991),
however, noted a female-biased sex ratio for woolly monkeys, with about
75% of kills being female. Woolly monkeys have not been well studied, but
populations in Venezuela show roughly equal sex ratios varying from 80 to
120 males per 100 females (Izawa, 1976, cited in Alvard and Kaplan, 1991).
Depending on the actual sex ratio of woolly monkey populations in the
Manu, it may also be the case that females are taken in greater proportion
than they occur in natural populations. The numbers as well as sex ratio of
prey species taken provide important data for determining the impact

and long-term sustainability of indigenous hunting practices on primate populations.

When female monkeys or other animals with young are killed, Matsigenka hunters recover the young and attempt to raise them as pets. Mortality appears to be high, especially among some species such as woolly monkeys. Spider monkeys, capuchins, squirrel monkeys, and night monkeys, as well as coatimundis, kinkajous, peccaries, parrots and macaws, mot-mots, nun birds, toucans, oropendulas, sun bitterns, many songbird species, and even tapirs, can be seen as pets in many Matsigenka households. Animal pets (*piratsi*) are cared for by women and children. Many pets are fed masticated food mouth-to-mouth, just like human children. Some animals are fed according to specific dietary needs: sun bitterns are provided with freshwater shrimp, toucans are fed fruits of Lauraceae, and insectivorous birds are given crickets. Wild animal pets receive much better treatment and nourishment than do hunting dogs. If pets survive to adulthood (mischievous behavior may lead them prematurely to the cooking pot!) they are eventually allowed to return to the forest. Pet monkeys released to the forest interact with wild troops, but also return to the vicinity of the house where they were raised. Their owners recognize them in the wild and do not hunt them because of their practical value as living decoys, and perhaps also for sentimental reasons. I once saw four spider monkeys swinging in the trees adjacent to a newly felled garden. Excited, I told the man of the household, 'Get your arrows! There are spider monkeys in the forest.' He laughed, and responded, 'Those aren't wild monkeys, they are old pets that escaped to the forest. I don't kill them. It would be like killing my own children.'

Spider and woolly monkeys are perhaps the most highly prized meats eaten by the Matsigenka. The meat is considered to be fatty and delicious during the rainy season of November through May, when many fruits are ripe. The marked dry season from June to October receives less than 100 mm of precipitation, only 5% of the total average annual precipitation of more than 2000 mm (Terborgh, 1990). During the dry season, fruit production by most tree species drops sharply. Frugivorous animals, including many monkey species, suffer nutritional stress and must turn to alternative, lower-quality foods such as nectar, leaves, pith, and arthropods (Terborgh, 1986). Though not formally prohibited, monkeys are generally not hunted during the dry season, as the meat is considered stringy and lean. The practice represents a form of saving, delaying harvest to increase the return on investment (i.e. time spent making arrows, time and energy spent in pursuit of prey) during a season when other resources, notably fish, are abundant. It also reduces total annual monkey mortality, particularly during the hungry dry season when they are most nutritionally stressed.

Primates in mythology and ethnozoology

Matsigenka mythology, like that of other indigenous Amazonian peoples, serves as an encyclopedia of ecological and taxonomic information about the plants, animals, and habitats of the rainforest environment (Berlin, 1977; Posey, 1983; Shepard, 1999b). A large share of Matsigenka mythology is dedicated to describing the origins of animal species. The theory of evolution implicit in these tales shares some similarities with eighteenth century Lamarckian notions equating organic evolution with progress. But for the Matsigenka, the Great Chain of Being began with mankind and proceeded by a 'devolutionary' principle such that animals descended from humans, and not vice versa. In ancient times, only human beings inhabited the earth. The first shaman, *Yavireri*, used his magical breath to transform humans into the great diversity of animal species that exist today. Often, these transformations represent a form of punishment for foolish behavior. Salient features of animals such as diet, habitat, social behavior, or unusual physical attributes are explained by analogy with human foibles, quirks and eccentricities. Taxonomic relationships between similar organisms are explained in terms of human kinship.

Yavireri's debut performance was to turn two groups of impolite guests at a party into spider monkeys and woolly monkeys. The first two animal species to be created were primates, suggesting that the Matsigenka recognize the close taxonomic relationship between humans and primates. Matsigenka hunters frequently remark how similar monkeys are to humans, with their opposable thumbs, facial expressions, and almost human way of carrying infants.

Comical and insightful tales describe the origins of other primate species. *Yaniri*, the howler monkey, was once a lazy shaman who liked to do nothing other than take hallucinogenic ayahuasca and sing ritual songs. Rather than raise his own crops, Yaniri continually borrowed beans from his brother-in-law, *Osheto*, the spider monkey. Rather than planting them, Yaniri simply boiled and ate them, and went back to borrow more beans once his supply was exhausted. He made up lies to fool his kinsman. 'Brother-in-law,' whined Yaniri, 'the beans you gave me to plant did not grow. Please give me some more.' Once again, Yaniri boiled and ate the beans, and a few days later came back to ask for more. Finally, Osheto discovered his brother-in-law's laziness and dishonesty. In anger, he punched Yaniri in the throat, which became swollen like the howler monkey's enlarged larynx. As punishment, Yaniri was transformed into the vociferous, slow-moving howler monkey, which still sings its shamanic songs in the forest. Matsigenka hunters frequently comment, upon encountering howler monkeys, 'They are *seripigari*, shamans.'

In another story, two second-rate shamans set out to steal fire-making technology from a fierce, all-female tribe, the *Maimeroite*. Both failed in their Promethean quest. One singed the hair off his face and was turned into the brown capuchin monkey, with its curious, flat-topped frontal outline, charcoal-black brow, and perpetually surprised facial expression. The other got drunk and fell head first into the women's toilet, becoming the white-fronted capuchin with its distinctive dark brown cap. The generic name for capuchins, *koshiri*, means 'thief' in Matsigenka, referring to this comical tale.

The tiny pygmy marmoset is known in Matsigenka as *tsigeriniro*, 'mother of squirrel monkey', *tampianiro*, 'mother of wind'[3], or *tampiashitsa*, 'wind tail'. They are considered to be magical and somewhat dangerous creatures. They may lead a hunter astray, flitting from tree to tree until the hunter is lost, and then vanish in a sudden gust of wind. Though they may be found in large, fairly noisy groups, these small animals are indeed difficult to spot in the slightest breeze.

In the rugged Andean foothills, there resides a creature known to the Matsigenka as *oshetoniro*, 'mother of spider monkey'. They are said to be like large monkeys, but are equipped with demonic powers and gigantic penises. They can summon wind and darkness, cause panic and confusion, and are said to rape and kill human victims[4]. Terborgh (1983) describes in almost breathless wonder the idyllic scenery of the Cocha Cashu Biological Station in Manu Park, where spider monkeys live in their natural state, completely unafraid of people, sometimes descending from the canopy to act out territorial displays just a few meters from human intruders. For the Matsigenka, the same scene is not idyllic, but rather demonic. Walking in the vicinity of Cocha Cashu with two Matsigenka friends in 1996, I encountered a large troop of spider monkeys, completely unafraid of our party. My Matsigenka companions, though respecting the informal prohibition of hunting around Cocha Cashu, pantomimed aiming and shooting imaginary arrows, and commented on the boldness and fearlessness of the monkeys. From their perspective, the encounter was unnatural, indeed uncanny: monkeys are animals of prey, they are supposed to escape from human predators. The bold behavior of the monkeys was reminiscent of the demonic *oshetoniro* found in remote areas uninhabited by humans. Both men suffered from nightmares that evening and blamed them on the harrowing encounter.

The metaphor of hunting is important in Matsigenka illness concepts and shamanistic healing. Just as humans hunt and kill animals for food, so do certain dangerous animals, demons, and other spirits look upon humans as animals of prey. One of the primary functions of the Matsigenka shaman as healer is to call back the souls of sick children, stolen by vengeful animal spirits or demons. Adult male spider monkeys and the demonic *oshetoniro* are among these dangerous beings. Why do monkeys have such a bad reputation? The almost human antics of pet spider monkeys are a source of daily amusement for many Matsigenka families, but this closeness to humanity is also disturbing. The dying struggle of a hunted spider monkey is perhaps an unconscious reminder of human mortality at the hands of more powerful forces.

In addition to their importance in mythology and folklore, primates also stand apart in Matsigenka ethnozoology. There is no Matsigenka word referring to the general category of monkey, but there is clear evidence from Matsigenka numeral classification that monkeys are treated as a distinct group within mammals (Shepard, 1997). Thus monkeys appear to constitute an unnamed or 'covert' (Berlin, Breedlove and Raven, 1968) intermediate category in Matsigenka ethnozoological classification. The Matsigenka language allows a speaker to distinguish or classify the kind of object being counted, somewhat like mass nouns in English ('one drop of water' as opposed to 'one cup of water'). Matsigenka numeral classifiers may be applied to both animate and inanimate nouns. For animals, numeral classifiers often create an analogy between some salient aspect of the animal's morphology, and some inanimate (in the sense of non-locomotive) object, typically a plant or plant part. For example, the classifier *-tsa*, for woody lianas, can be used in the animate form to refer to snakes. The classifier *-poa-*, referring to tubers, tree trunks, and other cylindrical objects, can be used in the animate form to refer to otters, catfish, caimans, and other animals with a sleek, cylindrical trunk and relatively small limbs. When counting mammals, Matsigenka speakers may distinguish as follows (see Table 6.2): 'one large-bodied animal of tapir' (classifier *-kana-*, also for chili peppers and banana bunches), 'one cylindrical animal of otter' (classifier *-poa-*, also for tubers and tree trunks), 'one small animal of mouse' (classifier *-kitso-*, also for seeds and other small, round things). Monkeys take two classifiers, based on the salient morphological distinction of prehensile vs. non-prehensile tails: 'one prehensile-tailed spider monkey' (classifier *-empe-*, also for tree branches), is contrasted with 'one non-prehensile-tailed squirrel monkey' (classifier *-shitsa-*, also for vines and string). The grammatical distinction

Table 6.2. *Matsigenka noun classifiers for mammals*

Classifier	Inanimate category	Animate category	Examples
-empe-	branched things (tree branches)	monkeys with prehensile tails	spider monkey, woolly monkey, howler monkey, capuchins
-shitsa-	stringy things (vines, string)	monkeys with non-prehensile tails	squirrel monkey, tamarins, dusky titis
-kana-	bodies, torsos (chili peppers, banana bunches)	large mammals	tapir, deer, dog
-kitso-	small, round things (seeds, pebbles)	small animals	mice (also small birds, insects)
-poa-	cylindrical things (logs, tubers)	cylindrical animals	river otters (also caiman, catfish)

Source: data from Shepard (1997).

also has taxonomic and practical relevance. All monkeys included in the class *-empe-*, 'prehensile-tailed', belong to the scientific family Atelidae (see Rosenberger, 1977; Schneider *et al.*, 1993), and most are important game species. Most monkeys included in the class *-shitsa-*, 'stringy-tailed', belong to the family Cebidae and none are hunted or consumed with great frequency.

These examples from mythology and ethnozoology demonstrate that Matsigenka knowledge and appreciation of biological diversity goes beyond utilitarian facts. The Matsigenka, like other indigenous peoples, are astute observers of nature, and like all good naturalists they have been able to synthesize their observations into an admirable taxonomic system that is both practical and reflective of natural relationships. Myths concerning the origins of biodiversity are a source of both entertainment and philosophical speculation. Animal species are anthropomorphized in folk tales, spiritual beliefs, and personal anecdotes. Metaphors drawn from ecological processes are used to understand illness, death, and other aspects of the human condition, while human intentions and emotions are projected into the ecological sphere (Shepard, 1999a, b).

Monkeys, manhood and medicine

Monkey meat is highly valued and seasonally scarce, and there also appears to be a prestige value associated with it. Quick, acrobatic, intelligent and often inaccessible in the forest canopy, monkeys represent a challenge even to skilled hunters. Being a good hunter for the Matsigenka is synonymous with being a good monkey hunter. Hunting, and specifically the hunting of large monkey species, is intimately associated with masculinity for the Matsigenka. Hunting is done exclusively by men. Matsigenka boys learn to craft and shoot their own bows and arrows when they are scarcely old enough to walk. From tiny toy contraptions used for impaling butterflies and frogs, a Matsigenka boy at about eight years old graduates to a half-size palm wood bow for hunting small animals. A boy's progress through various stages of adolescence towards adulthood is marked according to the animals he is capable of hunting: first small birds, then small monkeys such as the dusky titi and squirrel monkeys, then medium-sized capuchins, and finally woolly and spider monkeys. For the Matsigenka, monkey hunting is fundamental in the social construction of manhood.

Do Matsigenka men develop the aim and skill required to hunt monkeys through years of daily training with bows and arrows since boyhood? Or do some men inherit natural talents from a father with exceptional vision or athletic abilities? If you ask a Matsigenka, the answer is no. There is no such thing as good practice, or good luck, or good genes. There are only good hunting medicines. For the Matsigenka, hunting ability is acquired solely by the use of special plants that sharpen a hunter's visual acuity, aim, sense of smell, stamina, and luck. Dependent as they are on wild game and fish for virtually all of their protein requirements, hunting medicines are a crucial aspect of the Matsigenka pharmacopoeia. More than fifty species, nearly a quarter of the total number of medicinal plant species known to the Matsigenka of the Manu, are used as hunting medicines (Shepard, 1999a). These include plants taken by men to improve their own aim as well as plants given to hunting dogs to hone their sense of smell for specific animals. Medicines relevant to primate hunting can be divided into four categories: eyedrop medicines, purgatives, cultivated sedges, and hallucinogenic plants.

Kaokirontsi: *eyedrop medicines*

Matsigenka men apply the leaf juice of numerous plant species to their eyes in order to clarify vision and instill the hunter with the soul of *Pakitsa*, the harpy eagle (*Harpia harpyja*), the greatest hunter of the forest. Medicinal eyedrops (*kaokirontsi*) induce several minutes of intense stinging and virtual blindness. When the pain wears off, the hunter is thought to have 'good aim' (*ikovintsatake*) for several days. The Matsigenka view the eyes as the portholes of the soul; from a pharmacological perspective, active agents applied to the eyes are rapidly and efficiently absorbed into the bloodstream. Hunting eyedrops are taken only during the rainy season, specifically to improve a man's aim for hunting monkeys.

Most eyedrop remedies used by the Matsigenka for hunting come from the Rubiaceae, the botanical family of coffee and quinine. It is likely that these herbal medicines are active as stimulants or mild psychoactives on the central nervous system, providing a pharmacological basis for their use as hunting medicines. Different species of Rubiaceae carry the names of different game animals. The larger-leafed medicinal species are named after the larger game animals, and the smaller species are named after smaller animals. The most commonly used species are *oshetoshi*, 'spider monkey leaf' and *komaginaroshi*, 'woolly monkey leaf', both from the genus *Psychotria*, known to contain the psychoactive compound dimethyl-tryptamine and other bioactive compounds (Schultes and Raffauf, 1990).

Purgatives and emetics

Spiritual and bodily purity are important for the Matsigenka hunter. Eating spoiled or improperly cooked meat is a sure way of losing one's aim. If a woman allows a pot of meat to boil over and spill into the flames, the meat should be thrown away. If possible, a man should avoid carrying the animal he has killed. A companion, typically a brother-in-law or teenage boy, carries it for him. A man should never eat the head of the animal he has killed. Instead, the head is often given away to another family in the extended kin group. When a young man kills his first few large monkeys, he is not allowed to eat any of the meat. Unlike sportsmen in the United States, Matsigenka men never brag about their hunting abilities or the size of their kill. On the contrary, they tend to understate their catch. These practices instill a sense of humility, while ensuring that meat is shared with other families in the social group. Those who disobey these norms lose their hunting skill.

Any contact with menstrual blood or menstruating women, said to smell like carrion or raw meat (*janigarienka*), also takes away a man's aim. Men refer to sex in slang with metaphors from hunting, for example 'I shot her with an arrow' (*nokentakero*), i.e. had intercourse. Like a football coach warning his players about sex the weekend before the big game, Matsigenka men avoid sexual relations the night before hunting. A promiscuous woman is described as 'ruining the aim' of all the men in the village. After a sizable kill, couples tend to leave the children behind and go to distant garden sites to 'do some weeding'.

Sexual and dietary taboos associated with hunting establish an ethic of proper conduct between husband and wife. A good wife will be attentive to her cooking and not let the meat spoil, burn, or remain undercooked, and will not let the pot boil over. She will be careful during her menstruation to clean herself and not contaminate other people with her blood. A good husband will be skilled enough as a hunter that if meat does get spoiled, or if the pot boils over, he will discard it and not feel forced to eat it out of desperation. Such taboos also establish a balance between reproductive and productive responsibilities of the family, between sexual pleasure and the serious business of raising a family.

Inevitably, a man violates these norms, and thus loses his aim. A string of bad hunting days is a sure sign this has happened. The man's body is said to reek with the carrion smell of spoiled meat or raw blood, and his soul becomes possessed with the spirit of the vulture. A number of species of purgative and emetic plants are taken by hunters to clean themselves of sexual, dietary, and ritual impurities. Older men prepare their adolescent sons for manhood and monkey hunting through a rigorous regime of purgative plants belonging to a wide range of botanical families. Most are very bitter and all induce fits of vomiting or diarrhea of various degrees of severity. The more bitter and the more extreme the purgative effect, the better the medicine. The idea behind purgative remedies is to clean out the body of the spoiled meat and carrion-eating vulture spirit and replace it with the harpy eagle's hunting spirit. *Pakitsa*, the harpy eagle, is the epitome of hunting prowess for the Matsigenka. With its huge wingspan, powerful talons and acute vision, the harpy eagle is a formidable predator which, like the Matsigenka, feeds on monkeys in the forest canopy. The Matsigenka consider the feathers of eagles and hawks to be the best fletching for arrows. Matsigenka tales describe how the harpy eagle spirit, walking the earth in human form long ago, once taught Matsigenka shamans the secrets of its own hunting skill: special toxic plants for sharpening vision, cleansing the body and purifying the soul.

Medicinal sedges

Cultivated sedges, *Cyperus* spp., known as *ivenkiki* in Matsigenka, come in a bewildering number of varieties, each one with a distinct medicinal use. There are sedge varieties for treating fevers and headaches, for dispelling nightmares, and for healing arrow wounds. Women cultivate their own sedges for protecting babies from animal spirits, facilitating childbirth, improving their skill in spinning and weaving cotton, and resolving domestic disputes. Men cultivate sedge varieties used as hunting medicines. Each variety carries the name of the animal species they are used for. The most common hunting sedges are *oshetovenki* for spider monkeys, *komaginarovenki* for woolly monkeys and *shimavenki* for boquechico fish (*Prochilodus nigricans*). The importance of the three animal species in ethnobotanical nomenclature of sedges reflects the two salient hunting seasons for the Matsigenka: the dry season, characterized by fishing for boquechico, and the rainy season, characterized by hunting woolly monkeys and spider monkeys.

Every Matsigenka man cultivates his own sedge varieties, dispersed around his garden in discrete clumps. Many of the different sedge varieties used by the Matsigenka appear to be botanically identical plants, and only their owners are able to distinguish one variety from another (cf. Brown, 1978, p. 132). Cultivated sedges are typically infected by ergot-producing fungi of the family Clavicipitaceae, the same fungi from which lysergic acid was first isolated. The fungus infects the entire plant, destroying the fruits and flowers and infusing the roots with its alkaloid content. Ergot alkaloids are apparently responsible for the medicinal activity of sedges (Plowman *et al.*, 1990).

Since cultivated sedges are incapable of sexual reproduction, genetic diversity cannot be maintained through hybridization and selective breeding. This is why large numbers of cultivars must be maintained. But where do new cultivars come from? According to the Matsigenka, shamans ascend into the heavens in hallucinogenic trance to receive new sedge varieties from the guardian spirits. If shamans did not continually introduce genetic diversity into medicinal sedges and other cultigens, the varieties would lose their vigor and eventually die out (Shepard, 1999b). Without shamans, the Matsigenka would lose both their crop diversity and their hunting ability.

Unlike other hunting medicines, sedges are carried along on the hunt and ingested on the spot, just before shooting an arrow. The hunter chews the root bulb, which has a bitter, aromatic taste, like a mixture of ginger and juniper berries. He then spits the masticated bulb in a fine mist onto his hands, on the bow, on the arrow, and up towards the monkey, as he mumbles, 'Arrow! Fly straight and fast, straight to the heart, hit no branches. Monkey! Just sit still.'

The sedge is thought to infuse the hunter's body and weapons with its power, while mesmerizing the animal. When a Matsigenka hunter chews on a bit of sedge root in the forest, telling his arrow to fly straight to the monkey's heart, he is not engaging in idle superstition. He is giving time for the psychoactive alkaloids of the plant to take effect, while focusing his mind on the most important details of the scene in a Zen-like state of contemplation.

Hallucinogens and other psychoactive plants

The hallucinogenic vine *Banisteriopsis* (ayahuasca) and the solanaceous narcotics *Brugmansia, Brunfelsia, Solandra* and tobacco are important in Matsigenka shamanism and are also powerful medicines for the hunter. In fact, shamanism for the Matsigenka has as much to do with hunting as with healing. Taking ayahuasca is generally not allowed during the dry season, because of fires said to burn in the spirit world. Once the rains begin, the fires are put out, trees come into fruit, the monkeys get fat, and the ayahuasca season is ushered in. When you ask Matsigenka men why they take ayahuasca, their first answer is usually, 'to have good aim for monkey hunting'. The purgative effects of ayahuasca contribute to its power in purifying the hunter's soul. Most importantly, ayahuasca and other psychoactive plants bring the hunter into direct contact with the spirits who control the natural and supernatural spheres.

The *Saangariite*, 'invisible ones', are benevolent beings, the Matsigenka's principal allies in the spirit world. The Saangariite raise as pets all the game animals eaten by the Matsigenka (Baer, 1984). The curassows are the chickens of the Saangariite, the peccaries are their guinea pigs, the jaguar is their watchdog. They release these pets from their invisible villages for people to eat. By taking hallucinogenic plants, the Matsigenka shaman develops a relationship with a twin brother among the Saangariite. When the shaman goes into trance, the twins switch places: the spirit twin comes to sit, sing and heal among the humans gathered at the shamanistic ceremony, while the shaman visits the invisible village of the Saangariite. There, the shaman may make deals with the Saangariite, so that they release more of their pets for the Matsigenka to hunt and eat. The Saangariite may also provide the shaman with new varieties of crops and medicines, including the sedges used for hunting. This fusion of shamanistic religion with hunting magic is widespread in Amazonia (Reichel-Dolmatoff, 1971, 1976), and brings to mind Paleolithic cave paintings, which some authors say represent an effort to exert magical control over game animals (Lewis-Williams and Dowson, 1988).

Local faunal extinctions are said to occur when the Saangariite guardian spirits become upset that so many of their pets are being killed, hiding the animals in distant hillsides to be released at a later date (cf. Reichel-

Use Category

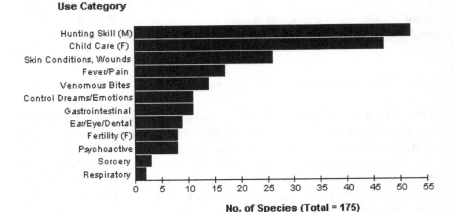

Figure 6.2. Use categories of medicinal plants. M, men's specialization; F, women's specialization.

Dolmatoff, 1978). As long as human populations remain relatively small, widely dispersed and somewhat nomadic, the spiritual explanation reflects ecological reality. The Matsigenka say that arrows are silent, and therefore do not frighten the invisible owners of the animals; shotguns, on the other hand, make a loud noise which makes the guardians pack up and move out quickly. This belief, too, reflects reality accurately.

Gender roles and ecological balance

Among the Matsigenka, hunting medicine is an area of men's ethnobotanical specialization, complementing women's specialized knowledge of plants used in caring for newborn infants. Whereas men use various toxic plants to gain skill for hunting specific prey, women use aromatic and succulent plants to defend their babies from the vengeful spirits of game animals. For every cultivated sedge used by men to improve their hunting skill, there is a sedge used by women to bathe babies. Men's hunting plants and women's childcare plants are the two most significant categories of botanical medicines. Together they account for more than a hundred species (Shepard, 1999a), roughly half of all medicinal plants used by the Matsigenka of the Manu (see Fig. 6.2). Male and female ethnobotanical knowledge reflects a complementary construction of gender roles while also describing an ecological and cosmological feedback loop between humans and their animal prey.

Matsigenka mothers give newborn babies warm baths of aromatic and succulent plants one or more times a day. The fragrance of these plants is thought to dispel certain rank-smelling animal spirits. If the mother does not know the proper plants, or if her husband kills animals he should not, animals can 'take revenge' (*ipugatakeri*) on a baby. The same verb is used when people get even or settle scores. Vengeful animals are said to 'steal the baby's soul' (*yagasuretakeri*), causing a wide range of illnesses including sleeplessness and crying, skin rashes, nosebleeds, upset stomach, and even crib death. The notion is widespread in the Peruvian Amazon and is known in mestizo Spanish as *cutipado*, derived from a Quechua word meaning 'to return, exchange' (Regan, 1993). In some cases, the parents must eat the meat for their child to be *cutipado*, in other cases merely 'messing with' (*yantakeri*) the animal is sufficient to result in vengeance upon a newborn child. During his wife's pregnancy and the child's early infancy, a man must be careful to avoid spiritually dangerous animals, and he must stay fairly close to home. Palm heart is harvested frequently, and preferably only bland fish and small game animals that are plentiful near the village are consumed. Even after the delicate stage of early infancy has passed, parents must be cautious in their consumption of meats while raising young children, and should use the proper medicines to avoid game animal revenge.

Gesneriaceae, the African violet family, contains a number of colorful, succulent and fragrant species that are used by women to protect babies from the vengefulness of primate species. As in men's hunting medicines of the Rubiaceae, each Gesneriad species bears the name of a monkey species, the spirit of which is thought to attack young children: *koshirishi*, 'capuchin leaf', is used to protect children from the capuchin monkey's spirit, *tsigerishi* for squirrel monkeys, *pitonishi* for night monkeys, and so on. Most dangerous for young children are the aggressive and powerful adult males of spider and woolly monkey troops. The most common Gesneriad species used to bathe infants are *oshetoshi*, 'spider monkey leaf', and *komaginaroshi*, 'woolly monkey leaf', virtually identical to the names of the most important Rubiaceae eyedrop medicines and *Cyperus* sedges used by men to improve their monkey-hunting skill. Thus there is a direct opposition between men's hunting medicines, used for killing certain monkey species, and women's baby-bathing medicines, used to defend children from the monkeys' vengeful spirits.

There may be an empirical basis to the observed 'revenge' caused by the meat of certain animals, since nursing babies can have allergic reactions to foods eaten by the mother. Given the high rates of infant mortality

(Shepard, 1999a), these beliefs reflect the general anxiety of parents about the health of their children, especially newborns. The warm baths of aromatic and succulent herbs keep babies clean and may help prevent skin infections. On another level, such practices establish norms of ideal behavior for parents, and create a balance between male and female roles in subsistence and child-rearing. Finally, these beliefs imply a system of ecological checks and balances in which the role of predator and prey may become reversed as Nature settles its scores.

A number of beliefs and practices surrounding bamboo ethnobotany likewise reflect the precarious balance between humans and prey species. Matsigenka men manufacture arrow points using the bamboo *Guadua weberbaueri*, known as *kapiro*. A man with a pregnant wife or a newborn child is prohibited from manufacturing arrows, lest the urticating hairs of the bamboo cause hemorrhages in the woman or rashes in the child. Furthermore, the glue-like resin used to fletch arrows is believed to cause the placenta to adhere to the womb and thus be retained, a potentially fatal complication of childbirth. Together with the concept of animal spirit revenge, these beliefs would appear to discourage and limit the hunting activities of young fathers. *Kapiro* bamboo stands flower, fruit, and die synchronously on long cycles of 15–30 years (Nelson, 1994). After a die-back event, arrow-making material is locally scarce for a period of one to two years during which the bamboo grows back. A number of alternate bamboo species of similar stem size to *kapiro* are available, for example *yaivero*, *Guadua glomerata*. However, the Matsigenka consider this species inappropriate as material for arrow points owing to spiritual considerations. It is said that if one kills monkeys or other animals with arrow points made from *yaivero*, the Saangariite spirits become angry and send game animal populations far away. This belief may have a basis in empirical observations. Perhaps *yaivero* bamboo points are less effective at killing prey, leaving more wounded animals that die later. Massive *kapiro* bamboo fruiting and die-back could alter the behavior and territorial distribution of game animals. The coincidence between such ecological alterations and the necessity of using substitute bamboos might have led to these beliefs. Furthermore, the die-back of local *kapiro* bamboo stands combined with a prohibition of alternate bamboo species might contribute to periodic migrations. In any case, even if they have no basis in empirical fact, these beliefs reflect a principle of ecological homeostasis that pervades Matsigenka cosmology and subsistence practices.

Figure 6.3. Large primates including woolly monkeys are among the most prized
game animals hunted by the Matsigenka.

Indigenous people and conservation: boon or bane?

Goethe once wrote that 'we are shaped and formed by that which we love.'
The same could be said of academics and that which they study. In the
debate over whether indigenous peoples are a boon or a bane to biodiversity conservation, anthropologists and ethnobiologists initially favored
indigenous people as stewards of nature. Perhaps the most outspoken and
influential ethnobiologist *cum* indigenous advocate is Darrell Posey, largely responsible for formulating the Declaration of Belém (1988) which first
proclaimed the role of indigenous peoples as stewards of the world's
biodiversity, and which formed the basis for more recent affirmations of
indigenous rights to traditional resources and cultural property (Posey,
1994). Posey and other scholars have written on the relevance of

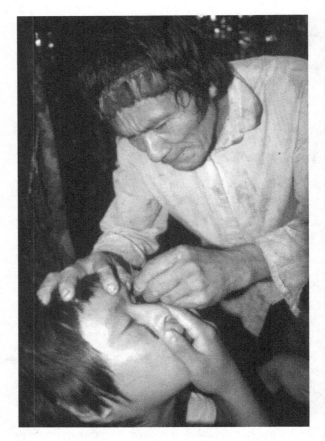

Figure 6.4. Matsigenka men use dozens of species of caustic or stimulating plants as eyedrops to improve their hunting skills.

indigenous knowledge to the study and conservation of biological diversity in the New World Tropics (Nations, 1980; Norgaard, 1981; Parker *et al.*, 1983; Posey, 1983, 1984; Berlin, 1984; Boster, Berlin and O'Neill 1986; Plotkin, 1988; Posey and Balée, 1989; Balée, 1994; Chernela, 1994; Shepard *et al.* 2001; Shepard, Yu and Nelson, 2002; Fleck and Harder, 2000).

Some tropical biologists, having witnessed the destruction of their beloved organisms or ecosystems at the hand of humanity, tend to view indigenous people, and especially acculturated populations with access to Western technology, as a threat to conservation. Though popular in the press, the concept of indigenous people as 'ecologically noble savages' has come under increasing criticism among biologists as well as ecological anthropologists (Johnson, 1989; Redford, 1991; Alvard, 1993; Robinson,

Figure 6.5. Female monkeys with young are taken disproportionately by Matsigenka hunters. Orphaned monkeys are raised as pets by women and children. If they survive to adolescence, monkey pets may be allowed to return to the forest.

1993; Brosius, 1997). John Terborgh, in his recent book *Requiem for Nature* (1999), dedicates an entire chapter, ominously entitled, 'The Danger Within', to a discussion of the threats posed by the Matsigenka to the future of Manu Park. He closes the chapter with the suggestion that the only way to guarantee conservation in this and other rainforest parks is by enticing acculturated indigenous communities to relocate to titled lands outside the park's boundaries. Given Terborgh's assertion that top predators 'are an essential component of healthy ecosystems' (1990, p. 57), it seems ironic that he calls for the removal of indigenous people, top predators *par excellence*, from parks. Kay (1998, p. 484), discussing North American ecosystems, notes that

Figure 6.6. Matsigenka women bathe newborn babies daily with warm, aromatic herbal infusions to protect them from the vengeful spirits of game animals.

Native Americans were the ultimate keystone predator and the ultimate keystone species... Until the importance of aboriginal land management is recognized and modern management practices change accordingly, our ecosystems will continue to lose the biological diversity and ecological integrity they once had, even in national parks and other protected areas.

Representations of indigenous peoples by academic researchers, policy makers, and the press have important practical implications. On the one hand, scientific affirmations of indigenous people as stewards of biodiversity have shaped and continue to influence popular opinion and the political will, creating a favorable atmosphere in some tropical countries for the affirmation of traditional land rights and the demarcation of indigenous reserves. On the other hand, projecting Western-style environmentalist or conservationist attitudes on indigenous world views can be misleading and inappropriate (see Brosius, 1997), and could backfire if indigenous people do not live up to the idealistic vision. Recent negative portrayals of indigenous people in the Brazilian press, for example, have been associated with a political movement backed by the agrarian sector to reduce the size and number of indigenous reserves, and therefore free land for logging, cattle ranching, and mechanized agriculture. Given the political reality, arguments such as Alvard's (1993), namely that native hunters maximize short-term returns and hence are not conservationists, seem narrow and distorted, ignoring important aspects of indigenous lifeways that indeed contribute to biodiversity conservation. Whether or not one chooses to define indigenous people as 'conservationists' in an absolute or technical sense, there seems little doubt that they are better conservationists, relatively speaking, than the logging companies, cattle ranchers, and agribusiness that eagerly await access to their well-conserved forests. Whether hard-core preservationists like it or not, indigenous areas account for 54% of all reserves by acreage in the nine Amazonian countries, and cover roughly 20% of the Brazilian Amazon (Peres, 1993). Thus indigenous societies are, in a very literal sense, stewards of half of the Amazon's protected biodiversity. The real question is not whether indigenous people are or are not conservationists, but rather how they can better participate in and benefit from conservation policies and projects for their lands.

Indigenous communities themselves have a strong vested interest in conserving game resources, which are crucial to their livelihood and world view. Though not exactly straightforward, sustainability of hunting practices is mostly a mathematical issue: if hunters kill more individuals than the prey population is capable of replacing, the population goes extinct. Anthropologists and biologists alike have addressed the question of hunting sustainability in tropical South America with a variety of quantitative methods and applied research techniques (Hames, 1980; Vickers, 1980, 1991; Hames and Vickers, 1982; Bodmer, Fang and Ibanez, 1988; Bodmer et al., 1994; Robinson and Redford, 1994; Bodmer, 1994, 1995; Bodmer, Eisenberg and Redford, 1997). Participatory, community-based research on questions of resource use is an important means of promoting dialogue between social scientists, biologists, and indigenous populations about conservation issues. It is time to move beyond 'noble savage' debates and begin thinking seriously about what it means for human societies – indigenous and non-indigenous, local and global – to co-exist with the tens of millions of other organisms inhabiting the planet.

Conclusion: the future of primate conservation in Manu

This chapter has pointed out a number of aspects of Matsigenka belief and practice that are relevant to primate conservation in Matsigenka communities of Manu Biosphere Reserve and elsewhere. In the Manu, traditional bow hunting practices appear to be sustainable, at least for the time being. Only two species, woolly and spider monkeys, are hunted intensely, mostly during the six-month rainy season. Both species are vulnerable to local extinction because of hunting pressure; however, the low human population density and large, non-hunted areas that surround human settlements have ensured a relative stability of primate populations to date. Little seems to have changed in terms of species preference and sex ratio of primate kills during the ten years since Alvard and Kaplan's (1991) more thorough study of Matsigenka hunting. However, the disproportionate taking of female monkeys (often carrying young) certainly has a deleterious effect on primate populations, especially if hunting pressure were to increase. The Matsigenka place a high prestige value on spider monkey and woolly monkey meat, and would seem to have a clear interest in preserving these important resources for future generations. Matsigenka beliefs about the revenge of game animal spirits, and how the Saangariite spirits hide animals when over-hunting occurs, reflect a form of ecological feedback between predators and prey. Such ideological elements could be incorporated within an explicit conservation strategy for primates and other game species.

However, Matsigenka culture and economics are now rapidly changing, perhaps more so than at any time since missionary contact. The most profound source of change is the GTZ-funded Matsigenka ecotourism lodge project, inaugurated in 1998. Though still operating on a preliminary basis, the lodge will eventually provide a substantial cash income. Whether the Matsigenka will use this income to buy (illegal) shotguns, thereby increasing hunting pressure, or to buy supplementary food, decreasing the pressure, is unknown. Schools, health posts, and other centralized services are steadily being improved by government and non-governmental organizations, representing a strong centripetal force that concentrates hunting pressure in a relatively small area. However, population growth may lead to the formation of new villages throughout a wider territory. In addition, population pressure, resource depletion, and the recent re-initiation of petrochemical exploration in areas adjacent to Manu, have begun to provoke waves of migration of both acculturated and uncontacted native people toward the safe haven of Manu (Shepard and Chichon, 2001; Shepard and Rummenhöller, 2000). Increasing exposure to Western goods and lifestyles is certain to change how the Matsigenka value different aspects of their environment and culture, in ways that can hardly be anticipated. Of particular concern is whether increasing Westernization will erode the traditional beliefs and practices that currently serve to reduce hunting pressure.

Published quantitative studies of Matsigenka hunting and resource use are available (Johnson, 1983, 1989; Johnson and Baksh, 1987; Alvard and Kaplan, 1991; Baksh, 1995), as well as an entire generation of research done at Cocha Cashu on primate populations (Terborgh, 1983) and other facets of the region's ecology. Anthropological and biological research needs to be synthesized and shared with local communities in applied, participatory projects such as have been carried out elsewhere in Peru (Bodmer, 1994; Bodmer *et al.*, 1997). The Matsigenka ecotourism lodge project includes a component of participatory environmental monitoring in Matsigenka communities. I contributed to the formulation of the monitoring plan (Shepard, 1998; APECO, 2000) and hope to remain in consultation about the project. These monitoring activities could benefit from periodic, quantitative reevaluation of actual hunting returns and faunal densities (da Silva *et al.*, in preparation). As the human population grows, and as indigenous populations both inside and outside Manu gain greater access to Western goods and services, community-based management of hunting and resource use will become increasingly important.

Primates, so obviously related to the human species, provide a particular fascination for cultures around the world. Many primate species are threatened with extinction due to changing technologies, habitat destruction and the unprecedented growth of the human population over the past century. In the New World tropics, large primate species often become locally extinct where their habitat overlaps with that of human hunters, especially where firearms are used (Bodmer *et al.*, 1988; Peres, 1990, 1991; Bodmer, Eisenberg and Redford, 1997). Primates often share their habitats with indigenous peoples who are, ironically, also threatened with cultural (if not biological) extinction due to many of the same forces. In some cases, indigenous hunters pose a direct threat to primate survival. At the same time, indigenous ecological practices and world view include certain mechanisms of natural and supernatural balance between predator and prey species. Understanding how local people interact with primate species is fundamental for developing conservation strategies, and contributes to our general knowledge of cultural and biological diversity. Although often at odds in day-to-day interactions, the destinies of primate species and indigenous cultures in the Amazon are ultimately intertwined.

Acknowledgements

The information presented in this paper was gathered and synthesised over many seasons of field work, funded by several generous fellowships and organizations including the Labouisse Prize Fellowship of Princeton University (1987), the Wenner-Gren Foundation (1995, 1997), the National Science Foundation (1996) and the A.L. Green Fund administered by E.O. Wilson of Harvard University (1999). Institutional support has been provided by Asociación Peruana para la Conservación de la Naturleza (APECO) and by the herbaria of Universidad Nacional San Marcos, Lima, and Universidad Nacional San Antonio Abad, Cuzco. A special word of appreciation goes to my parents, Glenn Sr. and Jean Shepard, who have supported me when no one else did. Thanks are also due to Manuel Lizarralde and Leslie Sponsel for contacting me about presenting a paper on this subject, and to Agustín Fuentes for organizing the invited session on ethnoprimatology where a preliminary paper was first presented. I thank Daniel Hoffman and Patricia Lyon for many useful suggestions on a preliminary draft of the paper, and Douglas Yu and Maria N.F. da Silva for comments on more recent drafts. I thank John Terborgh for introducing me to the Manu many years ago. Finally, thanks to Merino Matsipanko, Maura Vicente, Mariano Vicente, Elias Matsipanko, Mateo Italiano, Alejandra Araos, and many other Matsigenka companions, *Noshaninkaige Matsigenkaige*. I dedicate this paper to the memory of Marcial Vicente, a dearly missed friend.

Notes

[1] The event was funded by the Discovery Channel (Washington, D.C.), SuperFlow Productions (Colorado Springs, CO) and Friends of the Peruvian Rainforest (Philadelphia, PA) in recognition of the important contributions of Matsigenka native communities to the Discovery/SuperFlow film, 'Spirits of the Rainforest'. The event was organized by the pro-indigenous organization, Centro para el Desarrollo del Indígena Amazónico (CEDIA), and included the participation of the Peruvian National Institute of Natural Resources (INRENA), the directorship of Manu National Park, and Fundación para la Conservación de la Naturaleza (FPCN).

[2] At the Cocha Cashu Biological Station, spider monkeys will sometimes deliberately drop small branches near or on biologists walking the trails below! (D. Yu, personal communication).

[3] The suffix *-niro*, 'mother', is used in Matsigenka ethnobiology to indicate an organism that has a morphological or spiritual affinity to some other organism,

without necessarily implying a taxonomic affinity. *Tsigeriniro*, 'mother of squirrel monkey', is not a kind of squirrel monkey but rather a mischievous and magical creature that is small and sometimes accompanies squirrel monkeys. Likewise, *potsotiniro*, 'mother of annatto (Bixa orellana)' is not a kind of annatto, but rather an unrelated tree with orange cortex that, like annatto, can be used as face paint.

[4] Perhaps a variant of the *Mapinguari* of Brazilian folklore, these creatures may represent a folkloric memory of the recently extinct giant ground sloth (see Oren, 1993).

References

Alcorn, J. (1993). Indigenous peoples and conservation. *Conservation Biology* 7, 424–6.

Alvard, M.J. (1993). Testing the "ecologically noble savage" hypothesis: Inter-specific prey choice by Piro hunters of Amazonian Peru. *Human Ecology* 21, 335–87.

Alvard, M.J. and Kaplan, H. (1991). Procurement technology and prey mortality among indigenous neotropical hunters. In *Human Predators and Prey Mortality*, ed. M.C. Stines, pp. 79–104. Boulder, Colorado: Westview Press.

Alvard, M.J., Robinson, J.G., Redford, K.H. and Kaplan, H. (1997). The sustainability of subsistence hunting in the Neotropics. *Conservation Biology* 11, 977–82.

APECO (Asociación Peruana para la Conservación de la Naturaleza) (2000). Plan de Monitoreo Socio-Económico-Cultural del Proyecto Albergue Matsiguenka. Lima: APECO.

Baer, G. (1984). *Die Religion der Matsigenka, Ost Peru*. Basel: Wepf & Co. AG Verlag.

Baksh, M. (1995). Change in resource use patterns in a Machiguenga community. In *Indigenous Peoples and the Future of Amazonia: An Ecological Anthropology of an Endangered World*, ed. L. Sponsel, pp. 0. (Arizona Studies in Human Ecology.) Tucson: University of Arizona Press.

Balée, W. (1994). *Footprints of the Forest: Ka'apor Ethnobotany: The Historical Ecology of Plant Utilization by an Amazonian People*. (Biology and Resource Management in the Tropics.) New York: Columbia University Press.

Berlin, B. (1977). *Bases Empíricas de la Cosmología Aguaruna Jívaro, Amazónas, Perú*. (Studies in Aguaruna Jívaro Ethnobiology report no. 3.) Berkeley: Behavior Research Laboratory, University of California at Berkeley.

Berlin, B. (1984). Contributions of Native American collectors to the ethnobotany of the neotropics. *Advances in Economic Botany* 1, 24–33.

Berlin, B., Breedlove, D. and Raven, P. (1968). Covert categories and folk taxonomies. *American Anthropologist* 70, 290–9.

Bodmer, R.E. (1994). Managing wildlife with local communities: The case of the Reserva Communal Tamshiyacu-Tahuayo. In *Natural Connections*, ed. D.

Western, M. Wright and S. Strum, pp. 113–34. Washington D.C.: Island Press.

Bodmer, R.E. (1995). Priorities for the conservation of mammals in the Peruvian Amazon. *Oryx* **29**, 23–8.

Bodmer, R.E., Aquina, R., Puertas, P., Reyes, C., Fang, T. and Gottdenker, N. (1997). *Manejo y Uso Sustentable de Pecaríes en la Amazonía Peruana.* (Occasional Paper of the IUCN Species Survival Comission, No. 18.) Quito, Ecuador and Geneva: International Union for the Conservation of Nature (IUCN)/CITES.

Bodmer, R.E., Eisenberg, J.F., and Redford, K.H. (1997). Hunting and the likelihood of extinction of Amazonian mammals. *Conservation Biology* **11**, 460–6.

Bodmer, R.E., Fang, T.G. and Ibanez, L.M. (1988). Primates and ungulates: A comparison of susceptibility to hunting. *Primate Conservation* **9**, 79–83.

Bodmer, R.E., Fang, T.G., Luis, L. and Gill, R. (1994). Managing wildlife to conserve Amazonian forests: Population biology and economic considerations of game hunting. *Biological Conservation* **67**, 29–35.

Boster, J., Berlin, B. and O'Neill, J. (1986). The correspondence of Jivaroan to scientific ornithology. *American Anthropologist* **88**, 569–83.

Brosius, J.P. (1997). Endangered forest, endangered people: Environmentalist representations of indigenous knowledge. *Human Ecology* **25**, 47–69.

Brown, M.F. (1978). From hero's bones: Three Aguaruna hallucinogens and their uses. In *The Nature and Status of Ethnobotany*, (Anthropological Papers Vol. 67), ed. R.I. Ford, pp. 119–36. Ann Arbor: Anthropology Museum, University of Michigan.

Camino, A. (1977). Trueque, correrías e Intercambio entre los Quechas Andinos y los Piros y Machiguenga de la montaña peruana. *Amazonia Peruana* **1**, 123–40.

Chernela, J. (1994). Tukanoan know-how: The importance of the forested river margin to neotropical fishing populations. *National Geographic Research and Exploration* **10**, 440–57.

d'Ans, A.M. (1972). Les tribus indigènes du Parc Nacional du Manu. *Actas y Memorias, XXXIX Congreso Internacional de Americanistas* (Lima, 2–9 Agosto, 1970). Lima: Instituto de Estudios Peruanas, **2**(4), 14–19.

d'Ans, A.M. (1981). Encounter in Peru. In *Is God an American? An Anthropological Perspective on the Missionary Work of the Summer Institute of Linguistics*, ed. S. Hvalkopf and P. Aaby, pp. 145–62. Copenhagen: International Working Group for Indigenous Affairs (IWGIA).

da Silva, M.N.F., Yu D.W. and Shepard, G.H. Jr. Matsigenka primate hunting practices and conservation implications in Manu National Park, Peru. (Manuscript in preparation for submission to *Neotropical Primates*.)

Emmons, L.H. (1990). *Neotropical Rainforest Mammals*, illustrated by Françoise Feer. Chicago: University of Chicago Press.

Farabee, W.C. (1922). *The Tribes of Eastern Peru*. (Papers of the Peabody Museum of American Archaeology and Ethnology, vol. X.) Cambridge, Massachusetts: Harvard University.

Fleck, D.W. and Harder, J.D. (2000). Matses Indian rainforest habitat classification and mammalian diversity in Amazonian Peru. *Journal of Ethnobiology* **20**(1), 1–36.

Gentry, A.H. (1990). *Four Neotropical Rainforests*. New Haven: Yale University Press.

Hames, R.B. (1980). Resource depletion and hunting zone rotation among the Ye'kwana and Yanomamo of Amazonas, Venezuela. In *Working Papers on South American Indians*, No. 2: *Studies in Hunting and Fishing in the Neotropics*, ed. R.B. Hames, pp. 31–66. Bennington, Vermont: Bennington College Press.

Hames, R.B. and Vickers, W.T. (1982). Optimal diet breadth theory as a model to explain variability in Amazonian hunting. *American Ethnologist* **9**, 358–78.

Hill, K. and Kaplan, H. (1990). The Yora of Peru. *AnthroQuest* **41**, 1–9.

Izawa, J. (1976). Group sizes and compositions of monkeys in the upper Amazon basin. *Primates* **17**, 367–99.

Janson, C.H. and Emmons, L.H. (1990). Ecological structure of the nonflying mammal community at Cocha Cashu Biological Station, Manu National Park, Peru. In *Four Neotropical Rainforests*, ed. A.H. Gentry, pp. 314–38. New Haven: Yale University Press.

Johnson, A. (1983). Machiguenga gardens. In *Adaptive Responses of Native Amazonians*, ed. R. Hames and W. Vickers, pp. 29–63. New York: Academic Press.

Johnson, A. (1989). How the Machiguenga manage resources: Conservation or exploitation of nature? In *Resource Management in Amazonia: Indigenous and Folk Strategies*, ed. D.A. Posey and W. Balée. *Advances in Economic Botany* **7**, 213–22.

Johnson, A. and Baksh, M. (1987). Ecological and structural influences on the proportions of wild foods in the diets of two Machiguenga communities. In *Food and Evolution: Toward a Theory of Human Food Habits*, ed. M. Harris and E.B. Ross, pp. 387–405. Philadelphia: Temple University Press.

Kaplan, H. and Hill, K. (1984). The Mashco-Piro nomads of Peru. *AnthroQuest* **29**, 1–16.

Kay, C.E. (1998). Are ecosystems structured from the top-down or bottom-up: A new look at an old debate. *Wildlife Society Bulletin* **26**, 484–93.

Lewis-Williams, J.D. and Dowson, T.A. (1988). The signs of all times: Entoptic phenomena in upper Palaeolithic art. *Current Anthropology* **29**, 201–17.

McFarland-Symington, M. (1987). Sex ratio and maternal rank in wild spider monkeys: When daughters disperse. *Behavioral Ecology and Sociobiology* **20**, 421–5.

Milton, K. (1991). Comparative aspects of diet in Amazonian forest-dwellers. *Philosophical Transactions of the Royal Society of London* B **334**, 253–63.

Nations, J. (1980). The evolutionary potential of Lacandon Maya sustained-yield tropical forest agriculture. *Journal of Anthropological Research* **36**, 1–30.

Nelson, B.W. (1994). Natural forest disturbance and change in the Brazilian Amazon. *Remote Sensing Review* **10**, 105–25.

Norgaard, R.B. (1981). Sociosystem and ecosystem coevolution in the Amazon. *Journal of Environmental Economics and Management* **8**, 238–54.

134   G.H. Shepard

Oren, D.C. (1993). Did ground sloths survive to recent times in the Amazon region? *Goeldiana Zoologia* **19**, 1–11.

Parker, E., Posey, D., Frechione, J. and Francelino da Silva, L. (1983). Resource exploitation in Amazonia: Ethnoecological examples from four populations. *Annals of the Carnegie Museum of Natural History* **52**, 163–203.

Peres, C.A. (1990). Effects of hunting on western Amazonian primate communities. *Biological Conservation* **54**, 47–59.

Peres, C.A. (1991). Humboldt's woolly monkeys decimated by hunting in Amazonia. *Oryx* **25**, 89–95.

Peres, C.A. (1993). Indigenous reserves and nature conservation in Amazonian forests. *Conservation Biology* **8**, 586–8.

Plotkin, M.J. (1988). The outlook for new agricultural and industrial products from the tropics. In *Biodiversity*, ed. E.O. Wilson and F.M. Peter, pp. 106–16. Washington, D.C.: National Academy Press.

Plowman, T.C., Leuchtmann, A., Blaney, C. and Clay, K. (1990). Significance of the fungus *Balansia cyperi* infecting medicinal species of *Cyperus* (Cyperaceae) from Amazonia. *Economic Botany* **44**, 452–62.

Posey, D.A. (1983). Indigenous ecological knowledge and development of the Amazon. In *The Dilemma of Amazonian Development*, ed. E.F. Moran, pp. 225–58. Westview Special Studies on Latin America and the Caribbean. Boulder, Colorado: Westview Press.

Posey, D.A. (1984). A preliminary report on diversified management of tropical forest by the Kayapo Indians of the Brazilian Amazon. *Advances in Economic Botany* **1**, 112–26.

Posey, D.A. (1994). International agreements and intellectual property right protection for indigenous peoples. In *Intellectual Property Rights for Indigenous Peoples: A Sourcebook*, ed. T. Greaves, pp. 223–52. Oklahoma City: Society for Applied Anthropology.

Posey, D.A. and Balée, W. (eds) (1989). *Resource Management in Amazonia: Indigenous and Folk Strategies, Advances in Economic Botany*, vol. 7. New York Botanical Gardens.

Redford, K. (1991). The ecologically noble savage. *Orion* **9**, 24–9.

Redford, K.H. and Stearman, A.M. (1993). Forest-dwelling native Amazonians and the conservation of biodiversity: Interests in common or in collision? *Conservation Biology* **7**, 248–55.

Regan, J.S.J. (1993). *Hacia la Tierra sin Mal: La religión del pueblo en la Amazonia*. Iquitos: CETA.

Reichel-Dolmatoff, G. (1971). *Amazonian Cosmos: The Sexual and Religious Symbolism of the Tukano Indians*. Chicago: University of Chicago Press.

Reichel-Dolmatoff, G. (1976). Cosmology as ecological analysis: A view from the rain forest. *Man* **11**, 307–18.

Reichel-Dolmatoff, G. (1978). Desana animal categories, food restrictions, and the concept of color energies. *Journal of Latin American Lore* **4**, 243–91.

Reyna, E. (1941). *Fitzcarraldo, el Rey del Caucho*. Lima: P. Barrantes.

Robinson, J.G. (1993). The limits to caring: Sustainable living and the loss of biodiversity. *Conservation Biology* **7**, 20–8.

Robinson, J.G. and Redford, K.H. (1994). Measuring the sustainability of hunting in tropical forests. *Oryx* **28**, 249–56.

Rosenberger, G. (1977). *Xenothrix* and Ceboid phylogeny. *Journal of Human Evolution* **6**, 461–81.

Ross, E.B. (1978). Food taboos, diet, and hunting strategy: The adaptation to animals in Amazon cultural ecology. *Current Anthropology* **19**, 1–16.

Schneider, H., Schneider, M.P.C., Sampaio, L., Harada, M.L., Stanhope, M., Czelusniak, J. and Goodman, M. (1993). Molecular phylogeny of the New World primates (Platyrrhini, Primates). *Molecular Phylogenetics and Evolution* **2**, 225–42.

Schultes, R.E. and Raffauf, R.F. (1990). *The Healing Forest: Medicinal and Toxic Plants of the Northwest Amazon.* Portland, OR: Dioscorides Press.

Shepard, G.H.Jr. (1997). Noun classification and ethnozoological classification in Machiguenga, an Arawakan language of the Peruvian Amazon. In *The Journal of Amazonian Languages*, Vol. 1, ed. D.L. Everett, pp. 29–57. Pittsburgh: Department of Linguistics, University of Pittsburgh.

Shepard, G.H.Jr. (1998). Evaluación del Proyecto Albergue Matsiguenka, Fase 1: Realización de objetivos, efectividad de capacitación y impactos socio-culturales. Report submitted to Instituto Nacional de Recursos Naturales (INRENA) and Gessellschaft für Technische Zusammenarbeit (GTZ), Lima, December 1998.

Shepard, G.H.Jr. (1999a). Pharmacognosy and the Senses in two Amazonian Societies. Ph.D. thesis, Department of Anthropology, University of California at Berkeley.

Shepard, G.H.Jr. (1999b). Shamanism and diversity: A Matsigenka perspective. In *Cultural and Spiritual Values of Biodiversity*, *U.N.E.P. Global Biodiversity Assessment* Vol. 2, ed. D.A. Posey, pp. 93–5. London: United Nations Environmental Programme and Intermediate Technology Publications.

Shepard, G.H.Jr. (In preparation). Uncontacted native groups and petrochemical exploration in Peru. In *El Dorado Revisited: Mining, Oil, Environment, and Human Rights in the Amazon*, ed. L. Sponsel. Washington, D.C.: AAA Committee on Human Rights.

Shepard, G.H.Jr. and Chichon, A. (2001). Resource use and ecology of the Matsigenka of the eastern slopes of the Cordillera Vilcabamba. In *Biological and Social Assessments of the Cordillera de Vilcabamba, Peru, RAP Working Papers No. 12*, ed. A. E. Alonso, A. Alonso, T. S. Schulenberg and F. Dallmeier, pp. 164–74. Washington, D.C.: Conservation International. (In press.)

Shepard, G.H.Jr. and Rummenhöller, K. (2000). Paraiso para quem? Populaçes indígenas e o Parque Nacional do Manu (Peru). Paper presented at the annual meetings of the Associação Brasileira de Antropologia, Brasilia, July, 2000. [Paper available on the internet at < ftp://ftp.unb.br/pub/download/dan/ F.3-22RBA/sessao2/ shepardrummenhoeller.rtf > , and in press in *Terra das Aguas*, Brasilia: Núcleo de Estudos Amazônicos, Universidade de Brasilia (UnB).]

Shepard, G.H.Jr., Yu, D.W., Lizarralde, M. and Italiano, M. (2001). Rain forest habitat classification among the Matsigenka of the Peruvian Amazon. *Journal of Ethnobiology* **21** (1), 1–38.

Shepard, G.H.Jr., Yu, D.W. and Nelson, B. (2002). Ethnobotanical ground-truthing and forest diversity in the Western Amazon. *Advances in Economic Botany.* (In press.)

Terborgh, J. (1983). *Five New World Primates: A Study in Comparative Ecology.* Princeton: Princeton University Press.

Terborgh, J. (1986). Keystone plant resources in the tropical forest. In *Conservation Biology: The Science of Scarcity and Diversity*, ed. M.E. Soulé, pp. 330–44. Sunderland, MA: Sinauer.

Terborgh, J. (1990). An overview of research at Cocha Cashu Biological Station. In *Four Neotropical Rainforests*, ed. A.H. Gentry, pp. 48–59. New Haven: Yale University Press.

Terborgh, J. (1999). *Requiem for Nature.* Washington, D.C.: Island Press.

Townsley, G. (1987). The outside overwhelms: Yaminahua dual organization and its decline. In *Natives and Neighbors in South America : Anthropological Essays, Etnologiska Studier*, Vol. 38, ed. H.O. Skar and F. Salomon, pp. 355–76. Goteborg: Goteborgs Etnografiska Museum.

Vickers, W.T. (1980). An analysis of Amazonian hunting yields as a function of settlement age. In *Working Papers on South American Indians, (No. 2): Studies in Hunting and Fishing in the Neotropics*, ed. R.B. Hames, pp. 7–29. Bennington, Vermont: Bennington College Press.

Vickers, W.T. (1991). Hunting yield and game composition over ten years in an Amazon Indian territory. In *Neotropical Wildlife Use and Conservation*, ed. J.G. Robinson and K.H. Redford, pp. 53–81. Chicago: University of Chicago Press.

Zarzar, A. (1987). Radiografía de un contacto: Los Nahua y la sociedad nacional. *Amazonia Peruana* **8**, 91–114.

7   Monkey King in China: basis for a conservation policy?

FRANCES D. BURTON

Overview

All peoples construct 'nature'. The division of human FROM nature is such a construct; the development of humanity as controller of nature is enshrined in the myths, legends and religious tales from all over the world. Nowhere, perhaps, is this theme more clearly presented than in the myth of Monkey King, Sun M'Hong (Cantonese) or Sun WuKong (Mandarin). Although his character and some of his exploits may derive from the great Hindu being Hanuman, his development throughout the ancient novel *The Journey to the West* metaphorically addresses the development of humanity in its conscious striving towards a higher state. The obvious physical similarity between monkey and human also serves as reminder of the fragile distance between humans and nature. Traditional respect for monkeys owes much to the place of Monkey King in cultural thinking.

In 1984, when I first surveyed the hybrid macaques of Hong Kong, many of the traditional values were very much in evidence. Gifts to Sun WuKong were made through his mortal manifestations; Buddhism, in its devotion of *Fang Sheng*, encouraged the release of monkeys into the forest and courtesies towards them in the form of gifts of fruits and vegetables. Times change, and the ambience that fosters appreciation for wildlife amongst humans redefines its values in terms of competition for scarce resources, especially land. What was sacred becomes profane; what was honored is debased; pet becomes pest. Personal policy culturally endorsed is forgotten (Zent, 1999) and legislation becomes required in substitution (Yang, Head and Liu, 1994). The problem becomes to develop a set of laws that includes the values, attitudes and beliefs that once guided personal policy when each individual within a group acted in cultural accord with the other members, and to realistically integrate these traditional views with current needs.

Conservation policy needs to develop from a people's value system and knowledge (Kellert, 1993; Gragson and Blount, 1999). The more ingrained the values, the more leverage they might have in evoking the understanding, acceptance and even sacrifice of conservation. Lynn White (1967, p. 1205) put it thus: 'What people *do* about their ecology depends on what

137

they *think* about themselves in relation to things around them.' Religious systems seem like a natural location to find such values. Yet action does not conform to formal religious strictures. Participation in religion does not require adherence to its tenets, hence does not reflect on behaviour. If it did, there would already be conservation as naturally as there would be 'brotherly love'.

There are, however, deeper sources of a people's values. These are reflected in myths and legends which deal with primal questions: life, death, immortality, consciousness, place in nature, social relations and ultimately even purpose, but which are not systematized beyond cosmogony, nor hierarchialized with authority vested in a priesthood. These embedded values, in the Chinese context, come closest to what Julia Ching (1993, p. 206) calls 'popular religion' whose 'basic ideas ... pervade the culture as a whole'.

My suggestion is that folktales may provide the effectual basis for the development of conservation policy by profoundly residing in a people's cultural essence. This fundamental belief system is inculcated in children by way of such stories. The tale of Monkey King as developed in the *Journey to the West* is one that forms a test case. It is very old, although its current form was only written in the sixteenth century CE (Liu, 1964). It mirrors the interests of the syncretism of Buddhism, Daoism, and Confucianism developed in the Ming Dynasty when 'national culture' was deliberately recreated (Ronan, 1992), while presenting the dilemmas and paradoxes of early consciousness. Indeed, the references to the I-ching, the five elements, yin–yang are not part of the three schools, yet pervade the novel (Plaks, 1987), while the characters and events originate from ancient times (Dudbridge, 1970).

Background

The personage of Monkey King pervades Chinese culture. His image is everywhere. He is prominent as a children's icon (Guo Jian, 1986) as he is the subject of cartoon films, as well as children's books. What Monkey King 'means' is the subject here. The goal is to examine a facet of a world-view that behaves like a holograph so that reflecting on this narrow surface permits a vision of the whole. Monkey King continues to reside in the national spirit although relegated nowadays mostly to the domain of children. He has been transformed into a light-hearted being, used to promote everything from political figures to ice-cream. This transformation accords with the different purposes assigned to him over the several thousand years of his history. During the Cultural Revolution, for example, he was used as the main character in a film in a complex crypt-analogy for MaoTseTung. The evocation of his meaning recast Mao to the populace at a time when Mao's image was tarnished (Wagner, 1990). He remains a popular icon. Hence he continues to function as a link between the human-plane and nature, and as such, is important to consider in the development of conservation policy towards corporeal monkeys, which are considered his minions.

Literary criticism as well as religious analyses have illuminated the construction of this being as a device. Here I want to examine Sun WuKong from an anthropological perspective, the impetus for which came from observations of the monkeys of Kowloon and their relationship to humans. These hybrid macaques are descended from five or six interbreeding species and live in protected forest and beside roadways as well (Burton, 1989; Burton and Chan, 1987, 1989, 1996). Their synanthropic relationship developed from human self-interest. The need to build reservoirs meant planting forests to secure a water supply, which in turn permitted the increase of monkeys (Burton, 1989). They have a complex relationship with people, not the least reason for which is Monkey King. Buddhism, Daoism, and Confucianism have contributed to the increase of the monkey population by encouraging the release of monkeys held as pets or bought for this purpose, and by donations of food. At the same time, they are teased or made to perform, and occasionally disciplined and even injured. The interplay of social, political and economic phenomena make any modern story complex beyond understanding. The analysis of Monkey King permits a microscopic evaluation of this complexity.

Human nature: Sun WuKong

The personage of Monkey King engages the debate on 'what is human nature'. Since ancient times, the insights of hunters, farmers, religious leaders, and ultimately philosophers into the workings of the world were the results of empirical observation. The minute and the distant were equally unknowable in the absence of microscopes and telescopes. Technology in itself admits of humanity's physical nature and its limitations. Encountering the very small and dwarfed by the very large, humanity has been forced to revise assumptions about its place in nature: a humbling experience. Over the centuries we have increasingly recognized our kinship with all other living matter, while in fact our dominion over them has distanced us from them.

What is human nature depends on the construct 'nature'. Humans see themselves as animals within the environment, as part of nature. Equally, humans see themselves as progressing out of nature to a plane of existence above the other elements on Earth: the land itself, its waters, the wildlife in, on, and above the Earth. Biological science, especially post-Darwin, has fostered the first view, religion the second. Nature has been a deity and nature has been construed as a matrix, the mother in which all else resides. Nature has been interpreted as friendly, wild, the original, unadorned state; the true essence within all beings, honest, naïve, ingenuous. Humans in the natural state are sometimes pure, and guileless. Conversely, humans in the natural state are unprincipled, devious, disobedient, without laws (i.e. anomic).

Consciousness of our place in nature, of our power beyond nature, has a history as old as the first tracings of human manipulation of the world around. When humankind began to see itself as separate from 'nature' is assumed, however, to be associated with domestication, since the explicit attempt to control life and death of plants and animals marks the turning point (see, for example, Oelschlaeger, 1991). Folktales and religious cosmologies as well as popular performances actualize and enshrine this transformation. The 'dancing monkey' in Japan demonstrates the imposition of cultural control on movement. The performing monkey, forced to a bipedal posture – the human mode – becomes the quintessential metaphor contrasting culture and nature (Ohnuki-Tierney, 1983, 1990, 1995).

Folktales from ancient times resonate with the ambiguity of natural humans. The Sumerian tale of Gilgamesh is a prime example. In this story, Enkidu is wild or natural man, able to communicate with the other animals. He is confronted by Gilgamesh, the great hero who is 'civilized'. Conflict between them is resolved after Enkidu is brought to a temple vestal and, experiencing a woman's 'gift', is transformed (Ferry, 1992). Become conscious of himself, Enkidu can no longer communicate as an equal with the other animals. Some channel from inarticulate being to inarticulate being is closed. It is this *consciousness* that humans have tried to overcome, with chemical or spiritual means (McKenna, 1992). Tu Wei-Ming (1989:76) notes:

> The uniqueness of being human, however, is not simply that we are made of the same psychophysiological stuff that rocks, trees, and animals are also made of. It is our consciousness of being human that enables and impels us to probe the transcendental anchorage of our nature.

The corollary is that animals are not conscious, and live without benefit of laws. Such an existence, presuming only action and reaction, would justify human domination (and exploitation) of them. Here is where legendary figures, acknowledging our basic unity, may assist in conservation policy development. It is the transformation from mere actor to conscious doer that lies at the base of the Monkey King tale.

The novel is one of four great classics (Plaks, 1987). The best-known English translation was by Arthur Waley in 1943 (Waley, 1987), but a newer version, including all 100 chapters of the original, was published by Anthony Yu, beginning in 1977. The story, like all great tales, has an element of truth as it recounts the adventures of the Buddhist monk XuanZang (600–664 CE) to India to retrieve important texts of Mahayan Buddhism (Levy, 1984–85) (the Chinese Tripitaka). In the novel, the pilgrimage to the West is long, arduous and occasionally even dangerous. GuanYin, the Goddess of Mercy, intervenes to assure that XuanZang will complete his task and ensures that he meets several beings to assist him. Each of the four has sinned or erred such that the Jade Emperor, the deity of deities, has had him punished. GuanYin, in her mercy and to satisfy her plan, convinces the Jade Emperor to commute their punishments by assigning them to assist the Monk on his journey:

1 'Pigsy', Chu Pa Chieh (Chu WuNeng), a creature who is half-pig, half-human, originates in the Daoist pantheon where he was a naval commander. He was banished to earth for getting drunk and insulting the goddess of the moon. He is venal, gluttonous, slothful and lascivious. Although he is XuanZang's favourite, he is envious of Monkey King.

2 'Sandy', the monk of the sand (Liu, 1993), or Sha monk, is the least defined being in the story. Converted from cannibalism by GuanYin, he is a gloomy person who wears a necklace of skulls of the nine Chinese deputies sent to find the Buddhist canon, whom he devoured (Werner, 1934).

3 Pai Ma, the white horse, was originally a dragon. Because he ate the horse that the Emperor had given to XuanZang, he was obliged by GuanYin to transform himself into a horse. The horse figures prominently and significantly throughout the tale and is closely interwoven with Monkey King.

4 XuanZang originally lived in Heaven as the Elder Gold Cicada. However, because he fell asleep during one of Buddha's lectures, he was exiled to earth to endure 81 ordeals (Nienhauser *et al.*, 1986).

5 The fifth being, completing the quintessential Buddhist cipher (5), is Sun WuKong, (Sun M'Hong) Monkey King himself.

Figure 7.1. The birth of Monkey King.

It is important to note that the numbers throughout the text are deeply significant. Numbers not only correspond to early cosmogony, where for example, 8 corresponds to wood, 9 to metal, 5 to earth, etc. (Bodde, 1963) but multiples of 8 are Buddhist whereas those of 9 are Daoist. Thus 81 ordeals; 10 8000 leagues; and the interesting error, 9 apertures for *both* male and female.

Synopsis

The history of Monkey King is developed in the first seven chapters. His birth from an agate egg (less valuable than jade) is described (Fig. 7.1), as is his loyalty to his minions, the monkeys, his acquisition of magical powers

Figure 7.2. Monkey King and the Buddha's slaves.

and special attributes. These include his telescopic rod and, his ability to leap prodigious lengths, to splinter himself into armies of monkeys, and to grow, as well as to shrink. He is appointed a stable boy to the Royal Stables in Heaven, challenges Buddha, and steals the sacred peaches. These ultimately result in his capture and subjugation under the Mountain of Five Phases, Wu Xing Shan. His magical powers constitute an alchemy of form, but in one plane. Monkey does not transcend his monkey-ness, despite his prodigious abilities and his acquired immortal status. He may be a great and magical monkey; he may perceive himself as equal to Heaven; he may challenge Buddha himself, but ultimately, he is but a base form. To the Daoists, the ability to fly, to travel on clouds or from one world to another, is the second of three spiritual states where one is freed from all laws of matter, although not yet pure spirit. Higher than this is the 'saint', those

Figure 7.3. Monkey King steals for his master.

who have attained to extraordinary intelligence and virtue; lower than this is the 'immortal', the superhuman who may leave his earthly body as the grasshopper does its sheath (Werner, 1934). Monkey is immortal and can fly, but it is the Journey that will enlighten and liberate him.

The plot of the story is presented in chapter 8, where Buddha sends the Goddess of Mercy, GuanYin, to find a suitable pilgrim for the arduous task of going to India to receive the Sacred Texts. XuanZang and his disciples are identified and introduced. The following three chapters (9–12) explain XuanZang's mission to fetch the sacred scripture, and give the background to this project. The pilgrimage itself, its 81 ordeals, the return and reward occupy chapters 13–97. Chapters 98–100 fulfil the sub-text: the Disciples are presented to Buddha as they return to China with the sacred texts, and they are rewarded.

Delivered from their mortal flesh and bone,
A primal spirit of mutual love has grown.
Their work done, they become Buddhas this day,
Free of their former six-six senses' sway.

(Yu, 1983, p. 384).

Sun WuKong earns the title 'Buddha victorious in strife', a just acknowledgement of his loyalty, enterprise and devotion. It is he who is the focus of
the tale. His personal growth, his reaching beyond his early exploits of
mischief and self-aggrandizement to become the central force enabling the
mission to succeed, his comprehension of himself as 'Aware of Emptiness'
– the meaning of his name – convey the major theme of the novel.
Developed by Nagarjuna (*ca.* 100–200), this doctrine is stated:

Nothing comes into being,
Nor does anything disappear
Nothing is eternal,
Nor has anything an end.
Nothing is identical,
Or differentiated,
Nothing moves higher,
Nor moves anything thither.

(Ching, 1993, p. 133).

The *meaning* of the Journey

In the early part of the twentieth century, scholars like Hu Shih (1942)
considered *Journey to the West* (JW) to be structured as a satiric view of life
and the world. In his preface to Waley's translation of JW, he notes that it
is 'A book of ... profound nonsense ...' (1984, p. 5). At the other end of the
spectrum, the meaning of the Journey has been assigned to its role in
explicating one or another of China's great religions (see, for example,
Bantly, 1989). Scholars have debated for which of the Chinese religions the
Journey is an allegory, as the elements clearly presented are drawn from all
three major religions, despite the theme of a pilgrimage to obtain Buddhist
scripture. The fusion of Daoist, Buddhist and Confucianist thought was
the explicit intention of Ming Dynasty scholars following the twelfth
century origin of the San Chiao Kuei-I (Gernet, 1970; Ching, 1993). To Yu
(1977) and Plaks (1987), for example, the purpose of the *Journey* may have
been as a vehicle to present this possibility in the creation of a 'national
culture' (Ronan, 1992). Alternatively this syncretism is a by-product of the

period in which it was written, at a time when the dialectic and attempts at mutual extirpation of these three systems result in adoption of each other's themes (T.-Y. Liu, 1964; H. Liu, 1993). The greatness of the work lies in its construction as '...a self-consciously programmed allegorical composition' (Plaks, 1987, p. 224, ft. 118); the purpose of which is a 'pilgrimage in self-cultivation' (Yu, 1977, p. 37). The surname given to Monkey illustrates the Daoist 'Doctrine of the Baby', that is 'internal alchemy to designate the maturation of the holy embryo in one's body when immortality is reached' (Yu, 1977, ch. 14, p. 59). The main disciples gain immortality in Daoist fashion, through a process of self-cultivation. Their success in internal alchemy qualifies Monkey King, 'Pigsy' and 'Sandy' to be the guardians of the monk. The monk XuanZang is not at all like the real heroic figure. The real monk lived in the seventh century CE, and made the journey to India to fetch the sacred scriptures when he was 27 (Dudbridge, 1970). The mythologized monk of the *Journey* is cowardly and, despite his Buddhist vows, committed to bodily comforts.

The elements in the novel, and perhaps more cogently its history, reveal a level even more profound than this syncretic philosophy. The threads of this tale are deeply woven into folklore, illuminating the question of whether or not a conservation policy can be formulated which evokes this folk hero. It constitutes a 'pool of meaning' (Ohnuki-Tierney, 1987, in Kiefer, 1988, p. 645) emanating from 'popular religion' (Dudbridge, 1970, p. 166). The story line is so old, the motifs so ingrained, that the religious meaning has been dissipated. Ching identifies popular religion in China as crossing

> ... the lines of demarcation between Confucianism, Taoism, and Buddhism... There is no single sacred text or set of documents that can speak for it (1993, p. 206).

Identification of Monkey

'Monkey' figures as an important personage in early Chinese folklore. He is the ninth animal in the Twelve Terrestrial Branches (the zodiac). He is said to be born on the 23rd of the second month and to have general control over hobgoblins, witches, elves and the like. Monkey bestows health, protection and success (Williams, 1975). The appearance of this zodiacal 'monkey' is a generalized form, although certainly a macaque.

The archetype monkey king is Hanuman of India. He plays an important role on the side of good against an evil force that abducts the wife of

Rama and carries her to Sri Lanka. Hanuman's earthly representation is a langur, of the species *Semnopithecus* (formerly *Presbytis*) *entellus*. This species ranges throughout India, Nepal, and Bhutan to the border with China (Burton, 1995). Some authors take it for granted that Hanuman is the prototype for Monkey King. Ball (1969), for example, observes the similarity in the fiery eyes, the ability to transform, to leap vast distances, his relationship to various deities, and above all, his bravery and loyalty to his human master. Noting the presence of Tibetans in Sichuan, where so many of the monkey cult materials originate, Dudbridge (1970) argues that the Tibetan influence is paramount, especially since Tibetans trace their origins to monkey ancestors, and in one early folktale depict a monkey with attributes quite similar to Monkey King.

Perhaps, however, Monkey King is more appropriately identified as a gibbon. Scholars have confused gibbons with langurs, as when Dudbridge (1970) identifies 'the white ape' (Po-Yuan, Bao Yuan) as the silvery langur. Van Gulik (1967) examines the term 'nao', which refers to either monkey or gibbon. As drawings were often made from hearsay, they are not reliable sources for early periods either. Initially, the classifier for *nao* was an insect. The metaphor is clear: the insect changes its skin, metamorphoses, transforms. The use of this classifier, I believe, recognizes the historical transformation of primates to humans. Interestingly, when the term 'yuan' replaces 'nao' for the gibbon, the classifier becomes a quadruped (van Gulik, 1967). (Chinese ethnic minorities are also classified with this character.)

There are three species of gibbons in China today, concentrated in the south and southwest, including *Hylobates hoolock*, *H. concolor* and *H. lar* (Bartlett, 1999). The females of these three species are buff, tawny or white in colour, while the males are brown, or reddish brown or black as is the case with *H. lar*, with white patches along the cheek. In general body form, except for the absence of a tail, gibbons resemble the gracile leaf-eating monkeys of southeast Asia in their lean appearance and long limbs. The Chinese variants are black, or black with white heads (*Trachypithecus francoisi; T.f. leucocephalus*) and occupy southwest China, near the border with Vietnam.

The most abundant monkey genus in China, however, is the macaque. There are three species, *Macaca mulatta* (the common, rhesus monkey), *M. thibetana* (the bear, Pere David's, or Tibetan macaque) and the assam monkey (*M. assamenensis*). *M. mulatta* extends further to the east and to the south than the other two species. Macaques are red-bottomed and have tails that range from moderate to very short (*M. thibetana*). Fur colour ranges from dark brown or almost black (*M. thibetana*) to golden yellow, or beige (*M. assamensis*). Faces are pronounced, and their hands are proportioned like humans', with a highly opposable thumb. Their habitat ranges from rainforest to open forest, where they are just as agile in trees as they are on the ground. In China they live in the valleys as well as amongst the karst mountains, although there (e.g. LongZhou in Guangxi) they occupy lower levels than do the *Trachypithecus* monkeys (Burton *et al.*, 1995).

The text of the novel provides several physical descriptions of Monkey King. 'I have a round head sticking up to Heaven and square feet treading Earth,' said Monkey. 'I have nine apertures, four limbs, five upper and six lower internal organs, just like other people.'

'But,' said the Patriarch (a Daoist figure), 'you have much less jowl. For monkeys have hollow cheeks and pointed muzzles.' Monkey replies, 'I have my pouch, and that must be credited to my account, as something that ordinary humans haven't got.' Monkey King goes bare-headed and wears a red dress, with a yellow sash, and black shoes. This is a fitting description for *M. mulatta*, whose reddish fur on shoulders and back yields to lighter, more yellowish fur on the abdomen, and whose feet and hands are black. The 'pouch' to which Monkey King himself refers would likely be the cheek-pouches common in macaques, but not gibbons or leaf-eating monkeys. There is little mention of a tail, except in the episode where Monkey King attempts to avoid capture by changing his appearance to a wayside shrine, where his eyes are the windows, his mouth the door, and his long tail is disguised as a flag-pole. Macaques with such long tails do not live in China. They are members of *M. fascicularis*, from the more southerly reaches of southeast Asia, where, as in Bali, they are the representatives of Monkey King (Wheatley, 1999).

One of Monkey King's epithets is 'Fiery Eyes'. His red eyes – a characteristic he shares with the actual red-rimmed eyes of *M. mulatta* (and an attribute of Hanuman as well) – resulted from his punishment in a crucible. The smoke from the fire set to destroy him caused the condition from which he and his kind never recovered. Monkey King is also referred to as a 'cursed monkey', as well as a 'stinking ape', an epithet given him by Buddha. He is also 'a red-bottomed horse ape', which suggests *M. thibetana* or, again *M. mulatta*, both of which are large animals, with pronounced redness on the bottom, especially during reproductive phases. This association of monkey with horse is an important connection with multiple levels of meaning.

The origin and diffusion of Monkey King

The origins, significantly then, of this major character may well be a composite of nonhuman primate traits known for centuries to humans in contact with them. It is possible that Monkey King derives from other Asian sources. Originally a langur in India, with grey–white fur, he would have become a gibbon further north and east, looking superficially similar in colour, and general body conformation (except for the tail). As the legend entered China, he would have remained a gibbon given the ancient distribution of gibbons as noted in literary texts (van Gulik, 1967). Alternatively, he would have returned to being a leaf-eating monkey (perhaps *Trachypithecus*), this time retaining body form and agility, although not colour. Further east, with macaques more numerous, the final form of Monkey King emerged.

In the ancient world, the monkey was pet, worker, sacred being, and even mediator to the gods (Cammann, 1962; Ball, 1969; Morris, 1967; Langdon, 1990; Shaw, 1993; Ohnuki-Tierney, 1983, 1995). In the Ancient Aegean, (i.e. in the second millenium BCE), a variety of monkeys are shown working with or for humans, or more significantly, bestowing gifts on deities, as in the fresco in Thera where a goddess is offered the stigmas of crocuses which monkeys have helped to gather. Shaw's (1993) discussion of this European use of monkey as mediator for humans to the gods contrasts to the views of Asquith (1995) as well as Ohnuki-Tierney (1995). Although this role is important in Japan, they consider that the 'rational', European mind relegates 'monkey' to a status that precludes a role as mediator to the divine. The diffusion of monkey-gods and monkey-figures as early as Iron Age Sardinia and Italy, and as far distant as Phoenicia to the west and east to India (Langdon, 1990, p. 419), suggests yet another connection towards China, especially since trade routes from China to the west were established before the silk route, even in Shang times (Yetts, 1912).

Folk legend relates Chinese tales to those further to the west. The ancient story of a baby put in a basket among the rushes to save its life, best known in the tale of Moses, is repeated in the JW. The mother of the monk XuanZang left him to float on the river to protect him from a bandit, who had disposed of his father, usurped his role as official, and abducted his mother. Dennys (1876) notes some similarities to Arabic tales. Certainly contact between the two areas was frequent by the Neo-Confucianist period, around 1000 CE (Ronan, 1992).

Alternatively, the different forms and epithets attributed to Monkey King in the *Journey* may suggest more than one author (see, for example,

Liu, 1964) and this despite the general acknowledgement of Wu Ch'eng-en as the novelist in the 1600s. Plaks (1987) concurs with Liu (1964) that he may have compiled the novel, but was not the actual author, as the earliest extant edition dates from 1592 CE, and Wu Ch'eng-en had died 10 years earlier. The preface to that edition, however, does allude to an earlier version, as does a play on a similar theme. Liu (1964) considers the *Journey* one of four novelettes, deriving from 'prompt books' for story tellers. He concludes that the *Journey* was written in the Wan-li reign of the Ming Dynasty (1573–1619), derived from the *T'ang Tripitaka Master's pilgrimage to the West* compiled by Chu Ting-ch'en. Hence, Liu (1964, p. 63) concludes that Wu Ch'eng-en 'plagiarized' the story, and its prime character, Monkey, while adding some of the stories, injecting some of the meaning and adding his own embellishments. The current 'official' version, attributed to Wu Ch'eng-en, is actually adulterated by additions made by Wang Tan-I in the early Ch'ing (Liu, 1964).

Certainly the roots of the main character are ancient. Dudbridge (1970, p. 115) traces him back to around 25 CE, in the Han Dynasty, in a text '*Yi-Lin*'. However, the story Dudbridge (1970) considers antecedent to Monkey King in *Journey* is the Wu Chih Ch'i ape legend. The Ta Yu version of the story was known in the early ninth century CE (T'ang Dynasty). It concerned a black monkey with a white head, who was a female river demon (reminiscent of the white-headed leaf-monkey, *T. f. leucocephalus*, which lives along the MingJiang river that runs into Vietnam). The female river demon is Shui miu or Sheng mu and her brother is Hou Xing Che, the name of Monkey King before he receives his religious appellation (Sun WuKong). There are other early stories whose monkey character seems later to fuse with Monkey King. Dudbridge (1970) notes that the Daoist Cult of Ch'i t'ien Ta-sheng in Fukien celebrated a monkey or anthropomorphic deity with a monkey's head. This 'monkey spirit', like the later Monkey King, stole the peaches of Heaven in his quest for immortality. The younger brother of this character is Xing-che, another version of Hou Xing-che, or Monkey King before he receives his religious name. In Fuzhou, Sun Xing Che was worshipped as a household god. Fuzhou was also the location of another monkey story, thought to date from the twelfth century, that concerned a wounded monkey used as a cult object. This monkey was deemed to have malignant power and was controlled by a Buddhist, a significant theme in the *Journey* story, where Monkey King's behaviour causes him to be severely disciplined.

Mention has already been given to the white ape, Po-Yuan, the 'gibbon' (Van Gulik, 1967). This originally malevolent being was known in the T'ang Dynasty (eighth to tenth centuries CE), as an abductor (Dudbridge,

1970). Dudbridge does not consider the White Ape or Gibbon to have influenced the Monkey King story, although Shahar (1992) sees it as the major precursor. His argument rests on the fact that the antecedent story to the White Ape is the story of Huili, a Buddhist monk, and Sun Tuan, the simian. Around 330 CE, the real Huili founded a monastery at Lingyin Si, near Hanzhou. The monks actually reared 'monkeys', which became the source of several literary forms: tales and plays. They were fed at special locations, and poems were written about them. Shahar (1992) finds that the Lingyin Si story shares many features with JW. As in the story of XuanZang, Huili journeyed to India with Sun Tuan to get to the Western Paradise. Like Monkey King, the hero of this story lived in a cave on the slopes of a magic mountain, and while himself immortal, was the disciple of a mortal priest. Shahar notes that of the 60-odd poems written between the Tang and the Song Dynasties about the simians at this monastery, all but three identify them as gibbons. The exceptions note macaques (Shahar, 1992, p. 201). What was impressive to the Chinese about these lesser apes was their uncanny songs, their stature and taillessness, their ability to move bipedally, their monogamy, and their solitary habits. To the Chinese, this was quintessentially the behaviour of a gentleman (Van Gulik, 1967; Shahar, 1992). So the gibbon, too, has a dual role as a negative figure and yet as the very essence of civility (so dear to Confucius).

Whichever version may be antecedent to JW, all concern transformative change. This emphasis, so much at the core of Daoist thinking and of its studies in alchemy, is overlain with Buddhist themes and interspersed with Confucian ideals. The Daoist story *Pao Pu Tzu* relates the transformation of simians through several forms. It is virtually an evolutionary sequence in the modern, biological sense. 'Hou', the macaque (mandarin), when he is eight hundred years old, becomes an yuan, a gibbon, which after five hundred more years become known as a ch'ueh ch'üeh, which means either large monkey or large ape depending on the gloss (Dudbridge, 1970, p. 115), and segues after an additional thousand years into yet another form for an additional thousand years, only to finally assume the form of an old man (Ball, 1969, p. 117). The sequence is one of *transformations*. This story antedates an interesting Buddhist tale, the story of Yuan Sun, which name is a play on the word 'monkey' (Dudbridge, 1970, p. 116). The earliest version of this story has been traced back to 1420 CE. In it, Yuan Sun undergoes two transformations: (a) to Buddhism, and (b) when it is found dead, to a monkey, recalling the *Jatakamala* tale where Buddha himself was once a monkey (Batchelor, 1992; Aryasura, 1989). Yet Ball (1969) notes that the Buddhists were generally antipathetic towards monkeys, and classed them as one of the 'Three Senseless Animals'. Along with the Tiger, who was always angry, and the Deer, who was always love-sick, the Monkey was always greedy and covetous.

This Ming emphasis on Monkey, with his important role in human spiritual development, undoubtedly therefore owes a great deal to the influence of Daoism. Before contact with Buddhism (after 67 CE), Daoism had no knowledge of reincarnation, of life as suffering, or Nirvana. It was Buddhism that introduced the universe as atomistic, anarchic, and unorganized (Bodde, 1963). To Daoists, the natural state of humanity is like uncarved wood (Waley, 1984). Immanent in its transformation into other things, it is material not yet adulterated by knowledge or circumscribed by morality. This primeval chaos (i.e. undifferentiatedness) is when humanity was in a state of absolute harmony between himself and his surroundings. Then came the culture heroes, inventors of fire, agriculture and the like. Hence, Daoism looks upon the Sages, and their delivery of 'civilization' and knowledge, as misguided altruists (Waley, 1984).

Monkey of the mind

Chapter 14 of JW is titled *Monkey of the mind*: the most important metaphor of the novel. This chapter deals with 'bridling' Monkey of the mind and Horse of the will. Monkey of the mind is chaotic, undisciplined, unfocused. Monkey is the mind that comprehends and helps pilgrims out of problems. It is Monkey who kills the six robbers that come to attack the monk. These six robbers are the Buddhist metaphor for the six senses, which impede enlightenment. Only by going beyond the sense boundary can one arrive at true knowledge. Monkey earns the name 'WuKong', which means 'Awake to Vacuity'. He has superior knowledge of the Emptiness of All Things. It is this knowledge, derived from his experiences tempered by his character, that have enabled him to transcend his physical bounds, to *transform* himself much as an alchemist transforms base metal into nobler stuff. Monkey teaches not only that the true nature of experience is illusory, but that these experiences – however unreal – can become vehicles for enlightenment. Yu notes, 'Revealed in his person is the true attitude toward the doctrine of emptiness ... which signifies the ever-changing state of the phenomenal world, a constant flux of becoming' (chapter 14, p. 60). Note too, that Monkey does not *transmigrate* (Ronan, 1992): – that is, it is not his *soul* that reincarnates in a higher being: he is *already* an immortal (Daoist), and his transformation is recognized by his acts, hence his reward.

'Monkey of the mind' has several referents. Dudbridge considers the metaphor purely a Buddhist figure (1970). He traces its antiquity to the *Vimalakirti Sutra*, dating from the early 400s CE, where control of self is expressed by the association of Monkey with the unruly, undisciplined mind. By the T'ang dynasty, between the 700s and 800s CE, the metaphor is fixed as calming the monkey of the mind, and stilling the horse of the will or desire, but remains within the Buddhist context. Texts written in the thirteenth century indicate the increasing use of the metaphor as a 'stock allusion' (Dudbridge, 1970, p. 169), so that it is increasingly extended beyond Buddhism. Less concerned with strife and self-conquest than is Buddhism, the concept is nevertheless incorporated into its rival, Daoism (fourteenth century). Dudbridge (1970) emphasizes that the ongoing 'de-valuation' of this metaphor ultimately shifts its allusion from the spiritual to the sexual, and by the sixteenth century, when the novel is written down, these terms are commonly used in ribald association to desire and virility. The sexual allusions and play-on-words in the novel are rife. Plaks, (1987, p. 231) notes that the term 'firing time', huo-hou, is also the term for 'fire-monkey', i.e. Sun WuKong. Another reference is the 'dual cultivation of the physical and essential self', which by the Ming had come to refer to cults of sexual union.

The metaphor, however, arguably has precedence within Daoism not only as the natural state of humanity, but in association with the primeval state of chaos (that is the undifferentiated state) (Ronan, 1992), which exists before nature is affected by human power. Monkey is associated with the moon, whose reflection on water is but an illusion. In ancient times people will have observed that monkeys raise their arms above the head when they wish to be groomed on their abdomen or chest. The gesture was taken to be prayerful and associated with worship of the moon (Ball, 1969). The association is complex. The light of the moon casts reflections of itself on water: 'In this world of illusion, man is truly still the ape trying to seize on water, the reflection of the moon' (Ball, 1969, p. 119). To the Buddhists, Monkey must become conscious of himself in order to rise towards his Buddha state. To the Daoists, humanity is a transform of monkeyhood.

It is probably therefore that Monkey's punishment – and his crimes – are the most severe of all. He has committed hybris, stealing the sacred peaches of Heaven. He has dared to consider himself the equal of God and has named himself, 'The Sage Equal to Heaven'. For these sins and other similar acts of selfishness, arrogance, pride, egotism and conceit, and because he has become immortal and cannot be destroyed, he is punished. He is immobilized by means of a net of control causing incredible pain encircling his head, and he is buried beneath the Mountain of Wu Xing

Shan. Just as it was GuanYin who helped to capture him, it is she as the Goddess of Mercy who brings about his release in order that he might assist XuanZang on his Journey. Throughout the novel, it is Monkey whose bravery, kindness, wit, foresight and profound intelligence extricate the group from danger after danger.

Monkey's companion in the metaphor is 'Horse of the Will', who is white. A beast of burden in the story, rather than a disciple, he was originally a dragon prince (Pai Hu) (Ball, 1969). His relationship to Monkey is known from ancient times where monkeys are depicted riding horses. The complexity of the dual symbol includes sexuality, and also control. Like the Centaur of the ancient Greco-Roman world, the composite symbol of horse and monkey-rider reflects a powerful being, with the wit and agility of a monkey *cum* human, and the power and loyalty of a horse. The horse is one of the treasures of Buddhism, and is classed under the element 'fire' (Williams 1975). Pure white horses were considered, in the *Pen Ts'ao* (the Chinese *materia medica* published in 1596) as the best for medicinal use; the horse's eye is considered to reflect the human image (Williams, 1975), and most assuredly the monkey body too obviously reflects the human form. The horse is associated with the number eight: he is the eighth animal in the zodiac of the Twelve Terrestrial Branches.

The association between monkey and horse is an ancient and cosmopolitan one. Fibulae from tombs in Sardinia, dating from the Iron Age, depict horses with monkeys on their backs. Monkeys have long been associated with fecundity in the ancient Levant, where they were a 'potent symbol of fertility, a natural mirror for man' (Langdon, 1990, p. 423), whose emergent form was the Centaur. In the Chinese Zodiac, monkey, the ninth creature, follows the horse. Monkeys are the subject of small sculptures particularly associated with horses. These toggles for the bridle often depicted fertility. Cammann (1962) describes a 'notable' example depicting a monkey family of 18 young gathered around the leader monkey. This elder is seated in an armchair, which is being held up by other small monkeys. Its meaning is 'May your illustrious descendant have many descendants of his own'. Pictures or toggles depicting a monkey mounted on a horse were once very popular. The usual interpretation of this motif is 'ma-shang feng-hou', 'crazy monkey on horse's back' (Cammann, 1962, p. 125). But the Chinese language is eminently suited for double entendres, and Eberhard (1986, p. 192) gives its importance in the homonymic reading of this phrase: 'May you be straight away (ma-shang) elevated to the rank of count (feng-hou)'. Hence the association of 'crazy' and 'monkey' in Chinese, unlike its Japanese counterpart (Ohnuki-Tierney, 1990, 1995; Kiefer, 1988), does not deprecate the monkey or the human recipient of the designation.

Conservation

The intertwining of philosophies characterizes Chinese thought, and is purposefully syncretized in the Ming period. The three great philosophi-cal–religious systems, Daoism, Confucianism, and Buddhism, vied with each other, attempted at various times to violently extirpate each other (see, for example, Reischauer, 1955), influenced each other and so ulti-mately came to share values and cosmology. Buddhism advocates respect for the environment and the animals within it, since its cosmology recog-nizes that at one time all male animals were our father, and all female animals our mother. The Doctrine of Emptiness asserts that all things are unified; this is not to deny that they exist singularly, rather it emphasizes that all things are interdependent (Batchelor, 1992). The 'Jewel Net' of Indra (of the Hua-Yen school) presents the image of multi-facets each perfectly reflecting an image. Cook (1989, p. 225) explains the holographic image thus:

> If we know reality in the form of one phenomenon, then we know all of reality. It is for this reason that Hua-yen can make the seemingly outrageous claim that the whole universe is contained in a grain of sand.

Sun WuKong's epithet, 'Awakens to Vacuity', reflects this profound notion. It is the consciousness of this state that elevates Monkey, and it is in this process that one level of meaning of the *Journey* becomes clear: as humans are related to monkeys, they are like monkeys, at first without consciousness or conscience. In action both mischievous – even wicked – and good, the possibility of Enlightenment, of Awareness is his. Morowitz (1989, p. 48) clarifies the meaning of this when he compares the doctrine to modern thermodynamics:

> ... the structures out of which biological entities are made are transient, unstable entities with constantly changing molecules, dependent on a constant flow of energy from food in order to maintain form and structure. This description stands as a scientific statement of the Buddhist notion of the unreality of the individual. Among western thinkers, one might identify this kind of description with Heraclitus.

The major Daoist theme is transformation; an autogenerative, self-construed 'arising' (Ames, 1989, p. 125). The infinite potential for develop-ment is the way of the Dao: The Great Harmony. Alchemy, the transform-ation of one material form into another, is both a goal and a process. Since transformation is self-construed, an internal generation, it is *immanent*. It is the recognition of this that thematically runs through *Journey* in the personage of Monkey. The sacred as it attaches to Monkey, then, inheres in his being. And it is here that a reverence for life might be expected to develop. Cheng (1976, p. 18), identifies three principles common to Confucian and Daoist metaphysics:

1. the principle of holistic unity
2. the principle of internal life-movements [immanence] – all things are interrelated to form a network of interchange of processes, the transmission of moving force is conceived of as an exhibition of life activity . . .
3. the principle of organic balance and harmony

To which can be added the belief that the human spirit is liberated when it is one with natural surroundings.

Cheng (1976) concludes that Chinese thought is basically anti-atomistic and anti-mechanistic. But if this organic, holistic, interdependent view of living forms and the cosmos are so ingrained in Chinese thought and thematic in JW, how can one account for the massive environmental destruction that has occurred – and still occurs – in China despite recent efforts to the contrary (Harkness, 1998; Jahiel, 1998)?

Clearly, this deeply ingrained philosophy has not been sufficient to safeguard living forms or the environment (Niu and Harris, 1996). As Tuan (1968) demonstrated in his seminal work some decades ago, Chinese philosophical systems were *never* sufficient. Behavior towards wildlife is not constrained by sympathy for it. One of the earliest legends about monkeys concerns the self-sacrifice of a monkey king for his constituents. One day, a piece of unpicked fruit falls from the monkeys' special tree into the river and floats away. In a nearby kingdom the King finds the fruit so exceptional in perfume and taste that he organizes a gathering party. Finding monkeys in the fruit tree, the King orders his bowmen to shoot them. However, the monkey king makes a bridge of his body for the monkeys to pass over. As they run across him in panic, he is severely injured, and dies of this sacrifice (Cammann, 1962). The sensitive, compassionate empathy for the captive gibbon as reflected in genre poems does not stop people from capturing them and restraining them, albeit on a golden leash. Perhaps because the continuity between animal and human form is obvious, their treatment is not differentiated. Thus, when a King loses his pet gibbon, he has the entire forest cut down the better to find his beloved pet (Van Gulik, 1967).

The European romanticization of China had it that all negative activities towards the environment and wildlife were inspired by Western influence (Tuan, 1968). Yet human misuse of resources was already known in ancient times. The ingenious system of stacking steamers over woks, begun already in the Shang, was due to the difficulty *at that time* of obtaining fuel (Anderson, 1988). Tuan's (1968) list of ecological disasters notes that by the Song Dynasty (960–1279 CE) the demand for wood and charcoal could no longer be met. In the thirteenth century CE, the area around Hanzhou was totally deforested for timber to construct palaces. Indeed, it was often the religions themselves that inspired environmental destruction, as when Buddhism introduced cremation, with a resulting effect on local timber resources. Tuan quotes Schafer: 'Even before T'ang times, the ancient pines of the mountains of Shan-tung had been reduced to carbon, and now the busy brushes of the vast T'ang bureaucracy were rapidly bringing baldness to the T'a-hang mountains between Shansi and Hopei' (1968, p. 191) and this because of the requirement for black ink, which came from the soot of burnt pine. More recently, the ecological disasters of the Great Leaps Forward and the Cultural Revolution are certainly on a par (Chung and Poon, 1999).

The paradox of possibility

Perhaps, as with mythology itself, this *coincidentia oppositorum* (W. Lai, 1994, p. 33) is to be accepted: more paradox than contradiction. But is it workable? What elements could be drawn from Chinese popular religion and what Ching (1993) has termed 'Chinese thought', so clearly expressed in the personage of Monkey King and the novel JW? This 'archeology of metaphor' describes what we find here with Monkey King: an ancient theme profoundly attached to core values, opinions, and assumptions about how the world works. Clearly, there is much that can be evoked in his personage that would bring along with it a renewed interest in his associates – monkeys – and their habitat. His physical qualities are admirable. He is strong, lithe, athletic (can leap extraordinary lengths), well-built and well-proportioned. He therefore meets Chinese criteria of beauty and perfection of being. His personal abilities and attributes are, however, what makes him eligible to spearhead a campaign because they mark him as singular amongst mortals. He is capable of magic: sometimes because the objects he possesses are magical such as his rod; sometimes because he himself is magically endowed. He can fly. He can make prodigious leaps, as when he attempts to escape from Buddha's hand, and because he is a mischief, urinates and leaves graffiti on a 'pillar', only to discover that he has never left Buddha's presence: the defiled pillar was one of Buddha's fingers. His enormous intelligence is symptomatic in his ability to deceive, as when he stole the peaches of Heaven. More importantly perhaps, it is evident in his ability to understand, as in his sympathetic acceptance of the

Table 7.1. *Attributes of Monkey King*

1. Physical qualities	1. Abilities	2. Attributes
cute	magic	intelligence
handsome	trickery/deception	wisdom
lithe	understanding	humour
athletic		mercy
		knowledge
		leadership
		remorse
		repentance
		mischief
		childishness
		innocence
		loyalty

limitations of XuanZang, who would be at a loss were it not for Monkey King's skills in organizing and protecting the group. His attributes are virtually legion! This is no mere 'animal': this is a creature endowed with Buddha-essence, a Daoist transform. Already an Immortal at the beginning of his journey, it is his personal attributes of intelligence and character that enable him to transcend, to become fully conscious, and in his newly awakened state, he is justifiably rewarded with the title, 'Buddha victorious in strife'. His are the attributes of childhood: innocence, mischief, humor, open-heartedness and remorse. His also are the attributes of a consciousness matured: loyalty, wisdom, repentance and mercy, and hence leadership.

When commercial film-makers in Disney's studio produced *Bambi*, it was surely not their intent to virtually create a movement. Cartmill (1993) has shown how this story, and more importantly its commercial realization, brought a generation to a new appreciation for wildlife and a new attitude towards nature and its treatment. As mawkish as some of that may be, children matured carrying with them into adulthood a sympathy – even empathy – that ultimately affected conduct and legislation. Recently, Bernbaum (1998) has reflected on conservation projects that take their inspiration from 'enduring' religious and cultural traditions. Monkey King is one of these and remains a popular figure. He is the original 'super-hero', whose image has been exploited in mass marketing of food and toys. His exploits and character seem eminently suited to become the basis to evoke conservation.

References

Ames, R. (1989). Putting the Te back into Taoism. In *Nature in Asian Traditions of Thought: Essays in Environmental Philosophy*, ed. J.B. Callicott and R.T. Ames, pp. 120–35. Albany: State University of New York Press.

Anderson, E.N. (1988). *The Food of China*. New Haven: Yale University Press.

Aryasura (1989). *Once the Buddha was a monkey: Arya Sura's Jatakamala*. (Translated from the Sanskrit by Peter Khoroche.) Chicago: University of Chicago Press.

Asquith, P.J. (1995). Of monkeys and men: Cultural views in Japan and the West. In *Ape, Man, Apeman: Changing Views, 1600–2000*, ed. R. Corbey and B. Theunissen, pp. 309–25. Leiden: Royal Netherlands Academy of Arts and Sciences.

Ball, K. (1969). The Monkey. *Decorative Motives of Oriental Art*, chapter XVI, pp. 117–24. New York: Hacker Art Books.

Bantly, F.C. (1989). Buddhist Allegory in the Journey to the West. *Journal of Asian Studies* **48**, 512–24.

Bartlett, T. (1999).The gibbons. In *The Nonhuman Primates*, ed. P. Dolhinow and A. Fuentes, pp. 44–49. California: Mayfield Press.

Batchelor, M. (1992). Even the stones smile. In *Buddhism and Ecology*, ed. M. Batchelor and K. Brown, pp. 2–17. New York: Cassell.

Bernbaum, E. (1997). Sacred mountains. *The UNESCO Courier, Paris* **9**, 34–6.

Bodde, D. (1963). *China's Cultural Tradition, What and Whither?* New York: Rinehart.

Burton, F.D. (1989). *The Monkeys of Kowloon*. 30 min video. Scarborough: Omega Films.

Burton, F.D. (1995). *The Multimedia Guide to the Non-Human Primates*. Print version. Toronto: Prentice Hall.

Burton, F.D. and Chan, L.K.-W. (1987). Notes on the care of long-tail macaque (*Macaca fascicularis*) infants by stump-tail macaques (*Macaca thibetana*). *Canadian Journal of Zoology* **65**, 752–5.

Burton, F.D. and Chan, L.K.-W. (1989). Congenital limb malformations in the free-ranging macaques of Kowloon. *Journal of Medical Primatology* **18**, 397–403.

Burton, F.D. and Chan, L.K.-W. (1996). Behaviour of mixed species groups of macaques. In *Evolution and Ecology of Macaque Societies*, ed. J.M. Fa and D.G. Lindburg, pp. 389–412. Cambridge University Press.

Burton, F.D., Harrison, S.E. and Snarr, K.A. (1995). A preliminary report on *Trachypithecus francoisi leucocephalus*. *International Journal of Primatology* **16**, 311–27.

Cammann, S. (1962). *Substance and Symbol in Chinese Toggles*. Philadelphia: University of Pennsylvania Press.

Cartmill, M. (1993). The Bambi syndrome. *Natural History* **102**, 6–10.

Cheng, C.-Y. (1976). Model of causality in Chinese philosophy: a comparative study. *Philosophy East and West* **26**, 12,18.

Ching, J. (1993). *Chinese Religions*. Maryknoll, NY: Orbis Books.

Chung, S.S. and Poon, C.S. (1999). The attitudes of Guangzhou citizens on waste reduction and environmental issues. *Resources Conservation and Recycling* **25**(1), 35–59.

Cook, F.H. (1989). The jewel net of Indra. In *Nature in Asian Traditions of Thought: Essays in Environmental Philosophy*, ed. J.B. Callicott and R.T. Ames, pp. 213–27. Albany: State University of New York Press.

Dennys, N.B. (1971 [1876]). *The Folk-Lore of China, and its Affinities with that of the Aryan and Semitic Races*. Detroit: Tower Books.

Dudbridge, G. (1970). *The Hsi-yu Chi; a Study of Antecedents to the Sixteenth-Century Chinese Novel*. Cambridge University Press.

Eberhard, W. (1986). *Lexikon chinesischer Symbole. English. A Dictionary of Chinese Symbols*. Translated from the German by G.L.Campbell. London: Routledge.

Ferry, D. (1992). *Gilgamesh: A New Rendering in English Verse*. New York: Ferrar, Strauss and Giroux.

Gernet, J. (1970). *Daily Life in China: on the eve of the Mongol invasion, 1250–1276*. Translated from the French by H.M. Wright. London: Allen & Unwin.

Gragson, T.L. and Blount, B.G. (1999). In Introduction. *Ethnoecology: Knowledge, Resources and Rights*, ed. T.L. Gragson and B.G. Blount, pp. vii–xviii. Athens: University of Georgia Press.

Guo Jian (1986). The victorious monkey: Favorite figure in Chinese literature for children. In *Triumphs of the Spirit in Children's Literature*, ed. F. Butler and R. Rotert, pp. 161–3. Hamden, CT: Shoe String Press.

Harkness, J. (1998). Recent trends in forestry and conservation of biodiversity in China. *The China Quarterly* **156**, 911–45.

Jahiel, A.R. (1998). The organization of environmental protection in China. *The China Quarterly* **156**, 757–87.

Kellert, S. (1993). Attitudes, knowledge, and behavior toward wildlife among the industrial superpowers; United States, Japan, and Germany. *Journal of Social Issues* **49**, 53–69.

Kiefer, C. (1988). The monkey as mirror, Review of Emiko Ohnuki-Tierney. *Journal of Asian Studies* **47**, 645–6.

Lai, W. (1994). From protean ape to handsome saint: The monkey king. *Asian Folklore Studies* **59**, 29–65.

Langdon, S. (1990). From monkey to man: The evolution of a geometric sculptural type. *American Journal of Archeology* **94**, 407–24.

Levy, D. (1984–5). A quest of multiple senses: Anthony C. Yu's The Journey to the West. *Hudson Review* **37**, 507–15.

Liu, H. (1993). Chung-chou hsueh kon. A symbolic aesthetics reflection about the original (translated by D. Chan). *Zonghou* (n.s.), pp. 86–91.

Liu, T.-Y. (1964). The prototypes of monkey (Hsi Yu Chi). *Revue Internationale de Sociologie. Tung Pao* **51**, 55–71.

McKenna, T. (1992). *Food of the Gods: The Search for the Original Tree of Knowledge*. New York: Bantam Books.

Morowitz, H. (1989). Biology of a cosmological science. In *Nature in Asian Traditions of Thought: Essays in Environmental Philosophy*, ed. J.B. Callicott and R.T. Ames, pp. 47–54. Albany: State University of New York Press.

Morris, D. (1967). *Primate Ethology*. London: Weidenfeld & Nicolson.

Nienhauser, W.H., Hartman, C., Ma, Y.W. and West, S.H. (eds.) (1986). *The Indiana Companion to Traditional Chinese Literature*, pp. 413–18. Bloomington: Indiana University Press.

Niu, W.Y. and Harris, W.M. (1996). China: The forecast of its environmental situation in the 21st Century. *Journal of Environmental Management* **47**, 101–14.

Oelschlaeger, M. (1991). *The Idea of Wilderness*. New Haven, CT: Yale University Press.

Ohnuki-Tierney, E. (1983). Text, plan, and story: The construction and reconstruction of self and society. *Proceedings of the American Ethnological Society* **114**, 278–314.

Ohnuki-Tierney, E. (1990). Monkey as metaphor? Transformations of a polytropic symbol in Japanese culture. *Man* (n.s.) **25**, 89–107.

Ohnuki-Tierney, E. (1995). Representations of the monkey (Saru) in Japanese culture. In *Ape, Man, Apeman: Changing Views, 1600–2000*, ed. R. Corbey and B. Theunissen, pp. 297–307. Leiden: Royal Netherlands Academy of Arts and Sciences.

Plaks, A.H. (1987). *The Four Masterworks of the Ming Novel*. Princeton: Princeton University Press.

Reischauer, E.O. (1955). *Ennin's Travels in T'ang China*. New York: Ronald Press Co.

Ronan, C.A. (1992). *The Shorter Science and Civilisation in China. An abridgement of Joseph Needham's Original Text*. Vol. I. London: Cambridge University Press.

Shahar, M. (1992). The Lingyin Si Monkey disciples and the origins of Sun Wukong. *Harvard Journal of Asiatic Studies* **52**, 193–227.

Shaw, M.C. (1993). The Aegean garden. *American Journal of Archaeology* **97**, 661–85.

Shih, H. (1942). Introduction to the American Edition. *Monkey* by Wu Ch'eng-en, transl. by Arthur Waley. New York: Grove (1984; original copyright John Day 1943).

Tu, W.-M. (1989). The continuity of being: Chinese visions of nature. In *Nature in Asian Traditions of Thought: Essays in Environmental Philosophy*, ed. J. Callicott, J. Baird and R.T. Ames, pp. 67–79. Albany: State University of New York Press.

Tuan, Y.F. (1968). Discrepancies between environmental attitude and behaviour: examples from Europe and China. *Canadian Geographer* **12**(3), 176–91.

Van Gulik, R.H. (1967). *The Gibbon in China*. Leiden: E.J. Brill.

Wagner, R.G. (1990). Monkey king subdues the White-Bone Demon: A study in PRC mythology. *The Contemporary Chinese Historical Drama*, Chapter 3, pp. 139–234. Berkeley: University of California Press.

Waley, A. (1984). *Wu, Ch'eng-en, Monkey*. (Translated by Arthur Waley.) New York: Grove, original copyright John Day 1943.

Werner, E.T.C. (1934). *Myths and Legends of China*. London: George G. Harrap.

Wheatley, B.P. (1999). *The Sacred Monkeys of Bali*. Illinois: Waveland Press.

White, L. Jr. (1967). The historical roots of our ecologic crisis. *Science* **155**, 1203–7.

Williams, C.A.S. (1975). *Outlines of Chinese Symbolism and Art Motives*. Folcroft, PA: Folcroft Library Editions.

Xu, H.G., Wang, S.Q. and Xue, D.Y. (1999). Biodiversity conservation in China: Legislation, plans and measures. *Biodiversity and Conservation* **8**, 819–37.

Yang, C.X., Head, J.W. and Liu, S.R. (1994). China's treatment of crimes against the environment: Using criminal sanctions to fight environmental degradation in the PRC. *Journal of Chinese Law* **8**, 1–2. (http://www.columbia.edu/cu/chinalaw/backissues/issue82/head9.html)

Yetts, W.P. (1912). *Symbolism in Chinese Art*. Leiden: The China Society, Brill.

Yu, A. (1977–83). *Journey to the West*, vols. 1–4 (translated and edited by Anthony C. Yu: vol. 1. 1977; vol. 2. 1978; vol 3. 1980; vol. 4. 1983). Chicago: University of Chicago Press.

Zent, S. (1999). The quandary of conserving ethnoecological knowledge: A piaroa example. In *Ethnoecology: Knowledge, Resources and Rights*, ed. T.L. Gragson and B.G. Blount, pp. 90–124. Athens: University of Georgia Press.

8 Reflections on the concept of nature and gorillas in Rwanda: implications for conservation

PASCALE SICOTTE AND PROSPER UWENGELI

Introduction

The mountain gorilla (*Gorilla gorilla beringei*) of the Virunga region is endangered (Harcourt, 1996), and has been the focus of intense research and conservation efforts. Since 1959, work has been conducted on the socio-ecology of this ape, most of which is linked with the Karisoke Research Center (Robbins, Sicotte and Stewart, 2001). Even though conservationatists know more about mountain gorilla socio-ecology than 30 years ago, still relatively little is known about elements outside the forest that may ultimately affect the forest itself. For instance, research has looked at the people's attitude towards the forest, and at some of their practices influencing the forest (Weber, 1987, 1989; Plumptre and Williamson, 2001), but this socio-economic work around the Virunga is still in its infancy. The vision that people of Rwanda have of the forest and of wild spaces is also poorly documented. It seems, however, that it would be useful to better understand this relation to wild spaces, and more generally, the 'cultural construction of nature' in Rwanda to develop conservation programs better suited to the people of Rwanda.

This lack of understanding for the Rwandan cultural construction of nature provided the impetus for interviews, perfomed by P.S. in 1997–8, with 13 Bahutu living close to the Parc National des Volcans[1]. They were contacted with the help of members of the Karisoke staff. One of the criteria for selection was to interview men that were part of the forest-cutting effort that led to the settlement of the Kinigi area in the 1960s. At this time, these men were in their early twenties. Most of the men interviewed were thus over 60 years old. During these interviews, the aim was to document the vision that informants had of wild spaces, particularly of the forest and its inhabitants. For many cultures, apes have traits that allow them to cross the boundary between 'nature' and 'culture' (Haraway, 1989; Sicotte and Nisan, 1998), and can be useful subjects to investigate the 'nature' of this boundary. Gorillas were placed at the center of a constellation of 'others', and informants were asked to contrast these others with

163

the gorillas, by explaining to what extent they shared traits with or were different from them. These 'others' were monkeys living in the forest, humans living traditionally in the forest (the Batwa), and humans living outside of the forest. Informants were also asked to relate the story of their establishment in the Kinigi area. Finally, they were asked if they knew stories about wild and domestic animals. Thirteen interviews represent undoubtedly a small sample, but in association with historical and ethnographic literature on Rwanda, and several informal conversations with Rwandan friends, they provide material that suggests that there is a need for further research in this area.

The terms 'wild spaces', 'nature' and their opposite terms 'domesticated spaces', 'culture', which we use throughout this text, may have distinct meaning depending on the cultural context, or may encompass different elements. The danger here, which we may not have completely avoided, is for interviewer and informants to use terms that correspond to different realities. This is why in interviews, P.S. avoided using these terms, and simply used the terms 'forest', or 'park' to refer to the area where the gorillas are found.

In this chapter, we first give a brief description of Rwanda, its protected areas, and the conservation efforts surrounding the mountain gorillas. Following Weber (1989), we will argue that the people in power (whether in the government or in development circles) have given mixed messages to the population about the importance of preserving the protected areas. We will then argue that for most Rwandans (those who traditionally identify themselves as agriculturists and pastoralists), the forest is a place to be feared. However, the interviews suggest that the gorillas themselves are not associated with fear. In fact, most informants describe them in terms of their resemblance to humans, which contrasts to the way the forest-people of Rwanda (the Batwa) are being described, in opposition to humans. We will conclude by suggesting a few avenues to continue the investigation of Rwandan's relation to wild space, and how to integrate it into conservation programs.

The country and its traditional organization

Rwanda is located in East-Central Africa, in a region referred to as the African Great Lakes. It is a small land-locked country, characterized by a high altitude (above 1700 m) and a humid climate. The designation 'Pays des Mille Collines' (the country of the thousand hills) describes the physical landscape of this country well. Rwanda is a country of rolling hills, and the hill not only makes up the physical landscape of the country, but is also at the basis of the social and political organization, in a country that traditionally did not have villages or towns (Desforges, 1972; Lugan, 1997).

Three 'groups' form Rwandan society: the Bahutu, the Batutsi and the Batwa[2]. Together, they are the Banyarwanda, 'the people of Rwanda'. Banyarwanda share a common language (Kinyarwanda) and a system of cultural referents, but have different traditional subsistence activities. The Batwa, who are the least numerous group, are divided in two main groups. The 'Mpunyu' are hunter–gatherers that lived, until not long ago, in remote places such as forests and marshes. They are now generally confined to the periphery of these areas, areas that are now protected. The others are potters, who lived and still live in close assocation with the Bahutu and the Batutsi, but at the bottom of the social hierarchy (Seitz, 1993). The Bahutu, who constitute the majority of the population, are agriculturists, and the Batutsi are pastoralists. This being said, many Bahutu households possess one or two cows. Furthermore, Bahutus are often hired as *vachers* (i.e. cowherds). On the other hand, not all Batutsi households possess cattle, and certainly not all of them possess large numbers of cattle (Newbury and Newbury, 2000).

There has been considerable intermarriage between the groups, as evidenced by the fact that membership of a lineage does not coincide with 'ethnic' affiliation (Lugan, 1997; Jefremovas, 1997). 'Ethnic' boundaries could also be reorganized for Bahutu or Batwa to 'become' Batutsi (Desmarais, 1978; Seitz, 1993; Lugan, 1997).

The creation of Rwanda as a nation-state probably happened during the fourteenth century, under the impulse of a lineage of Tutsi, the Nyiginya. The Nyiginya progressively gained control of areas adjacent to the one they were occupying in Central Rwanda. The African Great Lakes region is characterized by the development of several states which were socially dominated by pastoralists. It is probably in Rwanda that this domination of a small group of pastoralists over the rest of the population reached its peak (Desmarais, 1978; Lugan, 1997). Rwanda actively maintained its isolation from the commercial circuits developed by traders throughout the nineteenth century (Lugan 1997). When the Germans arrived in Central Africa at the end of the nineteenth century, the territories in the center and in the south of modern Rwanda were well assimilated into the structure of the state. In other areas, particularly in the northwest, small autonomous kingdoms were more loosely connected to the central power and to the monarchy, and they sometimes rebelled against centralization (de Heusch, 1966; Vidal, 1985; Desforges, 1972;

Lugan, 1997; Jefremovas, 1997; Newbury and Newbury, 2000). With the Belgian protectorate after World War I, these regions became integrated into the state structure (Desforges, 1972).

The pressure for land and the forests of Rwanda

Rwanda knows – and probably has experienced for a long time – problems of shortage of land. Population and cattle density in the 'domesticated zones' is high. People, agriculturalists and pastoralists, have had a profound influence on the landscape, and there are virtually no relict forests anywhere. This absence of relict forests in regions inhabited by agriculturalists and pastoralists in Rwanda does not seem to be a recent consequence of increased human density. For instance, Weber (1989) cites historical sources that report how intensively the land was used in Rwanda at the end of the nineteenth and in the early twentieth century, for cultivation and pasture. As a result, in the late 1920s, the rate of deforestation was about 1 km per year on a wide front (Weber, 1989, p. 124). The first missionaries that settled in Rwanda in 1905 had to use thousands of porters to carry trees that were large enough to build their mission, and these trees came from a forest that was three days' walk from the site of the mission (Desforges, 1972). Finally Vidal (1984, p. 70), on the basis of photographs of central Rwanda taken in the early twentieth century, describes the 'slow work of clearing and erosion by men and cattle that had been necessary to strip the hills' prior to the arrival of the Europeans, to make the landscape look so 'naturally suited for cattle grazing'.

Land plots are usually divided between the sons after the death of the father (Vidal, 1984). When a plot became too small to support a family, migration was traditionally a strategy to acquire more land (Weber, 1989). The lack of land now makes this strategy impossible. In fact, many of the areas uninhabited until recently were so because they were marginal, either high in altitude, marshy, dry, or infested by tsetse flies, and as a result were not suitable for cultivation or for cattle. The very fact that they were marginal allowed these areas to be maintained as 'wild spaces'. Conservation policies introduced by the Belgians and maintained by the government of Rwanda after independence in 1962 have further helped protect these areas (Weber, 1989), despite the fact that land is in short supply.

The forests that remain are now islands of montane forest that cover a portion of the Zaire-Nile Divide (ZND). Figure 8.1 represents the situation in the mid-1980s. The islands of forests are 6, 2, 3 and 4; zone 5 is savanna environment. There are no connections between these islands of

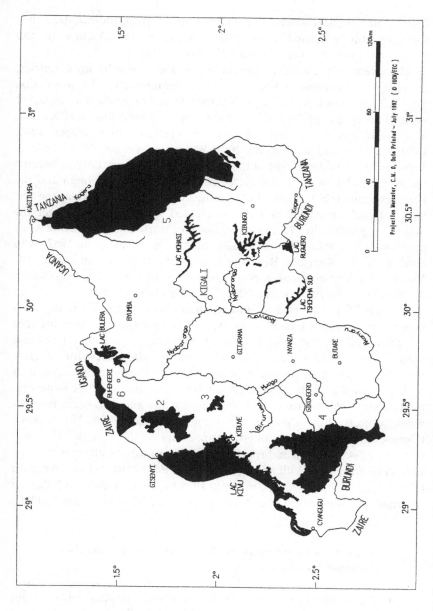

Figure 8.1. Protected areas of Rwanda in the mid-1980s. (Map reproduced with permission from "Environmental synopsis: Rwanda", IUCN 1993.) Zone 2, Gishwati Forest; Zone 3, Mukura Forest; Zone 4, Nyungwe Forest; Zone 5, Akagera Park and Mutara Hunting Reserve; Zone 6, Parc National des Volcans.

forest, but the short distances between some of them indicate that they were probably connected in the recent past. For instance, Virunga (6) and Gishwati (2) were connected until 1958 (Weber, 1989).

The forests of the ZND region (in Rwanda and neighboring countries) are recognized as being an important target for research and conservation efforts. They show a great deal of floral and faunal diversity. The appeal of some of the species under threat – among them the mountain gorillas – has also contributed to the research and conservation efforts (Weber, 1987, 1989; Plumptre and Williamson, 2001; Weber and Vedder, 2001). Indeed, 1926 saw the establishement of the Parc National Albert (later to become the Parc des Virungas in Zaire/Democratic Republic of Congo and the Parc des Volcans in Rwanda) by the Belgian Authorities. This was the first national park in Africa and it was created specifically to protect the mountain gorillas. In 1959, Georges Schaller started his pioneering work on mountain gorillas. He was followed by Dian Fossey in 1967, who established the Karisoke Research Center on the Rwandan side of the Virunga where research on gorillas and their ecosystem is ongoing (Schaller, 1963; Fossey, 1983; Stewart, Sicotte and Robbins, 2001).

The need for conservation and research efforts in the ZND region is even more tangible today, because the civil and political unrest in the area has provoked large-scale movement of populations (both internally and across borders). Many forests have been used as hiding places for refugees, for displacees and by armed opposition forces. Furthermore, several of these protected areas have suffered or have been reduced in size to accommodate both short-term population movements and long-term resettlement projects. For instance, the refugee camps in the North Kivu (Democratic Republic of Congo) put pressure on the Parc des Virungas between 1994 and 1996 and resulted in extensive wood-cutting within the park (Plumptre and Williamson, 2001). The Parc National de l'Akagera (Fig. 8.1; zone 5) was reduced in size in 1998 to accommodate both Batutsi returnees and their cattle from Uganda (Bouché, 1998; Williams, 1999).

The case of the mountain gorilla and the forest of the Parc National des Volcans

Only 650 mountain gorillas remain, in two distinct populations of about 300 each: the Bwindi population in Uganda (McNeilage *et al.*, 1998), and the Virunga population that spans over Rwanda, Democratic Republic of Congo and Uganda (Sholley, 1991). The conservation of gorillas and of their ecosystem involves work in research, ecotourism, anti-poaching, and conservation education. Research into the ecology and social organization of the mountain gorilla led to the identification of the factors that shape their habitat use and their social interactions (Robbins *et al.*, 2001). This knowledge allowed the development of a successful ecotourism program centered on viewing the mountain gorillas in their natural habitat (Weber and Vedder, 2001). This program brought millions of dollars to the country and was among the highest sources of foreign currency at the end of the 1980s, before the onset of the war (Weber, 1987, 1989; McNeilage, 1996). People from local communities have been employed by the conservation and research projects in many different capacities, thus providing direct benefits for the people living around the park (Weber, 1989; Plumptre and Williamson, 2001).

The poaching in the Parc National des Volcans (PNV) is now mainly directed at antelopes, but gorillas are vulnerable to snares set for antelopes. Anti-poaching patrols aim to remove the snares and prevent poachers' activities in the park. These activities give net results in terms of number of snares removed (Plumptre and Williamson, 2001). The education program, directed at the local population during the 1980s, and at the park guides and guards during the 1990s, focused on the utilitarian function of the forest, which is an area of convergence of interests between Western conservationists and Rwandans (Harcourt, 1986; Weber, 1989). This program seems to have increased conservation awareness in areas around the Parc National des Volcans. Interviews in the mid-1980s revealed that more people were able to associate a positive value with the park (e.g. tourism revenue and water regulation) than before the education programme was initiated (Weber, 1989).

However, Weber and Vedder (2001) note that the education program alone may not have been responsible for this change, as the benefits brought by the tourism program have probably also contributed to the change in attitude towards the park. It may, in fact, be impossible to measure the impact of the education program *per se*. Furthermore, it is unclear to what extent behaviors have changed, and how profound these changes are, if any. In a questionnaire survey administered by Rwandan research assistants among people living close to the PNV in 1996, 11% ($n = 181$) admitted to poaching in the PNV, and 33% to buying poached meat (Plumptre *et al.*, 1997). Poached meat could be found in the Ruhengeri market in 1993–94, and poachers and distribution points for poached meat were known by local people (Nzeyimana, personal communication). Thus, it does seem that some people still see the forest as a reservoir for resources (Weber, 1989). It is also not uncommon, at the level of the regional government, to hear voices demanding that areas of the PNV be excised from the park to allow for human settlement or cattle pasture. In recent years, the central government has resisted these demands (Plumptre and Williamson, 2001).

In the past, however, conservation policies did not prevent the central government, in association with international development organizations, from making several large excisions to the protected areas of Rwanda during each decade since independence in 1962. These excisions were made to allow the introduction of new cash crops (such as potato, which resists the cold climate of the high-altitude regions), to expand the cultivation of others (pyrethrum, tea, exotic trees), to promote the development of the dairy industry by providing more pasture land (Weber, 1987, 1989), and to allow the resettlement of returnees (Bouché, 1998; Williams, 1999; A. Plumptre, personal communication).

The area surrounding the remaining forest of the Parc National des Volcans, in particular, has been settled relatively recently. The last major forest cutting effort around the Parc des Volcans was organized in the 1960s as part of a project to develop the culture of a cash crop, pyrethrum, which is a flower that produces a natural insecticide. Forty percent of the forest remaining on the Rwandan side of the Virunga (ten thousand hectares) was cut down for this project. Five thousand families were given a plot of land excised from the forest (2 ha) that they had to clear, on the condition that they would devote a portion (40%) of this land to planting pyrethrum (Weber, 1989). This policy was maintained until the mid-1980s, despite the fact that the prices of pyrethrum had plummetted on the international market (Little and Horowitz, 1988).

If one compares Fig. 8.1, which represents the protected areas of Rwanda in the mid-1980s, with the current situation, the Gishwati forest and the Mukura forest have been cleared, and the Akagera park has been substantially reduced (by 40%).

Given these mixed messages from the authorities, it is not surprising that people see that opening up more land could be an option, and even be in the 'national interest', as it was often the argument invoked. In that sense, people that keep exploiting protected areas such as the PNV may in fact be simply 'borrowing' in advance from what they see as a reservoir that will eventually be open for exploitation anyway. Perhaps in all this, the surprising aspect is that the conservation message has been heard by some people, and that it is being put into practice.

For instance, our informants were people that came to the Kinigi area because they had too little or no land at all in their communes of origin. They all point out how successful their life has been since moving into the area; many own one or a few cows, several goats, a bicycle, and several point out that they were able to send their children to school. Clearly, these people and their families benefitted from using the land excised from the forest. That did not prevent them, at least on the surface and when talking to me, from sharing the discourse about conservation. They managed to reconcile the points of view between the importance of cutting the forest and preserving it. One informant met only two white people in the 1960s: 'One was involved with the pyrethrum project where parcels were created by cutting down the forest, the second was D. Fossey, who was trying to protect the gorillas and to prevent further cutting of the forest'.

Another tries to reconcile these two approaches: 'The minister at the time argued that to cut down the forest to cultivate pyrethrum was in the interest of the country'. When Fossey started her conservation work, 'she collaborated with the minister to convince him that wild animals were also part of the natural wealth of the country' (. . .) 'so there is a portion (of the forest) for the pyrethrum, and the rest is for the gorillas, because the two are important for our wealth.'

Informants try to integrate, to harmonize past practices. It seems doubtful, however, that they would offer resistance if there was a scheme to excise more land from the protected forest. Why? Partly, we think, because of the vision that Rwandans have of wild spaces.

Rwandans' relation to wild spaces

Traditional odes of the court describe Rwanda as a 'landscape entirely shaped by the activities of the herders', after warriors, men of the cow and of the spear, took control of the territory (Vidal, 1984, p. 68). Oral history collected in Central Rwanda by Claudine Vidal describes the pattern of settlement of areas at the periphery of the central kingdom, eight generations ago: Warriors mandated by the court defend and secure a hill; their descendants, who settle on the hill, 'kill the forest', and open the way for further exploitation of the land, for both cultivation and pasture (Vidal, 1984, pp. 68–70). The land ultimately remained the property of the king and of the central power. This ownership was maintained by a complex chain of clientships, where goods and services were being provided to the aristocracy and to the court by the people living on the land. This was the case in central Rwanda and also became the case in the northwest as it became more integrated into the sphere of the central power, although the forms and intensity of the clientships differed between regions (de Heusch, 1966; Desmarais, 1978; Newbury and Newbury, 2000).

Opening the land for cattle grazing and for cultivation brings humanness and order to an area. It thus should not come as a surprise that the forest is perceived by most Rwandans as a place of danger and of disorganization, distinct from the organized world of the *rugo*, the compound where the house is located and where the domestic life takes place. Danger and evil are found in 'chaotic' places, such as forests and marshes, caves, ravines and bushes (Gasarabwe, 1978). Even banana groves, hardly a 'wild space' by Western standards, can be feared because they are dark and shady. Shade and darkness are associated with fear and with the presence of ancestral spirits (Pottier and Nkundabashaka, 1992). The volcanic peaks of the Virungas, which can be seen from almost anywhere in Rwanda on a clear day, are considered to be the home of the ancestral spirits[3] (de Heusch, 1966). When ancestral spirits manifest themselves in Rwanda, it is always to torment the living (Vidal, 1984). Furthermore, forests and marshes have often given shelter to rebels and bandits (Desforges, 1972; Lugan, 1976; Vidal, 1985, Plumptre and Williamson, 2001). The fear, then, is not necessarily of the forest itself, but rather of what one can meet in the forest. 'In the forest, there is total darkness, so one walks without knowing who/what lays ahead and who/what is behind' (Uwengeli, 2000, p. 101).

This does not mean that people were not in contact with the forest. Many informants mentioned, for example, that people used paths through the forest to travel to neighboring countries. One informant recalls 'in times of famine, people have gone to get food in Zaire (. . .) the path (. . .) between Bisate and Sabyinyo was well known'. Another says that he knew the trail so well that he could have 'walked alone in the forest'. This is confirmed by Lugan (1976), who describes how people travelled long distances between regions, sometimes across forests, in order to get supplies in times of famine.

When my informants mentioned fear in relation to the forest, it was always because of the wild animals that one could encounter in the forest (leopards, lions, buffaloes, elephants are frequently cited). Several informants report either having been themselves attacked by wild animals, particularly buffaloes, or know people that have been attacked. Wild animals in the forest were also a concern for cattle. One informant says '(at the beginning, when we settled here) it was really dangerous, at 5pm you were already in the house and you (kept it) closed'. Some, on the other hand, say that fear was not an issue in light of the benefit gained by living on the new land. One, for instance, says 'I was not afraid because the field that (my family) had when I was a child was too small for me and my own family. I was not afraid (to be on the new land) even if there were animals in the forest.'

Many of my informants were *vachers* when they were young boys, and took cattle into the forest for grazing. Many of them also literally lived in the forest for a few weeks at the beginning of the forest cutting around Kinigi. In their contacts with the forest, we would argue that these men always brought 'orderly elements', or 'elements of humanness' that meant that they were conquering the forest. This is certainly the case in the phase of the cutting down of the forest, where they cut the trees, built houses, and rapidly planted their first crops. It can also be argued to be the case by bringing cattle into the forest, because the cow is the domesticated animal above all else in Rwanda, for both Bahutu and Batutsi. The cow is associated with beauty and purity, and belongs to the 'human' side of the world, in marked contrast with the Batwa (see below) (Smith, 1975; Desmarais, 1978). Alternatively, the cows might be offering some kind of protection to the humans: the Fulani believe their cows protect them from the dangers of the bush (van Beek and Banga, 1992).

The association between forest and fear came up slightly differently in casual conversations with educated conservation officers, however (some of them Bahutu, others Batutsi). One of them admitted, laughing at himself while reporting the story, that he had to restrain himself from running on the trail during his first visit in the forest in his new job, because he was so afraid of the dark. He eventually overcame his fear, but not by conquering the forest. We will return to this point later, in suggesting avenues for future work.

This association between fear and forest is probably found, under one form or another, in many agricultural and pastoral societies (as opposed to hunter–gatherers and horticulturalists, who live in the forest). Or if it is not fear, it may be that the forest is associated with undesirable values, and humans must transform this environment. At the risk of oversimplifying, here are a few examples from other areas in Africa. The Zafimaniry of Madagascar associate the forest with a lack of clarity; open and clear views are desirable (Bloch, 1995). The Dogon of Mali fear the spirits that live in the bush, despite the fact that the bush is also the source of life (van Beek and Banga, 1992). For the Mende of Sierra Leone, the forest is associated with danger and uncertainty, and the establishment of their villages and fields involves 'carving safe spaces out of the forest' (Leach, 1992, p. 80). The fear of the forest is certainly an element found in Western culture, and it is only recently that Nature has been seen as in need of protection from human disturbances and exploitation, in some special designated areas (Nash, 1968). In fact, in Western societies' recent history, the notion that human contact spoils the harmony of Nature has emerged.

Perception of the gorillas

Several factors influence the way informants talk about the gorillas. Before the conservation education program was put in place, most Rwandans were not able to physically describe a gorilla (Weber and Vedder, 2001). Gorillas are absent from traditional tales and stories, despite the fact that the Virungas and the spirits that inhabit them are the objects of stories, and that many informants spoke of tales that involved wild and domestic animals (mouse, hare, chimpanzee, dog, elephant, hyena . . .). This lack of presence of the gorillas in the symbolic world parallels what is found in published records on Rwandan oral literature (see, for example, Smith, 1975; Gasarabwe, 1988, 1991), and contrasts with the situation in West Africa (see, for example, Torday 1913: 'Why the gorilla does not speak'). That does not imply that people did not know about the existence of gorillas. After all, people used paths through the forest to travel to areas in Uganda and Congo, so some people must have seen gorillas. For instance, one informant stated that his grandfather had told him about the gorillas, but had described them as dangerous animals that would eat people. He said that he did not believe that such animals existed, until he saw them on his land. Indeed, the land that was cleared for pyrethrum cultivation was at an altitude where large areas of bamboos were found. Many people had left bamboo patches on their land, as bamboo is a useful building material. Gorillas rely heavily on bamboo shoots for feeding during the rainy season, and informants say that gorillas came back for a few seasons to what constituted a part of their range before the forest had been cleared, to feed on the bamboos. Some informants suggested that there are no stories about gorillas because they are quiet animals and they are generally afraid of humans.

Another factor that is likely to influence the way informants speak about gorillas is that primate meat, including gorilla, is not eaten tradition-ally in Rwanda. This contrasts with the situation in other African coun-tries, including Rwanda's neighbour, The Democratic Republic of Congo (Bowen-Jones, 1997–8). This may suggest that the close relatedness be-tween primates and humans is recognised, and that this recognition pre-vents 'cannibalism' from taking place (Uwengeri, 2000). As one informant points out, 'the chimpanzee tasted the potato, and thought that it was good. If potatoes are good for chimpanzees, they are good for humans', implying that chimpanzees and humans are alike. It seems, however, that in Uganda, primate meat is not eaten because it is considered 'dirty' (A. Plumptre, personal communication). Finally, it could be that bushmeat is generally less appreciated in Rwanda than domestic meat, a suggestion which is apparently unsubstantiated. P.U. reports in fact that people are nostalgic for an era where bushmeat was more easily accessible in markets.

The informants were also in contact to a certain extent with the work done at Karisoke on the gorillas, via personal contacts with gorilla workers or exposure to the education program of the Mountain Gorilla Project. They also knew that P.S. had worked with gorillas for many years, which certainly influenced their answers to her questions. For instance, all the informants said that they thought the gorillas were 'good' animals. This judgement is certainly facilitated by the fact that none of the crops used by the people living at the edge of the forest on the Rwandan side of the Virungas (potatoes and pyrethrum) attract gorillas. The situation is different on the Democratic Republic of Congo side of the Virungas, and in Bwindi (Uganda), where gorillas can cause substantial damage by crop raiding.

When asked to describe similarities between humans and gorillas, the informants speak about both physical and behavioral resemblances between the two species. Many informants said that they had gorillas on their land early after the clearing of the forest, and that when they saw a gorilla from a distance, they thought it was a human ('I saw somebody on my land, and wondered to myself, who is this person? (...) When I saw the hair, I realized it was not a muzungu (white person), it was not a Rwandan – I asked my neighbours, and they said it was a gorilla'). Informants state that gorillas resemble humans in body shape (shape of hands, fingers, face, eyes). They leave tracks that one could take for a human foot. They wear a 'jacket' (probably a reference either to the band of silver hair that males have on their back, or to the fact that they have a bare chest). When they eat, they peel their food and discard the 'hard parts'. Informants also reported that gorillas carry their infants like humans do (infant gorillas ride on their mother's back, which may be reminiscent of the way Rwandan women carry their babies). Informants also say that gorillas can be taught almost anything, from writing with a pen to driving a car. Finally, an informant who was closer to the work done at Karisoke stated that gorillas are able to control themselves, a quality intrinsic to what constitutes humanity for Rwandans ('gorillas do not mate in front of everybody in the group'). They can also control others ('they intervene in each other's fights to restore order').

In contrast, informants describe the Batwa as lacking the control over their lives and their environment which is expected of humans, and they clearly belong to the animal world. They do not cultivate ('even if they have a field, they go and collect the remains from other fields rather than cultivate themselves'). They do not build 'proper houses', houses that are solid ('even now that they are out of the forest, they build houses like the ones they used in the forest') merely 'temporary shelters'. This assimilation of the Batwa to the 'natural', 'animal' sphere by the other Rwandans has been amply documented in the ethnographic literature (see, for example, Smith, 1975; Desmarais, 1978; Seitz, 1993). These examples show that some categories (in this case gorillas and Batwa) can cross the boundary between nature and culture. Batwa are recognised as humans, but many of their traits make them closer to the animal world than to the human world. However, Batwa were sometimes, in exchange for exceptional services in traditional Rwanda, allowed to marry women of the aristocracy. They became Batutsi as a result. This confirms that boundaries can be crossed.

Informants know very well, of course, that gorillas are not humans. The fact that boundaries can be crossed, and that gorillas are described in terms of their resemblance to humans, does not make them *de facto* belong to the 'human' side of the nature–culture continuum. Some of our informants made intriguing comments regarding the relationship between cows and gorillas, and in light of the opposition between gorillas and Batwa (we already noted the opposition between cows and Batwa earlier), it is interesting to take a closer look at these comments. When talking about the work that they did as *vachers* when they were young boys, two informants mentioned that the gorillas follow the cows in the forest. One informant said that gorillas and cows could also rest together. He described one such case by saying that the cows were in the middle, with the gorillas and the *vachers* on both sides. He specifically said that the cows were a buffer, a protection against the gorillas, which supports the idea presented earlier that cows can be a means to enter the forest safely.

Another described the many tourists and researchers he saw heading for the forest to find the gorillas, and his exposure to the gorillas via films presented by the Mountain Gorilla Project. He said that the gorillas have 'guardians', people that stay with them and follow them all the time. By 'guardian', the informant probably meant the local Rwandans that work for Karisoke and for the tourism program. These people are employed so that white people (tourists and researchers) can have access to the gorillas. The function of 'guardian' may then be reminiscent of the work of a *vacher*: to keep the animals safe and to know where they are. Informants know that gorillas trigger interest in scientists and in white people in general. One form that this interest takes for researchers is that, like Rwandans with their cows, they know the history and the characteristics of individual gorillas. This is well known to Rwandan trackers and research assistants, whose task it is to report births, transfers, or wounds in the habituated groups, and who have to learn to recognize individual gorillas in order to be able to do so. These two elements (the notion of 'guardians' and the very specific type of interest) suggest that the people interviewed transposed some of the experiences of the most meaningful animal–human relationship for Rwandans – the one with the cow – onto the gorillas, and onto the relationship that they perceive between white people and gorillas.

If this is true, gorillas may thus have gone recently from having no 'symbolic presence' in Rwandans' understanding of the world, to a position where they are in opposition to the natural world, very much like the cow. This could suggest that a focus on gorillas' resemblances or connections to humans in an education program could project the gorillas more into the 'cultural' side of the world, which may not be desirable for conservation purposes, as one could argue that gorillas could survive without their habitat.

Conclusion

In Western culture, nature and culture are conceptualized as entities in opposition. For a long time, nature was viewed as needing to be conquered or tamed by humans. Now that conservation is an important issue in the West, nature is seen as in need of being protected from human action, as humans are seen as spoiling nature. In order to be able to relate to this Western discourse on nature, one has to accept the notion that nature is harmoniously organized. Not all cultures necessarily share this vision of nature. Where Westerners now see destruction and loss of harmony, others may see the beginning of order and harmony. For example for Japanese the opposing pole of 'nature' is not 'culture', but the continuum is rather between a 'cooked, domesticated' nature, which is considered pure, and a 'raw, wild' nature, which is associated with impurity and barbarism (Kalland and Asquith, 1997, p. 13). There seems to be enough evidence to suggest that, in Rwanda, 'nature' and 'forests' are not entities that seem to evoke order and harmony. We suggest that a better understanding of a Rwandan's vision of wild space is imperative, and that it should take into account the point of view of the multiple groups that compose Rwandan society: not an easy task. In other words, it is necessary to investigate the possible differences in views along the 'ethnic' axis, and along a geographical axis (north–south), taking into account gender and education.

In practice, some Rwandans seem to have experienced a change in their perception of the forest in recent decades. The most important suggestion that we could make at this point is to document how this transition occurred, and how it is formulated and reconstructed by the actors.

Furthermore, we need to ask these Rwandans to describe emotionally and intellectually the changes that they experienced in their perception of the forest and of their attitude towards it. A better understanding of this process could allow a reconstruction of the channels through which the conservation message is passed. This change in perception apparently did not only occur in a few educated 'Westernized' Rwandans: one of the Karisoke employees with whom P.S. had several conversations on this topic linked his commitment to his work to his father's strong conservation ethics. This transition, the elements that triggered it, and possibly the new construct of Nature and of the forest that it could produce, could form the basis of a new conservation curriculum specifically adapted to the Rwandan context.

Acknowledgements

This research was funded by the University of Calgary. The authors thank the Office Rwandais du Tourisme et des Parcs Nationaux for research permission, as well as the Dian Fossey Gorilla Fund International. Dr E. Williamson and Jean-Marie Gatana Wa Biso provided logistical help for which we are grateful. We also thank Profesor Deo Mbonyinekebe at the Université Nationale du Rwanda. J. Murphy, A. Plumptre and T. Wyman provided useful comments on earlier versions of this paper.

Notes

[1] These interviews were performed by P.S. in French and in Kynarwanda with the help of two research assistants who acted as translators (Jean Bosco Bizumuremyi and Emmanuel Hitayezu). Jeanne Musuku provided invaluable help in typing the interviews and clarifying some translations.

[2] These 'groups' have sometimes been referred to as 'ethnic groups', 'castes', 'social classifications', and are the main 'entities' that are usually mentioned in discussions about the inhabitants of Rwanda. This is certainly an oversimplification, in the light of the argument by Newbury and Newbury (2000) that there were many more 'regional variations' in Rwanda. They identify 'at least nine distinct cultural zones, with their own (politically autonomous) histories, before the eighteenth century , and often with important cultural connections reaching beyond current Rwandan state boundaries' (p. 851).

[3] This is not only put in evidence in the ethnographic literature, but is also heard from people. For instance, in conversation with P.U., a Karisoke tracker commented that he had always been afraid of Karisimbi, the highest peak,

because he believed that he would lose his voice if he went close to the summit. He attributed this to the presence of spirits.

References

van Beek, W.E.A. and Banga, P.E. (1992). The Dogon and their trees. In *Bush Base Forest Farm. Culture, Environment and Development*, ed. E. Croll and D. Parkin, pp. 57–75. London: Routledge.

Bloch, M. (1995). People into Places: Zafimaniry concepts of clarity. In *The Anthropology of Landscape: Perspectives on Place and Space*, ed. E. Hirsh and M. O'Hanlon, pp. 63–77. Oxford University Press.

Bouché, P. (1998). Les aires protégées du Rwanda dans la tourmente; évolution de la situation de 1990 à 1996. *Cahiers d'Ethologie* **18**, 175–86.

Bowen-Jones, E. (1997–8). A review of the commercial bushmeat trade with emphasis on Central/West Africa and the great apes. *African Primates* **3**, S1–S37.

de Heusch, L. (1966). *Le Rwanda et la Civilisation Interlacustre*. Université Libre de Bruxelles.

Desforges, A. (1972). Defeat is the only bad news: Rwanda under Musiinga, 1896–1931. Ph.D. dissertation, Yale University.

Desmarais, J.C. (1978). Le Rwanda des anthropologues: l'archéologie de l'idéologie raciale. *Anthropologie et Sociétés* **2**, 71–93.

Fossey, D. (1983). *Gorillas in the Mist*. Boston: Houghton Mifflin.

Gasarabwe, E. (1978). *Le Geste Rwanda*. Série 'La voix des autres', 10/18. Paris: Union Générales d'Editions.

Gasarabwe, E. (1988). *Contes du Rwanda: Soirées au Pays des Mille Collines*. Collection 'La légende des mondes'. Paris: L'Harmattan.

Gasarabwe, E. (1991). *Kibiribiri l'Oiseau de Pluie: Contes du Rwanda*. Collection 'La légende des mondes'. Paris: L'Harmattan.

Haraway, D. (1989). *Primate Visions*. New York: Routledge.

Harcourt, A.H. (1986). Gorilla conservation: Anatomy of a campaign. In *Primates, the Road to Self-sustaining Populations*, ed. K. Benirschke, pp. 31–46. Berlin: Springer-Verlag.

Harcourt, A.H. (1996). Is the gorilla a threatened species? How should we judge? *Biological Conservation* **75**, 165–76.

Jefremovas, V. (1997). Contested identities: Power and the fictions of ethnicity, ethnography and history in Rwanda. *Anthropologica* **39**, 91–104.

Kalland, A. and Asquith, P.J. (1997). Japanese perceptions of nature: ideals and illusions. In *Japanese Images of Nature: Cultural perspectives*, ed. R. Curzon, pp. 1–35. Richmond: Curzon Press.

Leach, M. (1992). Women's crop in women's spaces: gender relations in Mende rice farming. In *Bush Base Forest Farm. Culture, Environment and Development*, ed. E. Croll and D. Parkin, p. 146–68. London: Routledge.

Little, P.D. and Horowitz, M.M. (1988). Authors' Reply to 'Agricultural policy and practice in Rwanda: A response to Little and Horowitz'. *Human Organization* **47**, 271–3.

Lugan, B. (1976). Causes et effets de la famine 'rumanura' au Rwanda, 1916–1918. *Canadian Journal of African Studies* **10**, 347–56.

Lugan, B. (1997). *Histoire du Rwanda*. France: Bartillat.

McNeilage, A. (1996). Ecotourism and mountain gorillas in the Virungas. In *The Exploitation of Mammal Populations*, ed. V.J. Taylor and N. Dunstone, pp. 334–44. London: Chapman and Hall.

McNeilage, A., Plumptre, A.J., Brock-Doyle, A. and Vedder, A. (1998). Bwindi Impenetrable National Park, Uganda Gorilla and Large Mammal Census, 1997. Wildlife Conservation Society Working Paper 14.

Nash, R. (1968). *Wilderness and the American Mind*. New Haven: Yale University Press.

Newbury, D. and Newbury, C. (2000). Bringing the peasants back in: Agrarian themes in the construction and corrosion of status historiography in Rwanda. *American Historical Review* **105**, 832–77.

Plumptre, A.J. and Williamson, E.A. (2001). Conservation oriented research in the Virunga Region. In *Mountain Gorillas: Three Decades of Research at Karisoke*, ed. M.M. Robbins, P. Sicotte and K.J. Stewart, pp. 361–89. Cambridge University Press.

Plumptre, A.J., Bizumuremyi, J.B., Uwimana, F. and Ndaruhebeye, J.D. (1997). The effects of the Rwandan civil war on poaching of ungulates in the Parc National des Volcans. *Oryx* **31**, 265–73.

Pottier, J. and Nkundabashaka, A. (1992). Intolerable environments: towards a cultural reading of agrarian practice and policy in Rwanda. In *Bush Base Forest Farm. Culture, Environment and Development*, ed. E. Croll and D. Parkin, pp. 146–68. London: Routledge.

Robbins, M.M., Sicotte, P. and Stewart, K.J. (eds) (2001). *Mountain Gorillas: Three Decades of Research at Karisoke*. Cambridge University Press.

Schaller, G. (1963). *The Mountain Gorilla: Ecology and Behavior*. Chicago: University of Chicago Press.

Seitz, S. (1993). *Pygmées d'Afrique Centrale*. Collection Langues et Cultures Africaines. Louvain, Paris: Peeters.

Sholley, C.R. (1990). Conserving gorillas in the midst of guerrillas. *American Association of Zoological Parks and Aquaria Annual Proceedings*, pp. 30–7.

Sicotte, P. and Nisan, C. (1998). Femmes et empathie en primatologie: réflexions sur la représentation des femelles primates et des femmes primatologues. *Autrement* HS **106**, 77–102.

Smith, P. (1975). *Le Récit Populaire au Rwanda*. Collection 'Classiques Africains.' Paris: Armand Colin.

Stewart, K.J., Sicotte, P. and Robbins, M.M. (2001). Mountain gorillas of the Virungas: a short history. In *Mountain Gorillas: Three Decades of Research at Karisoke*, ed. M.M. Robbins, P. Sicotte and K.J. Stewart, pp. 1–26. Cambridge University Press.

Torday, E. (1913). *Camp and Tramp in African Wilds*. London: Seeley, Service & Co.

Uwengeli, P. (2000). Essai d'exploration des attitudes et conceptions des Rwandais vis à vis des espaces sauvages. Mémoire de Licence, Université Nationale du Rwanda.

Vidal, C. (1984). Enquêtes sur l'histoire et sur l'au-delà, Rwanda, 1800–1970. *L'Homme* **24**, 61–82.

Vidal, C. (1985). Situations ethniques au Rwanda. In *Au coeur de l'Ethnie: Ethnies, Tribalisme et État en Afrique*, ed. J.L. Amselle and E. M'bokolo, pp. 167–84. Paris: Editions La Découverte.

Weber, A.W. (1987). Socio-ecological factors in the conservation of afromontane forest reserves. In *Primate Conservation in the Tropical Rain Forest*, ed. C. Marsh and R. Mittermeier, pp. 205–29. New York: Alan R. Liss.

Weber, A.W. (1989). Conservation and development on the Zaire-Nile Divide: An analysis of value conflicts and convergence in the management of afromontane forests in Rwanda. Ph.D. dissertation, University of Wisconsin, Madison.

Weber, A.W. and Vedder, A. (2001). Mountain gorillas at the turn of the century. In *Mountain Gorillas: Three Decades of Research at Karisoke*, ed. M.M. Robbins, P. Sicotte and K.J. Stewart, pp. 413–23. Cambridge University Press.

Williams, S.D. (1999). Akagera: from research to conservation. In *La recherche: un outil pour la conservation et la gestion des aires protégées*, ed. E.A. Williamson, pp. 20–30. Unpublished report to Dian Fossey Gorilla Fund International.

Part 3
Conservation of nonhuman primates

A common theme that arises from every section of this book is that of complexity and the inextricable interconnections between ecology and human cultures. While the title of this compilation involves the word 'conservation' and all of the chapters touch on the subject, the chapters in this section focus on the nexus of culture and ecology in the light of conservation issues. There is no longer the possibility of studying a group, or population, of nonhuman primates without coming into contact with human interaction, manipulation, and/or habitat destruction. Ninety per cent of the world's primates are found in tropical forests and it is precisely these forests that are being converted to human use faster and more dramatically than any other habitats on earth. Fifty per cent of all primate species are a conservation concern to the Species Survival Commission (SSC) and the World Conservation Union (IUCN), and 20% are considered endangered or critically endangered. Today these numbers continue to rise in the face of continuing threats.

Unfortunately, the tropical areas that are home to most nonhuman primates are some of the most economically precarious and therefore politically unstable nations on earth. Human greed, lack of long-term foresight, global economies and colonial legacies contribute to dramatic food shortages among already impoverished people. Recent increases in the size of the human population and, in some cases urban people migrating to rural areas, leads to dramatic overuse and destruction of habitats in these countries. This destruction, often linked to people now having easy access to guns and ammunition, takes a heavy toll on the nonhuman primates, many of whom are traditional foods, readily accessible, large meat sources, and cost-effective prey items. The combination of threats facing both the human and nonhuman primates in these areas is daunting indeed. It is very possible that the twenty-first century will see not only the extinction of some (or many) nonhuman primate populations but also the continuing, or even worsening, socio-political and socio-ecological conditions for many human populations.

Can conservation activity slow the assault on the habitats and ecologies of these regions? Even the most strictly enforced park boundaries, structured use zones or resource management policies crumble in the wake of a civil war, famine or just plain old severe poverty and hunger. It has become readily apparent that no form of conservation action is possible without taking into account indigenous practices, local and global economies, and political realities in the utilization of local environments. Education, sustainable development programs, and a focus on solving the human primates' problems are probably the only permanent answers to the conservation dilemma.

In the first chapter in this section (Chapter 9) Agustín Fuentes proposes a complex web of interconnectivity as a model for attempting to understand the conservation issues on the Mentawai Islands, West Sumatra, Indonesia. By reviewing the nonhuman primate ecology, human culture and ecology, and the current economic and land use situations this chapter presents one view of a complex reality that researchers face whether they focus on nonhuman primates, human ecology and/or conservation issues.

Anthony Rose (Chapter 10) tackles what may be the most pressing human–nonhuman primate conservation issue in Africa today: the bushmeat crisis. This chapter proposes a biosynergistic approach, which ties any meaningful resolution of the bushmeat dilemma to a series of changes in the social and economic structures of the cultures and countries involved. This chapter attempts to draw an outline for a broad and far-reaching set of interconnections deeply rooted in values and social systems within the context of modern economies and political realities.

The chapter by Bruce Wheatley and colleagues (Chapter 11) presents a very specific conservation dilemma. Here the relationship between the introduced macaque monkeys of Ngeaur Island and the local indigenous people is reviewed. The monkeys are perceived as pests to the humans and were introduced by a colonial government earlier this century. What should the islanders do? This short chapter gives us a view into the problems of introduced species on islands, the complex issues surrounding extermination and re-location of monkeys, and into cultural aspects of tourism and taro gardens.

In Chapter 12, David Sprague examines the relationship between farming patterns and the use of crop-lands by monkeys in rural Japan. Here he ties a shift in cultural resource use and economic priorities to an increase in conflict between humans and monkeys.

The chapters in this section are far-ranging, but do not focus solely on the connections between economic development and conservation. If the reader is interested in the history and issues created by the linking legal issues, economic development, and conservation, we recommend recent publications by John F. Oates, (1999), Guruswamy and McNeely (1998), and articles by Conrad Kottack (1999) and Peter Brosius (1999).

References

Brosius, P.J. (1999). Green dots, pink hearts: displacing politics from the Malaysian rain forest. *American Anthropologist* **101**(1), 36–57.
Guruswamy, L.D. and McNeely, J.A. (eds) (1998). *Protection of Global Diversity: Converging Strategies.* Durham and London: Duke University Press.
Kottack, C.P. (1999). The New Ecological Anthropology. *American Anthropologist* **101**(1), 23–35.
Oates, J.F. 1999. *Myth and Reality in the Rain Forest.* Berkeley: University of California Press.

9   Monkeys, humans and politics in the Mentawai Islands: no simple solutions in a complex world

AGUSTÍN FUENTES

Introduction

Nonhuman primates are the focus of a wide array of investigations relating to human primates. Physiological models, evolutionary models, primate-wide trends in anatomy, and behavior are but a few of these foci. Many of these investigations are fruitful and rewarding to several branches of scientific knowledge. Here, however, I would like to take a slightly different direction and discuss nonhuman primates not as a model for humans but rather as sympatric, and evolutionarily similar, participants in a rapidly changing ecosystem. In many areas of the world, human and nonhuman primates are experiencing similar challenges brought about by changing ecological, social and economic realities. Although nonhuman primate behavior and ecology, human ecology, and conservation are all broad fields by themselves, it is my contention that the third (conservation) is most effective when the first two are merged and treated as a unified area of investigation. I hope to demonstrate here that by incorporating elements of human cultural ecology, nonhuman primate behavior and ecology into a broad picture of a dynamic ecosystem we may have a better chance of either successfully assessing ongoing conservation status or implementing new conservation programs.

Ecology is (grossly) the entire set of interactions between organisms and their environment. Although it can be studied as component pieces it does not actually occur as such. Conrad Kottack (1999) has recently revisited the concept of 'ethnoecology', which he defines as 'a society's traditional set of environmental perceptions' (p. 26) and cultural model of the environment and its relation to people and society. This set of cultural perceptions are critical to keep in mind when examining the ecology of human culture, especially in light of the dramatic and rapid alteration of environments that we are currently seeing around the world. Although humans frequently conceive of the 'environment' or 'the ecosystem' as external to themselves, this is not so. We are primates, we are mammals, and we respond to and interact with our physical environment using a wide array of biological and cultural adaptations. This broad conceptualization of the environment can be well understood by incorporating the concept of 'place'. As Miles Richardson pointed out in 1989 (in Biersack, 1999) 'We humans are flesh and blood primates and the world of the living and the dying is our home. As flesh and blood primates we occupy space, as creatures of symbol we transform it into place.' This 'Place', or human impacted/occupied ecosystem, can involve such disparate elements as local hunting and gathering, small-scale markets and garden crops, and the global market and trade in tropical hardwoods, coffee or other such commodities. One could see this as tied to a global ecology in the sense that forces (cultural or economic) geographically removed from our focus ecosystems have a dramatic effect on the aspects and quality of energy flow in and out of the system. This global ecology is part of the ecosystem in which our focal subjects (human and nonhuman) occur and interact. Given this broad reaching 'ecology' any discussion of a local or specific ecosystem should incorporate elements that affect it, even if they are geographically removed.

In many areas of the world (primarily circum-equatorial) humans and nonhuman primates share similar ecotypes and overlap heavily in biotic resources and space utilized (see Chapters 4, 5, 6, 14 and 15, this volume). One could say that in many cases humans and nonhuman primates (NHPs) share 'co-ecologies'. In light of this 'co-ecological' sympatry there is a recent surge of interest in what Leslie Sponsel has called 'ethnoprimatology' and Bruce Wheatley has termed 'cultural primatology', the study of the cultural, behavioral and ecological interrelations between human and nonhuman primates. This chapter is, in essence, an exercise in this area of investigation.

The Mentawai Islands

During 1989–90, 1992 and 1996, I conducted research with human and nonhuman primates on the Mentawai Islands off the west coast of Sumatra, Indonesia. This chapter summarizes my observations there and utilizes the published reports and communications of a wide variety of researchers working in the area since the early 1900s.

A chain of four islands in Westernmost Indonesia with a unique ecology and a long history of isolation, the Mentawais have interested anthropologists, primatologists, and conservationists alike. These islands are home to five species of primates who share and compete for the space and place of the Mentawais.

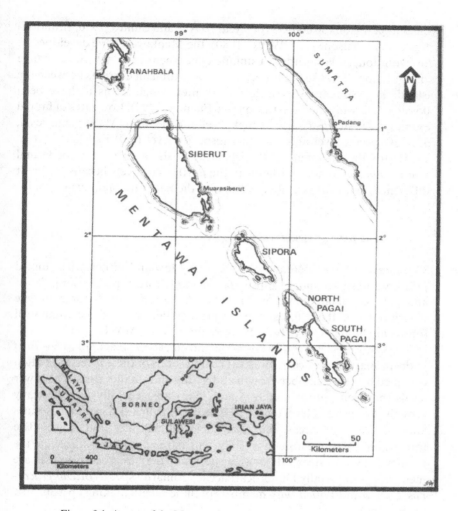

Figure 9.1. A map of the Mentawai Islands.

Location

The Mentawai Islands constitute some of the most unique and interesting places on earth. Lying 85–135 km off the west coast of central Sumatra between 0°55′S and 3°20′S and 98°31′E and 100°40′E, the four islands of Siberut (4000 km²), Sipora (850 km²) and North and South Pagai (1750 km²) (Fig. 9.1) are home to a wide variety of endemic flora and fauna. Owing to their long-term separation from Sumatra and the rest of the Sunda region (50 000–500 000 years) (Ave and Sunito, 1990; Brandon-Jones, 1978; Tilson, 1980; WWF, 1980), the Mentawais have developed as an evolutionary laboratory: a unique experiment in adaptation, change and isolation. Because of the deep trench and difficult currents between the islands and Sumatra very little faunal interchange seems to have been possible following their separation (see Fuentes, 1994). Low rates of faunal exchange are suggested by the high level of endemism (3% at genus level, 32% at species level and 32% at subspecies level (WWF, 1980; PHPA, 1995)) for the mammals of the islands. An alternative route for faunal interchange, via the Batu islands to the north, is unlikely because of great differences in faunal assemblages between the Batu and Mentawai islands.

The forests: structural ecology

The forests of the Mentawais are primary lowland dipterocarp, mixed primary forest, secondary forest, *Barringtonia* forest, peat swamp forest and palm-dominated swamp forest (see Whitmore, 1984, for complete descriptions of these forest types). Floristic analyses of the Mentawai forests are limited (see PHPA, 1995; Whitten, 1980; WWF, 1980).

The canopy is between 35 and 45 m and emergents reach up to 60–70 m in the primary dipterocarp forests. The mid-levels of these forests are often occupied by dense climber growth and young and smaller trees. The lower levels (between approximately 5 and 15 m) are generally free of heavy growth. The ground level tends to be heavy in palms and rattan in ravines and low-elevation areas, but fairly open on hillsides and ridges. The topography of these forests is hilly with streams moving through the ravines between ridges. Major tree species of the dipterocarp forests are members of the family Dipterocarpaceae, primarily species of *Dipterocarpus*, *Shorea* and *Vatica*, with members of these genera making up a 30–50% of the canopy trees.

Mixed primary forest is structurally similar to the dipterocarp forest; however, dipterocarps are not as dominant and the average climber load per tree is heavier. Common families are Bombaceae, Dipterocarpaceae, Euphorbiaceae, Myristicaceae, Sapindaceae and Sapotaceae. Dipterocarps are still the most common large trees in this forest type, making up 15–30% of all trees of greater than 15 cm diameter at breast height (dbh). Many palm families are also present on the ground and in the lower levels of this forest, including Arecoideae, Caryotoideae, Coryphoideae and Lepidocaryoideae (the rattans). Both types of primary forest on the Mentawai Islands have very few members of the family Leguminosidae.

Secondary forests are areas where some human disturbance had taken place in the past. The canopy is low, about 20–30 m high, with few emergents. Mid-levels are more crowded and the ground is often covered in low growth (for example, shrubs and saplings). Larger tree species tend to be similar to those of mixed primary forests.

Barringtonia forest is found primarily along the shoreline areas (within 1 km of the shore) and is dominated by *Barringtonia, Hibiscus* and *Pandanus* species (WWF, 1980). This forest is low, the canopy is under 20 m, and there are no emergents. The ground level is open and the soils are mixed or sandy.

Peat swamp forest is found in a few areas between primary mixed and palm-dominated swamp forests. These swamps are characterized by nutrient-poor peat soils covered by a mat of roots. The canopy is low, 15–20 m, and many of the trees belong to the genera *Artocarpus* and *Palaquium*.

Palm-dominated swamp forest is found both intermixed in the gullies and low areas of primary forest and between the primary forests and rivers or coastal forests. This forest can be flooded continuously or only occasionally depending on the distance from a river and the slope of the ground. Common families present are Coryphoideae, Lepidocaryoideae and many members of the genus *Nepenthes* (pitcher plants).

Other significant vegetation types are the mangrove swamps (on the west coasts), which are typified by the *Rhizophora* and *Brugueria* mangrove complexes, and the Nipa palm forests (*Nypa fruticans*) which cover much of the rivers as they move away from the coast. Additionally, large stands of sago (*Metroxylon sagu*) are found in many swampy areas and are heavily utilized by the local inhabitants. The tropical hardwood dominance of the Mentawai forests makes them a high priority target for logging concessions.

This structural description of the Mentawai forests is the primary backdrop to a discussion about habitat use and ecology of the nonhuman primates found on these islands. The generalized notion of a nonhuman primate ecosystem is described as a set of sympatric flora and nonhuman fauna which could, in turn, be impacted by humans.

The nonhuman primates

The Mentawai Islands have the highest endemism rate for primates of any island group: four endemic primates in a total area of 6550 km². The next highest primate endemism rate is in Sulawesi, with seven primates in 189037 km² (WWF, 1980). All four of the Mentawai primates are either threatened or endangered. All three monkey species are represented by two subspecies: a Siberut form and a form found on the three other islands.

Although these fascinating primates have been the focus of few studies, most researchers agree that these animals are in danger and that multiple factors are driving them towards extinction (Fuentes, 1998, Fuentes and Ray, 1996; Mitchell and Tilson, 1986; Tenaza, 1987, 1988, 1989). At present there are a few Mentawai macaques in zoos in Indonesia and Europe, but there are no captive populations or breeding programs for the other three Mentawai primates.

Of the four Mentawai primates the Kloss' gibbon, *Hylobates klossi*, the smallest of the gibbons, is the best known. As the subject of three dissertation studies (Tenaza, 1974; Tilson, 1980; Whitten, 1980) its ecology and vocal behavior are well documented. Of the gibbons *H. klossi* has the 'simplest' song and may be unique among its congeners in not displaying a male–female vocal duet (Kappeler, 1984; Whitten, 1982). Adult males and females sing separate songs, with the female long call lasting up to one half-hour. The Kloss' gibbon preferentially utilizes primary forests, using canopy emergents with heavy liana cover as primary sleeping sites (Whitten, 1980, 1982). Unlike many of the other gibbons, the Kloss' gibbon adults do not spend the majority of their time close to one another; both adults frequently sleep in different trees and do not display many of the pair bond behaviors associated with the genus *Hylobates* (Fuentes, 2000).

The Mentawai macaque is the least known of the four Mentawai primates and little has been published regarding its behavior and ecology. There appear to be two subspecies, *Macaca pagensis pagensis* of the Pagais and Sipora (Fig. 9.2), and *Macaca pagensis siberu* (Whitten and Whitten, 1982), found on the northern island of Siberut. The macaque groups range in size from five to 25 animals, and solitary males are also seen. Frequently the groups split up into smaller foraging units during the day, only seen en masse during feeding in coconut gardens or moving through the primary forest (Fuentes and Olson, 1995). The macaques feed on a variety of plant species both in the primary and secondary forests and in the gardens and coconut groves surrounding human villages. Like all of the Mentawai primates the macaques frequently vocalize in the early morning hours, but do not exhibit the multi-group vocal interactions common to the gibbons and langurs of the Indonesian archipelago. However, the macaques do interact with both species of langurs, moving and feeding on occasion with langur groups.

The monospecific pig-tailed langur, *Simias concolor*, is perhaps the most endangered of the Mentawai primates. An unusual macaque-convergent morphology, distinct vocalizations and interesting social organization set

Figure 9.2. Captive juvenile *Macaca pagensis (nemistina)* on North Pagai Island.

it apart from other colobines (Tenaza and Fuentes, 1995). Groups generally contain one adult male and one or more adult females, with ca. 50% of the groups having two or more adult females. Mean group size is 4.1 individuals and home ranges vary from 7 to 20 hectares. This species uses all levels of the forest, but mainly relies on the 25 m closest to the ground. *Simias* is also one of the few anthropoid primates to occasionally 'park' their infants as they feed (Fuentes and Tenaza, 1996). It has also been the subject of studies by Tilson (1977) and Watanabe (1981).

The Mentawai Island langur, *Presbytis potenziani*, is the only cercopithecoid primate found in two-adult groups throughout its range. *P. potenziani* resides in the upper canopy of primary and disturbed rainforests where it feeds mainly on climbers and epiphytes (Fuentes, 1996). The adults are monomorphic in relation to body size and there is no paternal care of the young (Fuentes, 1994). In most reports the adults do not duet. However, Tilson and Tenaza (1976) recorded some duetting on Siberut Island, Indonesia, and both adults vocalize in over 50% of intergroup encounters observed (Fuentes, 1994). Behavioral indications of territoriality, or range defense, such as intergroup conflict, or active boundary

Figure 9.3. Villagers of Desa Betumonga, North Pagai Island.

monitoring are rare in this species (Fuentes, 1994, 1996). Although the Mentawai Island langur does appear to live in two-adult plus offspring groups it displays very little, if any, of the standard monogamy behavior profile. The general behavioral profile of adult male and female Mentawai Island langurs appears to be quite similar to that of the other, polygynous, members of the genus *Presbytis*, which inhabit the Sunda region of Southeast Asia (Fuentes, 1994, 1999).

The human primates

The Northern Island of Siberut has been occupied by humans for at least the past 2000–3000 years. The colonization of Sipora and the Pagais came later, possibly as recently as 200–400 years ago (Loeb, 1928, 1929; Nooey-Palm, 1968; Schefold, 1972, 1991). The indigenous peoples of these islands are proto-Malayan and most likely descendants of people from the Nicobar Islands or via those islands and/or Nias (Figs. 9.3 and 9.4). According to their own traditions the people originated in Nias and moved southward to Siberut, Sipora, and eventually the Pagai islands. This pattern of dispersal is easily traceable through the clan names of the Sipora and Pagai

Figurre 9.4. Mentawai Islander children.

Islanders, as all the names come from the southern portion of Siberut (Nooey-Palm, 1968).

The traditional Mentawai culture, as far as can be reconstructed, consisted of small, generally egalitarian bands composed of 30–80 members (the *uma*) who occupied (and defended) discrete territories centered around a longhouse (also called *uma*). These groups cultivated taro, sago, bananas and coconuts, collected forest crops, and relied primarily on hunting monkeys and, secondarily, on seafood for protein. Although the *uma* had no full-time specialists, the *kerai* (a shaman) held an important role in the communication between the *tai* (general spirits), the *simagre* (human spirits) and the *badjou* (life force). The Mentawai Islanders maintained an elaborate taboo system based on a balance between the above mentioned supernatural forces. This taboo system regulated hunting and harvesting and was a dominant characteristic of everyday life (Loeb, 1928; Schefold, 1972, 1991).

Although sporadic trade contacts between the Mentawai Islanders and the Minangkabou and Batak ethnic groups of Sumatra existed from at least the seventeenth century, the colonial Dutch government never took much interest in the Mentawais. A few attempts at building a school or two and a few dirt roads in the early part of this century were the extent of their interests. However, German Protestant and Italian Catholic missionaries have been intensively active since the 1950s in all of the Mentawais. This intensive missionary activity has resulted in a 98% conversion rate. This religious conversion along with changes mandated by the Indonesian government in living patterns (*uma* to village) and political structure (village headmen and governmental representative) have disrupted much of the traditional lifeways of the Mentawai Islanders.

The current Mentawai population, as of 1996, is 56 000, of which 42 000 are indigenous Mentawai Islanders.

Human and nonhuman interconnections

The humans, monkeys and gibbons of the Mentawai Islands rely almost exclusively on the forest to satisfy their material, nutritional and spatial needs.

Habitat interconnections

All of the nonhuman primates of the Mentawais are tied to the forests of the islands. The four species avoid competition to an extent by focusing their foraging and ranging patterns in different layers of the forest (Fuentes, 1994). However, the Mentawai island langur and the Kloss' gibbon do compete for the same food sources and sleeping trees (Fuentes, 1994, 1996; Tilson and Tenaza, 1982). In this competition the gibbons win nearly every encounter. The two colobines (*P. potenziani* and *S. concolor*) are able to avoid conflict by exploiting very different niches. *P. potenziani* specializes in the upper reaches of the forests, feeding on climbers and epiphytes, whereas *S. concolor* exploits a wide range of leaves and fruits in the lower canopy (Fuentes, 1994, 1996; Tenaza and Fuentes, 1995).

The forests provide a wide array of support for the Mentawai Islanders. The trees provide the wood for houses and canoes. Nearly all of the long-distance travel in the Mentawais is by boat and critical intervillage links are maintained via ocean and river voyages. The forests are also home to the taro gardens, the mixed gardens and the stands of sago that provide the core nutrients for the islanders' diet. The taro and sago grow well in the swamps that fringe the lower-elevation areas and coastal edges of the primary forests. The mixed gardens are built directly in the forest, usually near the forest edge within a few hours' walk or canoe trip from the villages. These swidden plots can be used for up to 10 years before they are allowed to lie fallow. In the gardens a number of banana species are grown, mixed with other fruits and ground crops. The primary forest also provides fruit resources, such as durian and manggis, which are routinely collected in the appropriate seasons.

Since the 1980s the forests have also played an important role in the cash economy of the islanders. The mid-1980s see a heavy exploitation of *Aquilaria* tree species for the internal fungus that produced an economically valuable resin (kayu gaharu or eagle wood). However, once the majority of older *Aquilaria* trees were felled, the kayu gaharu became less economically favorable. Recently (1990s) a global upswing in the rattan market has led to an extreme exploitation of the diverse array of rattan species growing in the Mentawais. Many young men are able to generate substantial income via rattan harvesting throughout the year.

Both the human and nonhuman primates rely heavily on the Mentawai forests for their livelihood. Through this mutual reliance they are also interconnected to one another. The humans visit the forest to hunt all four species of nonhuman primates. The macaque is thought to be the least palatable ('asin' or sour) and many Mentawai clans have taboos against hunting the gibbons. However, both of the colobines are considered good food and *Simias concolor* is the favored food of the majority of islanders. A by-product of the hunting is a large number of pets. When adult females are killed they frequently leave behind an infant. It is common for the hunters to take the infant to the village and raise it until it is large enough to eat.

The gardens also link the human and nonhuman primates. All three monkey species regularly use the gardens as food sources and many groups of both macaques and Mentawai langurs appear to exploit the gardens as prime feeding patches. Losses can be as high as 35% of garden yields (A. Fuentes, unpublished data). The islanders frequently respond with series of traps ringing the gardens and considerable extra effort in monitoring the gardens (which can be some distance from the villages).

Cultural interconnections

The primates of the Mentawais are not only tied together by their shared habitat, they are also untied via central elements of the human primates' culture. Traditionally, the Mentawai Islanders saw the nonhuman primates as both 'cousins' and magical sources of *tai*, *badjou* (spirits/life forces) and nutrition. The monkeys of the Mentawais (and the gibbon for some *uma* of Siberut) played integral roles in the elaborate taboo system that patterned the Mentawai life cycle. Births, deaths, sickness, marriage, rites of passage all were marked by *punen*, or feast celebrations, for which certain monkeys were required as the main courses (Nooey-Palm, 1968, Schefold, 1991). The skulls of these monkeys were placed on the long beams of the *umas* to guarantee a strong house, fertility and successful future hunts. In fact, one of the earliest translations of a Mentawai chant was that of a Siberut *kerai* as he offered portions of the *punen* meal to the monkey skulls in his uma:

> This is for you, spirits of the monkeys which we shot. If we go hunting again summon your friends and let them be fat ones.

Habitat, resources, and culture unite the human and nonhuman primates of the Mentawais. They share and compete for elements of their forest environment and are all impacted by changes to the forest and coastal ecosystems.

Nonhuman primate and human primate ecologies: similar threats

Today the main threats to the Mentawai Islands and all of their inhabitants are the disruption of communal and traditional social structures, deforest-ation, and habitat conversion.

As many of the traditional beliefs and practices disappear or merge with Christian and Indonesian generalities, the monkeys of the islands continue to play an important role in the beliefs and diets of the Mentawai Islanders. Sunday celebrations, Christmas, and other such holidays have taken on many aspects of the *punen*, and the stories of hunts of the past are told and retold in many villages. However, the nonhuman primates of the Mentawais are now officially protected by the government of Indonesia and their hunting is illegal. Although few arrests have been made, the threat is there. Unfortunately, even if hunting is continued in a regulated fashion the decreasing numbers of monkeys and gibbons and the vanishing habitat paint a very bleak picture for the nonhuman primates of the Mentawai Islands.

Logging is rampant (Figs. 9.5 and 9.6). As of 1996 less than 17% of the original forest cover remains in the southern three islands (Fuentes, 1998). Since 1996 logging has continued and there are no current published estimates of remaining forest cover. The logging impacts both forest-dwelling animals and the Mentawai Islanders. Islanders must now compete with loggers for forest land in which to plant their extensive swidden gardens. Recently, it has become common practice for the logging companies to surpass their assigned quotas by obtaining rights to log village

Figure 9.5. Logging on South Pagai Island.

Figure 9.6. Post-logging beach transfer site on North Pagai Island.

lands. They are able to do this by purchasing the lands in exchange for televisions, satellite dishes and gas-powered generators. These items are costly to run and maintain, and have an increased need for fuel and parts. This increased need is met by either increasing the extraction of forest products or selling further parcels of land to the logging concessions.

Logging on Siberut was halted in 1994, which left nearly 50% of the island's forest cover. As of 1996 Siberut remained the only island in the chain where traditional cultural elements are present in high concentration amongst some of the remaining *uma* groups (now seen as clans). Siberut also has large, robust populations of all four Mentawai nonhuman primates (Fuentes, 1998) and the lowest ratio of religious conversion. There appears to be a tie between the occurrence of forested habitat and some degree of social stability for the primates of the Mentawais. This may be related to the set of ecological and behavioral adaptations by the nonhuman and human primates to the same complex forest environments. Unfortunately, in 1999 logging resumed in Siberut, with at least two companies logging land ahead of an expected transmigration of up to 300 families from Java.

As the forests shrink the monkeys and gibbon are forced to move to smaller forest patches and marginal habitats, where they are easy prey to hunting, disease and capture for the pet trade. This has also resulted in increased reliance by the nonhuman primates on the humans' gardens. So as the forest and available space for gardens decreases the value of the gardens go up for humans and nonhuman primates.

Transmigration and 'sawah' agriculture represent major stressors on the Mentawai ecosystem. The dramatic increase in population density brought about by government transmigration can have highly deleterious effects on small island habitats. At present Sipora has undergone the most substantial transmigration, but all three other Mentawai islands are scheduled to receive other Indonesians from the 'central' islands of Java and Sumatra (as many as 10 000 families) (Fuentes, 1998; Berber, 1995). There is also a push to get the Mentawai Islanders to reduce their use of swidden gardens and place more emphasis on wet rice production. Owing to the shallow soils, lack of rocks, and hilly topography of the Mentawais, wet rice agriculture can perform well for only a few years after which the area utilized is useless and frequently colonized by *Imperata* grasses (*alang-alang* or elephant grass). Although many of the nonhuman primates are beginning to rely on the swidden gardens for dietary needs (more so now owing to deforestation pressure), none of them have been able to utilize wet rice paddies in the same way.

Finally, the dramatic upsurge in the reliance on a cash economy and external elements in the global economy have directly affected all the primates in the Mentawais. The Mentawai Islanders are now nearly exclusively reliant on cash crops to purchase tobacco, alcohol, parts for church construction and maintenance, clothing, cooking utensils and other necessities. This reliance on cash crops spurred massive clove (and coconut) planting in the 1970s and 1980s, but the global crash in clove (and copra) prices of the 1990s have caused a turn towards other cash sources, namely forest products. The need for hard currency has led to habitat conversion and the removal, and subsequent ecological damage, of many forest products (rattan, eagle wood, etc.). This habitat conversion to clove production and the destruction via forest products extraction are both a nonsustainable use of resources for the humans and a severe blow to the nonhuman primates' spatial, nutritional and social needs.

Researchers, tourists, and capitalists

In addition to the impacts of the above mentioned habitat and economic challenges the primates of the Mentawais are also jointly affected by a distinct set of external forces via foreign academics, conservationists, profiteers and a range of international tourists.

Primatological research

Because of the unique cultural and evolutionary situation presented by the Mentawai Islands, they have been a site for sporadic research interest by primatologists and anthropologists. Although most of the cultural anthropological research has been conducted on the Northern Island of Siberut (Schefold, 1991), primatologists have been working on both Siberut and the Pagai Islands since the 1970s.

Since the 1980s research by primatologists Richard Tenaza, Monica Olsen, the author of this chapter, and, currently, Lisa Paciulli and Sasimar Sangchantr has had a pronounced impact on the southwestern region of North Pagai Island. This region, Betumonga, has been the focus of short- and long-term primate research and is currently recognized as a semi-protected site by the government of West Sumatra. Prior to the mid-1980s there were no foreign visitors to the Betumonga region, except occasional Catholic and Protestant missionaries. Lack of roads to the region and the western-facing beaches made travel by foot or boat very difficult and quite dangerous. This region had been relatively isolated since a brief period of Dutch interest (a road and a school) in the 1930s.

Logging from 1989 through 1999 has reduced the thousands of hectares of forested land owned by the villages of the Betumonga region to under 1000 ha, of which approximately 600 ha remains a primate research site with a high density of all four Mentawai nonhuman primate species. As early as 1989 researchers negotiated a cessation of hunting with two villages in the Betumonga region. However, hunting and land conversion for gardens does occur in the times between researchers' tenure at the site. The primary reason for the remnant forest and its high density of primate species lies with its nearly 12 year history as a research site and more specifically the occupation of the site and lobbying activity by Lisa Paciulli from 1996 through 2000.

With each researcher who came to study primates there also came a minor economic windfall to a few families within the associated village. While these benefits generally resulted in small amounts of cash, the associated prestige and subsequent disparity amongst families frequently led to bickering and a degree of social instability. As new researchers moved in to the region they occasionally began working with different families than the previous researchers. These transitions had impacts on the village structures and hierarchical relationships across villages. They also may have led to a range of 'interference', by villagers, with data collection and research projects. Some researchers also brought, and enforced, particular cultural and behavioral requirements for association with their projects. In a sense, researchers have had an impact of altering the local economy (via cash-based wage labor), introducing distinct cultural patterns, and providing an additional variable influencing the ranking of individual families within village hierarchies.

In this case we see foreign researchers, drawn to the Mentawais by the nonhuman primates, who then have significant impact on the lives and villages of the human primates while actively studying and ostensibly 'conserving' the nonhuman primates.

Surf wars, National Parks and ecotourism in Siberut

Although the Pagais do receive some surf tourists (more in the Betumonga region than other areas), it is the Northern island of Siberut where the majority of the small but growing Mentawai tourism trade occurs. There are two main draws for tourists in the Mentawais: surfing and 'culture'.

The surf tour industry in the Mentawais started as early as the late 1980s. By the mid-1990s there had been half a dozen surf videos and many more photo spreads and magazine stories about the 'untamed Mentawais' (Baker, 1998). As recently as June, 2000, a pro invitational surf contest was held based on a boat moored off of Siberut with a cash prize of US$67 000 for men and US$37 000 for women. Currently there are a few major companies vying for the lucrative surf market. All of these companies are non-Indonesian owned, and there are no Mentawai Islanders as managers for these companies. In 1995 Great Breaks International, a surfing tour company, was able to secure the rights to purchase a 100 ha region in Southern Siberut containing some of the best surf breaks. This company has proposed a Mentawai Sanctuary that will include a hotel and facilities, package surf tours and ecotours. These ecotours were to include both 'culture' visits and 'eco-friendly' animal experiences. There were even initial conversations about the inclusion of a primate conservation element in the initial project (A. Fuentes, personal observations). The resulting animosity between Great Breaks and other tour companies has resulted in a set of 'surf/turf wars' in Siberut (Baker, 1998; Cameron, unpublished). Although the majority of these elements impact primarily coastal areas of Siberut, the economic and political impact can have potentially significant effect on the human and nonhuman primates of the Mentawais.

As early as 1980 the northern island of Siberut's Teitei Batti wildlife sanctuary was established as a national park and a UNESCO Biosphere reserve (WWF, 1980). By 1994 the Asian Development Bank had designated 18 million US dollars to carry out the Siberut Integrated Conservation and Management Plan (ICMP). The plan called for Siberut to be managed through a zoning system that included sanctuary, traditional use, intensive use and park village areas. The conservation and management area included over 50% of Siberut's land mass. The project intended to maintain biological diversity through a partnership with local institutions. There was to be an establishment of village councils, an improvement of healthcare and schooling, and the promotion and institution of small-scale cash crops, subsistence crops, logging and entrepreneurial endeavors. The entrepreneurial endeavors were to include coordinated ecotourism as a main focus. All of this was to be done in conjunction with a strong research and conservation program focused around the four Mentawai nonhuman primates (PHPA, 1995).

As of 2000 all US$18 million had disappeared, the Asian Development Bank had pulled out of the project, most of the advisory staff had resigned and the only evidence of infrastructure were a new park headquarters, a research building, visitor center and lodge, and an undocumented length of newly cut trails (DeAnna Howell, personal communication; Sasimar Sang-chantr, personal communication). No long-term research has been established, nor have any surveys or park enforcement activities been continuously conducted.

No simple solutions

At the beginning of this chapter I defined ecology as the entire set of interactions between organisms and their environment. For the Mentawai

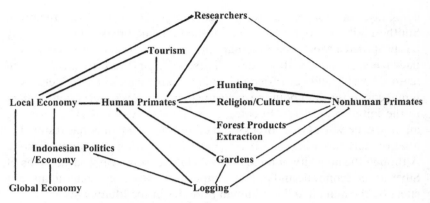

Figure 9.7. Interconnections diagram.

primates (human and nonhuman) the interactions within and between species and the human social, political and economic realities directly impact the environment in which all of the species exist. I suggest that the ecological 'space' (the Mentawai flora and fauna) is altered and can be viewed as the Mentawai 'place' (the current situation of the Mentawais as outlined in this chapter).

In order to study the nonhuman primates of the Mentawais we are faced with the reality of defining our ecosystem and realizing that much of the system itself is intertwined with human Mentawai culture, and the Indonesian and global economies. The forests are active areas of subsistence for the monkeys, ape, and humans of the Mentawais. They are also sources of economic gain for humans on and off the islands. Humans from other parts of the planet come to the forests and coastal regions to extract or utilise the flora for logging and forest products, and geological structures for surfing. The amount of forest extraction or modification and primate hunting practiced by the Mentawai Islanders is related to the external economic system and a cash economy in addition to dietary needs and socio-religious practices and patterns. Given these eco-complexities, how do we study the nonhuman primates or attempt conservation of these species? The concept of an ethnoprimatology, an envisioning the nonhuman primates as part of a 'place' or larger ecosystem that incorporates multiple levels of assessment beyond the traditional examination of ecological space, may be a partial solution to the situation presented here (see Fig. 9.7).

There are serious challenges to the Mentawai forests. The entire Mentawai ecosystem (however defined) is in a state of flux with multiple and competing impacts at many levels. I suggest that a current understanding and baseline assessment of Mentawai Island nonhuman primates, which is necessary for any conservation action, must be conducted in the context of the current Mentawai situation. Broad swathes of primary rainforest may no longer be the main forest environment available to the monkeys and ape. Garden environments, logged forests, and small forest patches could rapidly become the main ecotype on the Mentawais. These are human-altered, maintained and monitored 'places'. For example, any understanding of the nonhuman primates' forest use patterns and ranging must also examine the human primate planting and forest extraction patterns *and* the economic patterns that affect these. Changing patterns of human hunting result in differential selective environments for the monkeys and the gibbons. However, these changes are tied to cultural and economic alterations for the humans. All of the Mentawai primates must adapt to the current rapid and systematic decrease in available forest use area. This can relate to changes in diet, range use, group structure and social interactions for all of the primate species (human and nonhuman). By focusing on an interaction paradigm rather than a 'pristine' paradigm we may be better able to understand the range of behavior and ecological adaptations possible and thus be better prepared to initiate or assess potential plans for conservation at a variety of levels.

Obviously, this is a very complex way to attempt to envision an ecological system if one wishes to accurately model and assess conservation strategies. There are no simple solutions and it may be that this is too complex or too labile for current conservation modeling. However, the recent interest in undertaking ethnoprimatological research (see other chapters in this book) suggests the possibility of a growing awareness of the context that we currently find ourselves in. This entails a relaxing of the traditional boundaries between the 'science of primatology', the practice of conservation and the traditional ethnography and emerging political and cultural ecologies. By attempting to meld these different approaches we may be better armed to successfully provide complex potential solutions to areas where there are no simple ones.

Acknowledgements

I wish to thank the people of the Mentawai Islands for all of their assistance and commend their perseverance in the face of an uncertain future. I am especially indebted to the villages of Betumonga, Sioban, Sikakap, and Sinakak, Nusa Mentawai Tour and Aurelius Napitupulu, and the family of Pak James Huttagulung, for their assistance and friend-ship. I also thank the Government of Indonesia and Lembaga Ilmu Pengetahuan Indonesia (LIPI) for permission to conduct research in their country. Partial funding for this research was provided by Primate Conservation, Inc., the Center for Southeast Asian Studies at UC Berkeley, and Sigma Xi.

References

Ave, W. and Sunito, S. (1990). *Medicinal Plants of Siberut*. Switzerland: WWF International Nature Report.

Baker, T. (1998). Boat wars. *Surfer Magazine*, September, pp. 88–104.

Berber, C.V. (1995). *Tiger by the Tail: Reorienting Biodiversity Conservation in Indonesia*. Amsterdam: Elsevier Press.

Biersack, A. (1999). The mount kare python and his gold: Totemism and ecology in the Papua New Guinea highlands. *American Anthropologist* **100**, 68–87.

Brandon-Jones, D. (1978). The evolution of recent Asian Colobinae. In *Recent Advances in Primatology*, vol. 3 (Proceedings of the 6th congress of the International Primatological Society), pp. 323–5.

Fuentes, A. (1994). The Socio-ecology of the Mentawai Island Langur (*Presbytis potenziani*). Ph.D. thesis, University of California at Berkeley.

Fuentes, A. (1996). Feeding and ranging in the Mentawai Island langur (*Presbytis potenziani*). *International Journal of Primatology* **17**, 525–48.

Fuentes, A. (1998). Current status and future viability for the Mentawai Primates. *Primate Conservation* **17**, 111–16.

Fuentes, A. (1999). Re-evaluating primate monogamy. *American Anthropologist* **100**, 890–907.

Fuentes, A. (2000) Hylobatid communities: changing views on pair-bonding and social organization in hominoids. *Yearbook of Physical Anthropology* **43**, 33–60.

Fuentes, A. and Olson, M. (1995). Preliminary observations and status of the Pagai macaque (*Macaca pagensis*). *Asian Primates* **4**, 1–5.

Fuentes, A. and Ray, E. (1996). Humans, habitat loss and hunting: the status of the Mentawai primates on Sipora and the Pagai Islands. *Asian Primates* **5**, 5–9.

Fuentes, A. and Tenaza, R.R. (1996). Infant parking in pig-tailed langurs (*Simias concolor*). *Folia Primatologica* **65**, 172–3.

Kappeler, P.M. (1984). Vocal bouts and territorial maintenance in the moloch gibbon. In *The Lesser Apes, Evolutionary and Behavioral Biology*, ed. H. Prueschoft, D.J. Chivers, W.Y. Brockelman and N. Creel, pp. 376–89. Edinburgh University Press.

Kottack, C. (1999). The new ecological anthropology. *American Anthropologist* **100**, 23–35.

Loeb, E.M. (1928). Mentawei social organisation. *American Anthropologist* **30**, 408–33.

Loeb, E.M. (1929). Mentawai religious cult. *University of California Publications in American Archeology and Ethnography* **25**, 185–247.

Mitchell, A.H. and Tilson, R.L. (1986). Restoring the balance: traditional hunting and primate conservation in the Mentawai islands, Indonesia. In *Primate Ecology and Conservation*, ed. J.G. Else and P.C. Lee, pp. 249–60. London: Cambridge University Press.

Nooey-Palm, C.H.M. (1968). The culture of the Pagai Islands and Sipora, Mentawai. *Tropical Man* **1**, 153–241.

PHPA (Perlindungan Hutan dan Pelestariam Alam) (1995). *Siberut National Park Integrated Conservation and Development Management Plan* (1995–2000): Vols. 1–3.

Schefold, R. (1972). Divination in Mentawai. *Tropical Man* **3**, 10–87.

Schefold, R. (1991). *Mainan Bagi Roh: Kebudayan Mentawai.* Jakarta: Balai Pustaka.

Tenaza, R.R. (1974). I. Monogamy, territoriality and song among Kloss' Gibbons in Siberut Island, Indonesia. II. Kloss' Gibbons sleeping trees relative to human predation: implications for the socioecology of forest dwelling primates. Ph.D. thesis, University of California, Davis.

Tenaza, R.R. (1987). Studies of primates and their habitats in the Pagai Islands, Indonesia. *Primate Conservation* **8**, 104–10.

Tenaza, R.R. (1988). Status of primates in the Pagai Islands, Indonesia: a progress report. *Primate Conservation* **9**, 146–9.

Tenaza, R.R. (1989). Primates on a precarious limb. *Animal Kingdom* **92**(6), 146–9.

Tenaza, R.R. and Fuentes, A. (1995). Monandrous social organization of pig-tailed langurs (*Simias concolor*) in the Pagai Islands, Indonesia. *International Journal of Primatology* **16**, 195–210.

Tilson, R.L. (1977). Social organization of Simakobou monkeys (*Nasalis concolor*) in Siberut Island, Indonesia. *Journal of Mammalogy* **58**, 202–12.

Tilson, R.L. (1980). Monogamous mating systems of gibbons and langurs in the Mentawai Islands. Ph.D. thesis, University of California, Davis.

Tilson, R.L. and Tenaza, R.R. (1976). Monogamy and duetting in an Old World monkey. *Nature* **263**, 320–1.

Wantanabe, K. (1981). Variations in group composition and population density of the two sympatric Mentawaian leaf monkeys. *Primates* **22**, 145–60.

Whitmore, T.C. (1984). *Tropical Rain Forests of the Far East*, 2nd edn. Oxford: Clarendon Press.

Whitten, A.J. (1980). The Kloss Gibbon in Siberut rain forest. Ph.D. thesis, University of Cambridge.

Whitten, A.J. (1982). The ecology of singing in Kloss Gibbons (*Hylobates klossii*) on Siberut Island, Indonesia. *International Journal of Primatology* **3**, 33–51.

Whitten, A.J. and Whitten, J.E.J. (1982). Preliminary observations of the Mentawai macaque on Siberut Island, Indonesia. *International Journal of Primatology* **3**, 445–59.

WWF (1980). *Saving Siberut: A Conservation Master Plan.* Switzerland: World Wildlife Fund.

10 Conservation must pursue human–nature biosynergy in the era of social chaos and bushmeat commerce

ANTHONY L. ROSE

Introduction

While social chaos spreads across equatorial Africa, human predation has become the most urgent threat to African primates. The explosion of commercial bushmeat hunting is destroying primate populations faster than their habitat can be cut down. The failure of conservation in great ape range countries is due primarily to human crises: poverty, illness, war, commercial greed, political corruption, lawlessness. There is a revolution going on in equatorial Africa, and it is being perpetrated by international exploiters who are radically manipulating the social order and cultural values of the African people so as to serve their personal and corporate pursuit of power and money. Because of this assault on African lifeways and values, social change and turmoil are the rule. In this milieu the practice of conservation must become proficient at understanding the context for change and working with its problems, causes, and solutions.

In this complex, harsh context, conservation's first challenge is to help African people to restore and sustain their cultures, their economies, and their reverence for nature as the foundation for ensuring the viability, diversity, and synergy of life in the region, and ultimately across the planet. The leaders of the conservation movement come from fields and disciplines that don't address the causes of Africa's social chaos. Neither are they practiced in the social change applications needed to resolve it. The main focus of this chapter will be to examine the commercial bushmeat crisis that presents conservation's most urgent challenge. Some of the social factors that underpin the crisis will then be presented. Paradoxes in the search for solutions will be discussed, and a call for new approaches to conservation, with new strategies and new players, will be made. Throughout we shall examine how the varied and shifting relationships among humans and nonhuman primates are vital to the survival of apes and monkeys in the wild, and to the future of human society in Africa.

208

Table 10.1. *Complex and harsh realities underlie the destruction of nature in Africa*

Problems	Causes	Solutions
Over exploitation	Greed / corruption	Transparency / monitoring
Cultural destruction	Lost social control	Community development
Fatalism/lawlessness	Poverty and disease	Health and welfare support
Bushmeat market growth	Misdirected finance	Non-game protein provision
Rampant poaching	Inadequate capacity	Protection and governance
Great ape extinction	Conservator timidity	Leadership empowerment
Consumption of nature	Lost care/reverence	Global wildlife missions

Bushmeat crisis and social chaos dominate African primate conservation

Hunting and primates

Across the forest regions of west and central Africa a conflux of factors are making human predation a leading threat to the survival of most primates, including the great apes. Primate hunting is reported in 27 of the 44 primate study and conservation projects described in IUCN's latest status survey on African primates (Oates, 1996b). In 12 of these territories, human predation is said to be a severe threat to species survival. The latest IUCN *Red List of Threatened Animals* (IUCN, 1996) shows a big rise in threat status for mammal species, with primates being the large order most threatened by extinction. The situation is worse in the areas where most remaining apes and monkeys live, outside parks and reserves. In Africa hundreds of unique and never studied primate populations are being annihilated, and thousands will follow if current trends continue (Oates, 1996a; Rose, 1996f; Ape Alliance, 1998; Ammann, Pearce and Williams, 2000).

The risk level for different species and populations varies with their numbers, reproductive vigor, and distributional range. Past declines have correlated most closely with human population growth and the destruction of habitat. Eltringham (1984) wrote that 'Gorillas and chimps costing several thousand dollars each are captured for zoos and medical research centers, but the quantity killed for food dwarfs the number taken alive.' Although capture of live apes for research has mostly stopped, a growing body of evidence now shows that shifts in human social and economic practices in the forests of Africa have greatly increased the killing of primates for meat. Oates (1996a) concludes

> ... while the total removal of natural habitat is clearly a major threat to the survival of many African forest primates, an analysis of survey data suggests that human predation tends to have a greater negative impact on primate populations than does selective logging or low-intensity bush-fallow agriculture.

Commercial bushmeat trade is not sustainable and will result in rapid local extinction and eventual extirpation of large mammals from the world's rainforests (Robinson and Bennett, 2000). Estimates, now considered by some as conservative, place annual bushmeat out-take at 1 million tonnes in the Congo Basin (Wilkie and Carpenter, 1999). That is more than a billion dollar business. Some researchers hypothesize that traditional hunting with nets, spears, and arrows might have been sustainable in the past. I would argue that the conditions for sustainable hunting via indigenous methods no longer exist in tropical Africa, or perhaps anywhere on earth. Outside influences, no matter what the purpose, provide economic and social pressures that transform the hunt to a modern process that will ultimately destroy wildlife populations. This chapter begins with a review of some of the evidence that supports this argument, to give the reader a feel for the situation in African primate habitat.

Guns and commerce

Even in areas with no commercial pressure, demand for chimpanzee and gorilla meat can be substantial. Kano and Asato (1994) compared ape density and gun hunting pressure from 29 Aka and Bantu villages along the Motaba River area of the northeastern Congo Republic and projected a bleak future for the apes. They found that over 80% of their 173 Aka informants were willing to eat gorilla or chimpanzee meat. Among 120 Bantu informants, 70% were willing to eat gorilla meat and 57% would eat chimpanzee. Because more Aka were involved in ape hunting, 40% reported having eaten gorilla or chimpanzee meat in the previous year, while 27% of Bantu had eaten apes in the same period. Aka informants estimate that 34–60 successful 'subsistence' hunters slaughtered 49 gorillas and 103 chimpanzees in 1992. Bantu claimed 7–9 hunters took 13 gorillas and 28 chimpanzees that year. Based on population density, Kano and Asato assert that the survival of gorillas and chimpanzees is at serious risk in this territory 'unless a strong system can be established which combines effective protection with the provision of attractive substitutes for ape meat to the local people'.

Although this hunting is carried out to serve the villages and not for commercial export, it is not traditional subsistence hunting: it is modern hunting supported by guns and ammunition imported from urban manu-facturers. The finding that village hunting of apes in a large habitat area is unsustainable when guns are used is not surprising. What is disturbing is the lack of any further reports on ape hunting practices in the area. We cannot say whether the illegal and unsustainable hunting has been stop-ped, or if apes have been wiped out along the Motaba River. We can assert with confidence that if village gun hunting of apes is unsustainable, then surely such illegal activities catalyzed by expanding commercial bushmeat trade will demolish ape populations. As the taste for bushmeat continues to spread across equatorial Africa at an accelerated pace and hunting and meat export capacity improves, African primates along with other edible wildlife are being faced with severe threats of local extinction (BSI, 1996; Ape Alliance, 1998; BCTF, 2000).

South of the Motaba River, Hennessey (1995) studied bushmeat com-merce around the Congolese city of Ouesso by means of survey and undercover methods. He reports that 64% of the bushmeat in Ouesso comes 'from an 80 km road traveling southwest to a village called Liouesso'. There a hunter who specializes in apes was responsible for most of the 1.6 gorilla carcasses sold each week in the Ouesso marketplace. That is over 80 gorillas per year in one city. Hennessey projects that 50 elephants were killed annually for meat and ivory in this same study, and 19 chim-panzees. Similar Aka–Bantu hunting and long-distance commercial bush-meat trade is described by Wilkie, Sidle and Boundzanga (1992) in the Sangha region west of Ouesso. There, many hunters preferred trading their meat at Ouesso in order to get a higher price than at logging concessions, confirming the report of Stromayer and Ekobo (1991) that Ouesso and Brazzaville are ultimate sources of demand. Wilkie *et al.* (1992) describe monkey meat for sale, but say nothing about apes. They do recommend that wildlife conservation officers and biologists monitor and protect duiker, primates, and elephants to regulate 'the harvest of forest protein'.

Ammann and Pearce (1995) reported intense hunting of apes for bush-meat in south-eastern Cameroon, across the border west of Wilkie's study site.

> The hunters in the Kika, Moloundou and Mabale triangle in Cameroon estimate that around 25 guns are active on any given day and that successful gorilla hunts take place on about 10% of outings. This would result in an estimated kill of up to 800 gorillas a year.

These same hunters say they bring out chimpanzee too, half as many as gorillas in this location: up to 400 per year. Although some of this ape meat is sold to logging workers in these forests, most is shipped on logging lorries back to Bertoua and all the way to Yaounde and Douala where a better profit can be made. Ammann (1996b) confirmed Hennessey's (1995) findings that a small portion of Cameroon bushmeat crosses the border for sale in Ouesso.

Illegal bushmeat including gorilla, chimpanzee, and bonobo in villages near reserves like Lope, Ndoki, and Dja, and in city markets at Yaounde, Bangui, Kinshasa, Pt. Noire, and Libraville, has been seen and photo-graphed (see, for example, Ammann, 1996a,b, 1997, 1998a; McNeil and Ammann, 1999; McRae and Ammann, 1997). Traders interviewed in those areas affirm that the fresh meat comes from nearby forests, whereas smoked meat can be transported long distances. It is well known that the million people who inhabit the largely forested territory of Gabon have a strong palate for bushmeat. Steel (1994) found half the meat sold in Gabon city markets is bushmeat: an estimated US$50 million unpoliced trade. Twenty per cent of the bushmeat is primates. This study reported very few apes observed, likely because ape-meat is not displayed openly in public markets.

The importance of well-planned and well-conducted undercover investigations of illegal bushmeat activities cannot be over-emphasized. In the absence of complete and continual monitoring of hunting and the bushmeat trade, we can only guess the numbers of primates killed to feed the tens of millions of people living in equatorial Africa. There can be little doubt that many more apes are butchered for meat in the forests every year than live captive in all the world's zoos, laboratories and sanctuaries: Perhaps 5000–10 000 each year, or more.

During extensive discussion with field researchers and conservationists (Rose, 1996b,c,d; Rose and Ammann, 1996), this author found expert consensus predicting that 'if the present trend in forest exploitation continues without a radical shift in our approach to conservation, most edible wildlife in the equatorial forests of Africa will be butchered before the viable habitat is torn down' (Rose, 1996f). This conclusion had already been reached by Oates (1996a) and has been confirmed by consensus of the wildlife protection and conservation nongovernmental organizations (NGOs) participating in the EU Ape Alliance (1998) as well as the more recently formed Bushmeat Crisis Task Force in the USA (BCTF, 2000). To stop this slaughter will require understanding the root causes of the crisis and building capacity to address them both in immediate emergencies and over the very long term.

Logging and anarchy

Ammann's (1993, 1996c, 1998b, McNeil and Ammann, 1999) wide-ranging ten-year investigation of hunting pressures in and outside the IUCN (1996) surveyed areas strongly indicates that unprotected and unstudied primates, especially those within 30 km of the expanding network of logging roads and towns, are being devastated by a burgeoning commercial bushmeat trade. The catalyst of this devastation is growth of the timber industry (Ammann and Pearce, 1995; Ammann, 1996b; Rose, 1996c, g; Ape Alliance, 1998).

Timber prices and profits are tied to provision of subsidized bushmeat to migrant workers. Every logging town has its modern hunting camp, supplied with European-made guns, internationally made ammunition, and men and women who come from distant towns and cities, hoping to make a living in the forests. With indigenous forest dwellers hired as guides and hand servants, immigrant hunters comb the forests, shooting and trapping. Anything edible is fair game in a market that starts with the wood cutters, truck drivers and camp families who scrape together their meager wages for a porcupine stew (Rose, 1998a). From this captive market base the bushmeat trade stretches all the way to fine restaurants and private feasts in national capitals where more rare and expensive fare is available. Little is done to teach or enforce wildlife laws. Prices charged for bushmeat can rise tenfold as it moves from hunting camp to big city market (Auzel and Wilkie, 2000).

Most timber executives admit there is a problem and say they are powerless to stop it (Incha, 1996; Splaney, 1998). In the past, logging managers have been reluctant to let outsiders into their concessions, fearing that problems will be uncovered and business disrupted, with no solutions provided. Recently a few timber companies have opened their concessions to bushmeat researchers and conservationists in response to pressures from outside (see, for example, Ammann, 1998a; Auzel and Wilkie, 2000; Eves and Ruggiero, 2000). These small interventions are still rare and their impact unproven. It remains true that

> ... almost all the companies in the forestry sector are 'outside the law'. Despite good legislation, there is no effective overseeing of actual operations
>
> (Horta, 1992).

The timber industry's reliance on bushmeat to feed loggers and their inability to educate workers and govern their concessions leads to indiscriminate hunting that fosters not only the breaking of laws, but also the breaking of customs. People whose colonial and tribal cultures once enforced taboos against eating the meat of apes and monkeys are beginning to try it (Ammann, 1998a). Further, the lure of cash attracts some people to hunt animals that they would not eat, and sell the meat to others (Lahm, 1993).

More worrisome is the agreement among primatologists and other diverse observers that destructive outcomes of bushmeat commerce have taken crisis proportions (Rose, 1996c; BCTF, 1999). What makes this a crisis is not only the numbers, but the way they develop. Juste *et al.* (1995) crystallize the essence of the crisis:

> With the advent of modern firearms, and improved communications and transport, subsistence hunting has given way to anarchic exploitation of wildlife to supply the rapidly growing cities with game.

The key word here is *anarchic*. In the absence of an effective political authority, having no cohesive principle, common standard or purpose, the bushmeat trade has exploded into a rush for personal profit not unlike the gold rush that transformed the western portion of the United States in the nineteenth century.

A respected French timber executive in Cameroon compared bushmeat to 'found money' and suggested that poor Africans cannot resist hunting any more than they can leave a hundred franc note lying on the forest floor (Incha, 1996). This view of financial greed as overriding human values of honesty, community, and compassion is a perceptual framework that came to Africa with the people who imported competitive cash economy: the traders and developers of the middle east and Europe. The fact that this economic value structure has served the outsiders more than the Africans is accepted. The possibility of controlling this destructive penchant for commerce *über Alles* has been largely ignored.

But it is not francs and dollars that cut trees or kill gorillas. The rational constructs of supply and demand are not capable of wielding chainsaws and firing shotguns. It is people who destroy forests and wildlife. Individual loggers and hunters are manipulated by specific timber managers and bushmeat traders, who have in turn been seduced into exploitative enterprise by the exigencies of their personal situations in a region plagued with societal turmoil and rife with anarchic and tribal scrambles for self-survival and private profit. It is imperative that international political and financial pressures and incentives be brought to bear on these uncontrolled business activities and the resultant social anarchy. At the same time, work must begin in earnest to expand African people's values beyond the imported view of wildlife and wilderness as an exploitable natural resource.

War and chaos

During the mid and late 1990s a vast threat to nonhuman primate survival has spread across equatorial Africa. Ethnic, tribal, and international conflict and war have emerged in a maelstrom of economic and social dissolution and sent many millions of people into deep turmoil. Slaughter and displacement of people stimulates the destruction of wildlife. Within national boundaries warring factions take turns acting as in-place political leaders and outcast refugees. On the political level, concern for environmental protection vanishes when one's own position of power is continually at stake. Homeless refugees, numbering in the tens and hundreds of thousands, buy and barter for any scrap of protein they can get: ape, ant or antelope. Taboos fall away in the face of starvation. Alliances shift, and new social norms are devised to suit each social and environmental situation. Rule of the gun supersedes rule of law: the man with firearms and ammunition calls the shots. This means that armed forces control the people not only by threat of harm but also by incentive: they control the means to hunt bushmeat. Starvation and assassination are in military hands. So are apes and monkeys.

This element of the crisis has delivered a rising tide of human and wildlife emergencies. People and primates once considered safe and stable are now in grave danger. The movement of militias and refugees out of Rwanda, Uganda, and Burundi into the eastern part of The Republic of Congo (DRC) has wreaked havoc on local villagers and indigenous peoples. Conservation workers flee, return, and flee again. Eastern lowland gorillas are vanishing fast, the target of armed soldiers and war-ravaged refugees who make camp in the forest. The gorillas in the highlands of the Kahuzi-Biega reserve in eastern Congo, once studied and visited by scientists and tourists, have been hunted at such an alarming rate that they have become one of the most endangered groups of apes on the planet: fewer than 100 remain (A. Plumptre, personal communication, 2000).

Meanwhile, on the other side of the DRC, an influx of bonobo orphans into Kinshasa attests to the poaching of these most rare of apes in their Congo Basin habitat (C. Andre, personal communication, 2000). It is now accepted by the conservation community that local and indigenous taboos on bonobo eating have fallen away in the face of streams of invading armies, insurgents, and refugees. Ironically, while stepping up 'emergency hunting', war reduces logging and its concomitant bushmeat activities. Timber operations cease and new projects are put off under war conditions. When stability returns, however, extractive economics returns to a milieu where survival mentality has severely suppressed ecological morality. It is imperative to use the hiatus in logging to influence plans and policies of international exploiters and the financial institutions that support them. The next influx of natural resource exploitation must be morally committed and organizationally competent to promote wildlife conservation in the context of ecological and social stability and sustainability.

Intrusion and disease

Thus far we have been looking at the explosion of illegal bushmeat commerce as a wildlife crisis that emerges from human greed, conflict, and chaos. For the apes in particular, our work on the bushmeat crisis has been manifested as a fight against the extinction of our closest living kin. But that genetic kinship appears to be the source of a crisis that threatens the health of humankind. Medical scientists have uncovered evidence that chimpanzees (*Pan troglodytes*) are the original source of the viruses that have propagated AIDS. Bushmeat hunting along each new logging road is likely to bring out more than ape and monkey meat. Although cause and effect are now interlaced, I am convinced that the growth of human intrusion into forests for commercial hunting is a root cause of the AIDS epidemic and other threats to human health (Rose, Ammann and Melloh, 1999).

Virologists have presented their evidence in science journals (Gao *et al.*, 1999) and at major medical conferences (Hahn, 1999a,b). They have told the public two things. First, we must stop the hunting and butchering of wild chimpanzees in order to avoid transmission of new strains of SIV. Second, we must launch a program to study wild apes and hunters *in situ* (Weiss and Wrangham, 1999). Biomedical research and action to influence the etiology and management of these viruses in apes and in bushmeat hunters and traders may expose the keys to preventing spread of HIV and AIDS. These discoveries escalate the bushmeat crisis from a regional social and conservation challenge to a global health issue, and increase the complexity of the problem many times over.

Still, the chimpanzee virus factor has yet to alter conservation priorities. In the years since the relationship of bushmeat and AIDS was publicized, little has been done on the conservation front to promote control of ape hunting as a preventative of human disease. Virologists seek grants to study the processes of disease transmission from apes and monkeys to hunters and bushmeat consumers (Young, personal communication, 2000). Concern that Africans will destroy apes to protect themselves from AIDS have given way to reports that Africans are denying the link, claiming 'we have eaten apes for centuries and only got AIDS very recently'. Of course the difference between contained exposure to small indigenous pockets of SIV and explosive broadcast exposure to widespread reservoirs renders that argument spurious. But underlying African denial is a dire fatalism that grows from a terrible truth.

Africans face death on a daily basis from epidemic AIDS, malaria, and infectious diseases. Excessive interaction with biodiversity is producing these deadly effects. The Africans who enter rain forests and those expatriates who go with them, whether to protect or exploit biodiversity, are being exposed to severe danger and putting all humanity at risk. Because these threats are invisible makes the situation all the more dire. Loggers, field scientists, tourists and conservationists are all culpable. Similarly, organizations like the United Nations, International Red Cross, and others who support refugee populations in proximity to forests are operating with a blind eye. There is growing evidence that a moratorium should be placed on all non-indigenous human intrusion into biodiverse rain forest habitat for local and global health reasons. But until there is direct evidence of ape hunters having new SIV strains, virologists and conservation biologists will likely avoid the issue as it relates to apes. During the years required to conduct such studies, it will be left to conservation activists to address the global health implications of human consumption of apes and other primates.

Face the crisis

The impact of bushmeat commerce grows with invasion of the logging industry, the movement of armies and refugees, and the spread of epidemic AIDS and other forest-borne diseases. These dangerous phenomena are grounded in a chaotic flux of social change in the region, which is caused by revolutionary cycles of colonial and global pressures. If current trends continue, the next century will empty the forests of edible animals, fill the human community with waves of new plague, and displace and destroy most of the people and primates in the forests of central Africa. As parks and reserves become the only places left for people to hunt or escape, such sanctuaries will need to be defended by armies, or they will all vanish. But this destruction is not inevitable. We must face the crisis head on, and recognize that it is the excesses of humanity that require management, not the forests and wildlife.

Conservation moves from saving biodiversity to promoting biosynergy

Conservation in the twenty-first century will succeed only if it becomes the domain of experts in social change, economic development, and human spiritual transformation. The rest of this chapter will examine factors that can engender that success. To arrest the growing multi-billion-dollar annual trade in bushmeat and its consequent social, ecological, and economic collapse across west and central Africa requires big changes in the practice of conservation. More of the same will not work. Conservation strategy has been dichotomised in recent decades. Some attempt to protect biodiversity from people. Others try to help the people use biodiversity sustainably. Both these strategies are unidirectional in their methods and objectives. I have called for a paradigm shift that affirms the overriding importance of *interchange* among key elements of humanity and nature (Rose, 1998b). We need to focus our science, our strategic planning, and our innovative interventions and developments on the relationships among human and nonhuman factors. The aim is to understand and to influence *biosynergy*: the collaborative and mutually beneficial interaction of all living elements within regional ecosystems, which leads to individual, social, and ecological stability, longevity, and enrichment. With commitment to mutual benefit for all stakeholders, human and nonhuman, we stand on the highest ground of global ethics and ideal.

Will we achieve this ideal for all interactive elements? Perhaps not, but we must strive for it, expose failures and attempt to correct them, identify successes and try to replicate them. We must begin to look for synergistic relationships as the main datum, with polarized outcomes being secondary foci. Such a paradigm shift will be difficult, requiring new methods, measures, and modes of operation. The construction of a biosynergy-focused science will replace dichotomies with interactives. This is a task for the future, beyond the scope of this chapter. What we can explore here is theory and constructs that enlighten the human dimensions of conservation. As a modest beginning, I have created a schema differentiating the operational driving forces for certain key social factors in wilderness, rural, and urban populations in African bushmeat territories (Rose,

Table 10.2. *Driving forces and key social factors in African bushmeat territories*

These concepts were created with support from Conservation International's Center for Applied Biodiversity Science, as part of their Bushmeat Initiative for West Africa, 1999–2000.

Location:	wilderness	rural	urban
Social group:	family	community	organization
Constitution:	**myth**	**precept**	**law**
Legislation:	ritual	custom	regulation
Management:	**kinship**	**consensus**	**contract**
Adjudication:	elders	leaders	enforcers
Identification:	**nature**	**society**	**individual**
Theistic power:	intrinsic deities	spirit presence	distant god
Commerce:	**hunt**	**trade**	**market**
Situation:	environment	association	workplace
Wildlife values:	**theistic**	**conflicted**	**utilitarian**

2000a). There are ten continua in this schema. The five marked with **bold** print are critical action–research areas for the restoration of biosynergy among people and wildlife in equatorial Africa.

Common differentiation between wilderness, rural, and urban locations are made according to relative densities of humans and their developments, on the one hand, and nature on the other. For this treatise I will try to define these terms in accord with social and biosynergistic factors. *Wilderness* environments are those where nonhuman communities and ecosystems govern most ongoing life processes, with human–nature interaction being predominantly synergistic or nature-dominant. *Rural environments* are co-governed by humans and nature, with varied ratios of biosynergy, nature dominance, and human dominance prevailing. *Urban environments* are those constructed and governed by human communities where human dominance over nonhuman nature prevails. The buffers between these environments – mixed environments and transit corridors – are often the most critical territories for social control.

The terms in the left column label the ten continua. The next three columns identify concepts that describe the human social characteristics of wilderness, rural, and urban environments on each of the continua. Those continua in bold print are the ones I will discuss briefly below.

220   A.L. Rose

The wildlife values continuum: theistic – conflicted – utilitarian

Studies of the ways humans value wildlife by Kellert (1996) have set a standard for social science modeling. Unfortunately this tome of work is focused on northerners in industrial society. Application of Kellert's rigorous attitude scales in African settings requires adaptation. None the less, work by Mordi (1991) has provided attitude survey data on wildlife values in Botswana which Kellert uses to expand his theory to nonindustrial societies and to hunter–gatherer society. In much of central Africa 'a general pattern of apathy, fatalism, and materialism towards nature and wildlife' prevails (Kellert, 1996). Most contemporary Africans have lost their traditional 'theistic' reverence for wildlife and many have taken on a more harsh utilitarian view (Mordi, 1991). With the advent and spread of cash economy, colonial religion, and urbanized central government, 'tribal values of conserving and protecting nonhuman life are rendered spiritually inoperable, while new ecological and ethical foundations for sustaining nature have not emerged' (Kellert, 1996). Bushmeat commerce is the latest manifestation of the economic and moral values of international 'resourcism'. People in Africa are being manipulated by the architects of global consumerism into giving up their cultural values and treating wildlife as a material resource.

Wildlife values still reflect somewhat the traditional – colonial dichotomy for humans living in wilderness (theistic) and urban (utilitarian) environments. It appears that the prevailing wildlife value-sets in many rural milieus are a conflicted mix of traditional and modern. Whereas indifference towards wild animals might be expected among urban dwellers who do not interact with them, rural people who are affected by crop-raiding animals and are educated to stay out of dangerous nearby forests may be expected to report anti-wildlife values. This will be especially true where imported religious training has stripped the theistic value from wild animals, leaving them to be viewed as little more than pests, thieves, and thugs (Lawrence, 1993). On the other hand, reliance on bushmeat for protein in many rural African settings strengthens the utilitarian value of wildlife. Thus rural hunters and hunting subcultures can perceive wild animals positively: as useful resources, not spiritual gifts.

The practical question 'which values can best be developed in rural and urban populations so as to reduce demand for bushmeat' will require specific study of the diverse and changing human subgroups. Short-term manipulation of values through passive economic incentives to not hunt, active incentives to protect wildlife, and social and legal disincentives such as fines and incarceration, are typically proposed as face-valid interventions.

Although these approaches can work, the fact that such tactics rely on external material values renders them risky, especially if linked to a cash economy. To simply reinforce the pursuit of money can backfire whenever the money source vanishes, or when economic need or desire rises. Without other influential values at play, a purely utilitarian wildlife protector, for example, can be bought off by a patron offering more money to hunt for bushmeat. This is why the social values held by candidates for jobs as field assistants and tourist guides become crucial to the hiring decisions of scientists and conservationists (see, for example, Fossey, 1983; Owens and Owens, 1992). Expert assessment of theistic, kinship, and other non-utilitarian social values by applied social scientists will optimize selection and motivation of staff hired for bushmeat control programs.

In the long term, however, public efforts in wildlife values education will turn the tide. This will be accomplished, ironically, in the missions and schools that promoted Western materialism, and on the commercial media. Programs to engender empathy for apes and other wildlife as part of the creation, and thus as spiritual and evolutionary kin, are now being developed in central Africa (Rose, Bowman, and Patterson, 2001). These innovations are essential to counter the continuing promotion of material utilitarian values by international exploiters.

The constitution continuum: myth – precept – law

In places where spiritual myth and ritual still influence community attitudes and behaviors, the establishment of pervasive conservation values could be quick and long-lasting. Vabi and Allo (1998) detail the workings of community myth and ritual practices in relation to commercial bushmeat hunting in eastern Cameroon. In brief, they describe the replacement of effective internalized myth-based social controls with ineffective external law-based administrative mechanisms. Individuals whose community and clan share common belief in the intrinsic theistic value and power of wildlife and wilderness can be expected to relate in predictable, synergistic ways to the ecosystems in which they live. Transgressions are punished and proprieties are rewarded by personal self-assessment and public comment, automatically and reliably. Daily and continual reinforcement of the myth–belief system carries through in ritual practice, and helps to maintain the community institutional framework uncontested through succeeding generations. Where certain animals are totems, their habitat is protected and their hunting strictly controlled and typically performed in sustainable traditional ways. Taboo wildlife and ecosystems are avoided because strong personal and communal sanctions insist on it. On the face of it, this kind of system in isolation seems perfect for maintaining human–nature biosynergy.

But as outside factors impinge, myth-based community conservation practices unravel and collapse along with other social systems and controls. Introduction of foreign technology, economics, affiliation, and religion undercut and transform indigenous society. The proliferation of guns into rural societies destroys the power balance between hunter and prey, and can be expected to erode traditional myth and its ritual controls on hunting society. To study gun hunters as if they are traditional people (see, for example, Alvard, 1993) is absurd. More ubiquitously and importantly, any bushmeat trader, marketer, or consumer using government-issued money to sell and buy bushmeat transported on foreign-built trucks and roads is also basically modern. The assertion of traditional social control on commercial bushmeat traders will be psycho-socially ineffectual. This will be especially true in urban centers, where the overarching social and religious influence is modern, individualistic, external, and legalistic.

In rural areas the mix of community precept and central law can be responsive to normative influence. Villages and small towns are enclaves of traditional community and clan practice. Village chiefs may serve as mediators between the legalists and the theists, between law and myth. To the extent that local people accept the chief as empowered by their deities and ancestors and working within their community myth–ritual system, his precepts may be honored and community conservation may emerge on a strong, lasting traditional footing.

Intrusion by foreign exploiters and conservationists can undercut the mediating power of village and community or clan leaders. Whether it is the logger paying a chief for the right to cut trees or a scientist paying him for the right to study apes, both are substituting money incentive for the traditional theistic and community empowerment. More subtle and important, a community leader's affiliation with foreign emissaries of any kind alters status structures in the community and risks offending mythic tradition. Of critical concern is the transfer of northern individualistic norms onto leaders, so that social affiliation isolates the chief and clan leader, making them individual operatives, no longer an arm of gods, ancestors, and the community.

The most explicit destruction of community conservation myth has come from foreign religion. Key to this adverse effect is the externalization of deities, which strips wildlife of theistic power and renders once sacred ground empty of spirit and open for total material conquest. Ironically, a movement is afoot in North American religious institutions advocating ecological justice through 'care for the creation'. This could become a huge source of funding and energy needed to make conservation work in the bushmeat arena. But just as scientific reductionism can defeat African community conservation by denying the existence of god–spirit in forest and wildlife, so can religious externalization of deities reduce the effectiveness of this positive movement in Africa.

Solutions are available. There are many examples of foreign religious missionaries enabling the coexistence of local and global religious myth. A very open-minded, innovative approach to the support of local myth in the context of modern religious and spiritual concern for the living creation seems to me to be the most promising untried area for research and development. Without the capacity or will to create a fully endowed and socially supported enforcement and judicial system spreading from cities to rural and wilderness areas, social control of unsustainable bushmeat commerce will require the reconstruction and institution of spiritual myth that supports the synergistic interchange of human community and biodiversity.

The identification continuum: nature – society – individual

The development of psychological identity, or ego, is diverse as the cultures in which people grow and live. Modern power societies encourage a kind of egocentric identity which allows social institutions (schools, businesses, governments) to manipulate and manage individuals for their corporate benefit through person-focused incentive systems. This kind of individualistic identity pattern appears to prevail among the affluent and educated residents of African cities, as it does in Europe and North America. Urbanites tend to see wildlife as a resource for their individual use as private means to facilitate pursuit of personal goals.

In contrast to urban selfishness, traditional people who live in wilderness areas view themselves as elements of nature, asserting ecocentric identity. As part of nature, one identifies ecological health and stability with one's own well-being. The human–wildlife totem relationship deepens nature-connected identity. A forest-dwelling person who uses the term 'I' may be referring to a panoply of interlaced human and nonhuman identities, and 'we' may indicate any or all flora and fauna who co-inhabit the natural world, not merely human family or community.

Again in a pivotal position, rural villagers identify themselves anthropocentrically as members of human society with proscribed social responsibility and privilege relative to the natural environment. The shift from identification with nature to identification with human society marks the loss of ecological sensitivities. Living and working in human-constructed habitat on human social tasks erodes the sense of self as animal in nature. Although rural people are in closer contact with the wild than urban dwellers, their identity may be shaped by a 'man against nature' frontier ethos (Cartmill, 1993). Paradoxically, on the psychosocial level, the ego that is identified totally with humanity may be less able to evoke concern for nonhumans than the individually focused one.

This suggests that education about and empathy for endangered animals will develop differently in urban and rural settings. Urban individualists may respond to personal instruction and one-to-one bonding with apes and other wildlife in sanctuary settings, for example. Rural socialists might be better convinced to protect wildlife through interventions that link nature to the satisfaction of community needs, which are central to the person's communal identity. In attempting to establish a gorilla research and tourism project in bushmeat territory, for example, it appears that we must establish the community identity of 'gorilla protection society' in the common mind (Rose, 2000d). So long as individuals – chiefs, conservation workers, and select villagers – are identified as mainstays of the project, it will not succeed.

The management continuum: kinship – consensus – contract

Authority and power to manage social behavior is vested more in relationships than in individuals. Elders need youth, leaders need followers, employers need workers. But these social compounds vary from location to location. Social management in wilderness dwellers is empowered by family elders. Rural villagers often invest management responsibility in community-level consensus systems. Urban societies have a preponderance of nonhistoric and temporary relationships to manage; they do so by individualized contract.

Urbanites transplanted into rural and wilderness settings attempt to install the management processes they know best. The difficulty obtaining contract compliance with people adapted to management through rural-consensus and wilderness-kinship systems has led to the proliferation of urban-migrant contract workers in rural and wilderness development projects. Typically, conservation projects are also run by outside contractual managers. To the extent that villagers are obligated to follow community consensus, contractual agreements with them will be countermanded. Similarly, the elders' pronouncements can override agreements forest dwellers may make with outsiders. Rural men and immigrant-urban hunters have attempted to secure authority positions in wilderness cultures by marrying forest dwellers, with mixed success depending on ability at cross-cultural management.

Community development projects, whether extractive or conservative, are increasingly built on consensual management models imported from Euro-America. Participatory rural development tries to capitalize on the rural consensus capacity in building agreement. Problems of population size, intercommunity conflict, and stakeholder non-participation make these efforts difficult to manage. Wilderness dwelling people are generally unskilled in urban-style consensus building, so they avoid conflict resolution processes. Thus the human stakeholders with the most to lose often have no voice in decision-making. The absence or dysfunction of a crucial stakeholder group renders participatory consensus invalid. This must be remedied. Not only is it unethical to leave out the human forest dwellers, it is ineffectual and dooms projects to failure.

There are other stakeholders, however, with more to lose and less voice in development planning: the fauna and flora. Because these stakeholders cannot function at the planning table, humans try to talk for them. Unfortunately the range of surrogate voices is very narrow. Conservation scientists present their findings and mitigations and wildlife biologists and foresters argue for statistical sustainability of animals and trees in rural appraisals and other participatory programs. These outsiders rarely speak in ways that reflect the interactive vitality of human societies and natural ecosystems. To wilderness-dwelling people, the natural world is to be lived in reverentially, not managed contractually. Urbanites, whether scientists or loggers, rarely allow village consensus to overrule their corporate or professional pursuits and proclamations. The management continuum is a crucial dimension in the reduction of bushmeat commerce. If management processes are not synergistic for all stakeholders, biosynergy cannot be pursued.

The commerce continuum: hunt – trade – market

Human commerce hinges on human need. At the bottom of the hierarchy of human needs is survival, and we usually associate food needs at that level (Maslow, 1993). One must have a fairly full stomach, and food for the next meal, in order to be influenced by higher-order needs for security, status, identity, and actualization.

Food preferences, however, are only partly related to hunger and nourishment. Whether in rainforest or metropolis, the foods we gather, trade, sell and buy are determined by myriad social factors. Wilderness dwellers prefer smoked porcupine to fresh chicken because it lasts longer, and better satisfies food *security* needs. Young men in rural villages agree to take a gun and hunt larger game to satisfy *identity* needs in a shifting cultural milieu. Village chiefs and Provincial governors enhance their *status* by serving ceremonial meals with expensive wild game meat. Affluent urban citizens may actualize their personal sense of power and potential with traditional foods and medicines imported from the rainforest.

All these underlying needs drive behavior, which becomes habit. At that stage, consumers typically report that they buy bushmeat because it tastes better than chicken or beef. It seems frivolous to eat endangered gorillas and protected elephants for the taste sensation. But taste familiarity itself provides a sense of personal and social security which is profound in all cultures. And, like the holiday turkey that serves as an icon for 'the good life' in North America, special bushmeat on the platter in many African homes signals the celebration of community. Our nervous systems are hardwired to accept familiar flavors and aromas that prove safe, and to reject unusual tastes. Ritual feasts rely on visual and culinary consistency. Perceptual adaptation levels develop rather quickly, and are slow to change. Thus, once communities and families begin to include newly available game meats in their diets and ceremonies, it will be difficult to reduce the demand.

This is why we must be especially concerned about the spread of bushmeat supply from wilderness and rural areas to the cities. Reducing 'the taste for game meat' in smaller rural populations is a formidable challenge. Reversing bushmeat demand in the high-density urban markets will become even more difficult, owing to the individualistic and multicultural complexity of social factors and human needs. Already in some quarters of major west and central African cities, bushmeat has become a habitual and expected part of the diet. This demand will give incentive for opening new sources and routes of supply, and supply will expand demand.

Urban demand and rural supply are interactive. Social factors mediate the two-way relationship between supply and demand. Bushmeat hunted in wilderness, traded in rural areas, and marketed in cities will satisfy human social needs, support new consumatory habits, and stimulate an accelerating demand for bushmeat products. To reverse these trends *ad hoc* will be more difficult than to prevent them. But prevention and correction are both multi-locus and multi-factorial propositions. Our colleagues who study economic variables have advanced the understanding of the interactive effects of price, household income, availability of bushmeat and substitutes, and market trends (Wilkie and Carpenter, 1999). Although there are relatively stable theoretical models for these interactions, they will not help us to predict and control bushmeat supply and demand in real-life settings without integrated treatment of social variables. Values, constitution, identification, and management are four social dimensions that are crucial to the pursuit of biosynergistic conservation.

Recommendations for research and intervention

Programs must be designed to deal with urgent, fundamental, and sustaining solutions to major social problems that underlie the bushmeat crisis (Rose, 1998c). Fundamental to the effectiveness of all interventions will be the establishment of a worldwide alliance to stimulate and maintain public and political concern for the wildlife crisis. Most urgent is the need for multidisciplinary crisis intervention projects to stop endangered wildlife slaughter in locations where militias, refugees, and exploiters are invading critical habitat. To sustain the reduction of bushmeat commerce over the long term, methods must be developed to restore reverence and a sense of kinship with endangered wildlife across equatorial Africa.

No matter what the solution, it must be developed in the context of long-term action research programs (LTAR) that will optimize interventions cumulatively. The LTAR model is most effective in the ongoing improvement of social change and management programs in large and complex commercial service systems (Stebbins, Hawley and Rose, 1982). A fundamental difference between the LTAR model and traditional basic science is the explicit and continual pursuit of social problems and solutions. Success in LTAR is defined as (i) uncovering mistakes and making corrections and (ii) identifying achievements and sustaining them. Finding out why and how things happen is subordinated to making things happen. The implementers of LTAR programs must be multidisciplinary teams of professionals with process and content expertise in the social system being treated. Members of the social system are partnered with outsiders to develop the strategic intervention targets and design and implement social change projects. The best LTAR builds capacity within the social system for self-improvement, so that over a period of years the action research programs and processes are internalized.

The social systems connected with bushmeat commerce stretch from African forests and savannas to corporate boardrooms in Europe, Asia, and North America. This far-flung 'informal organization' requires far-flung formally organized processes to effect the changes that will keep it from destroying the remaining natural and cultural heritage of equatorial Africa. There are three strategic targets that seem ripe for intervention. They fit into generic categories that are interactive, but distinguishable: supply control, demand reduction, and alternative development. Each has complexity and paradox which require rigorous action-research efforts.

Bushmeat supply control

Many European and North American wildlife advocates and their public supporters argue that we must start with bushmeat control for endangered species. The direct approach from the urban armchair says interdict, arrest, fine, and jail poachers, traders, and marketers of endangered bushmeat. Conservation biologists often argue further that this should be limited to parks and reserves. A social science perspective that accounts for driving forces and key social factors shows why these prescriptions backfire and how they might be improved.

The law enforcement approach adds yet another urban social dynamic to the conflicted rural community. It signals that conservationists and central governments, along with their international supporters, do not respect the commons nor the community ethos. This approach typically ignores rural precepts and subordinates local leaders, undermines the consensus power of the village, displaces customary social control systems, asserts individual identity over society, resorts to distant gods and their emissaries, elevates the importance of utilitarian wildlife values, and reinforces the preeminence of the market over trading systems.

The affront of outside enforcement and judgment to wilderness dwellers can be more potent than that to rural communities. Direct legalistic intervention to interdict forest dwellers who support commercial bushmeat hunters can devastate whole families and clans. Indirect effects can be similarly destructive to social cohesion, as urban-style intervention interferes with kinship relationships between wilderness dwellers and rural peoples. Common understanding of this extreme adversity explains why outside enforcers rarely attempt to arrest wilderness dwellers. Vabi and Allo (1998) suggest that control measures must 'emerge from careful location-specific and culture specific analysis ... greater emphasis should be placed on an understanding of the normative and social practices at the grassroots level of society.' This good advice is easier said than done.

Control interventions must rebuild the power of the rural community to construct contemporary customs and precepts through societal consensus based on amalgamated myths and rituals that will re-instill theistic values for nature and its conservation. This kind of effort will require a cadre of social change agents trained to work behind the scenes facilitating societal, economic, and spiritual redevelopment. First choice locations for pilot projects would be those where commercial bushmeat hunting is about to encroach, and rural communities are still relatively intact. Each village and clan would develop its own community control mechanisms to prevent influx of hunters and market traders and thus protect local wildlife and their own societies and people.

In addition there is need for effective management systems in the immigrant populations that enter rural and wilderness areas for temporary and longer-term exploitation and development. Evidence is clear that major interlopers such as loggers and miners do not have the capacity to manage and control the urban workers and families they hire and attract to live and work in their concessions. Programs must be designed, funded, implemented, and monitored to develop the organizations, managers, supervisors, and performance systems that will control and replace bush-meat commerce in all settings where urban societies have been trans-planted into rural and wilderness environments. As mentioned above, these would be LTAR programs co-developed by outside professionals and inside managers, staff and other interlopers.

The interface between imported bushmeat control for urban interlopers and rural society redevelopment for forest dwellers and forest-edge vil-lagers is critical. Ultimately the rural and wilderness social control systems need to be protected and strengthened so they can maintain their own self-management and keep their hunting community-focused and non-commercial. A key is to keep rural and wilderness men and women from being enticed into the cash economy of urban exploiters. This requires the presence of intermediaries with allegiance to process, not subculture. The parallel links between the three cohabiting societies are vital, and must be facilitated by full-time independent outsiders skilled in inter-group cross-cultural relations and social systems monitoring and maintenance. This imperative will be resisted by both individualistic urban and kinship-based wilderness factions. Rural communities are more likely to recognize the value of outside consensus managers. Practitioners of community-based conservation projects have reported time and again how well-running efforts unravel into inter-group disputes and ultimate failure when outside conservationists leave the scene. The answer is, don't leave without leaving behind a replacement who can facilitate the community relations manage-ment function.

Here we see the irony. We cannot tell the local people or the interlopers what to do from our urban armchairs. But at the same time we must find outsiders (or streams of them) willing to leave their armchairs and live as neutral facilitators in situations where 'what to do' is complex and often unknown. To control bushmeat supply will be an innovator's challenge.

Bushmeat demand reduction

To reduce bushmeat demand appears to be a marketer's nightmare. On the one hand we may need to re-ignite theistic reverence for wildlife and draw on indigenous totem beliefs to foster human–animal kinship, which precludes secular consumption of endangered species. On the other hand we may choose to evoke negativistic avoidance of wildlife and draw on individual fears to foster human–animal repulsion, also to stop consumption of endangered species.

Let us consider the high-profile issue of great ape bushmeat. We may find ourselves encouraging wilderness dwellers and interlopers to respect apes because they are kin, and to avoid them because they carry dangerous diseases. And in the cities, we may build empathy for our ape cousins by exposing urbanites to their human-like qualities in hands-on educational settings, while insisting that wild apes must be shielded from human contact in order to survive. The messages will be mixed; as mixed as the cultural overlays and interactions of Africa, which cover the widest range of any in the world. I suspect this mixing will make sense to most of the people most of the time, so long as we outsiders live by the same codes and values as we ask of Africans.

The modes of influence for reducing bushmeat demand are many. Perhaps the most far-reaching medium is radio. Popular formats such as docudramas and talk shows can provide entertaining opportunities for many publics to explore issues of health, human welfare, cultural change, environmental safety, and nature conservation. To stimulate discussion and thought is crucial, and radio allows many voices to be heard at once, across all societies from urban to wilderness. I have sat in forest hunting camps and heard the battery-operated radio blast music and news through the air at the end of the day. Everyone listened. Because everyone listens, it is critical in such programming to assure that an ethos of tolerance for different cultural norms is ever-present. Listeners must be exposed to entertainment and advertising that reflect their own beliefs, yet put them in larger and fuller context.

Although radio messages can create a climate for change, physical interventions at key nodes of the bushmeat commerce will be needed to modify behavior. Three critical spots come to mind: the market, the restaurant, and the home kitchen. To convince restaurateurs to forego the attraction and profit gleaned from bushmeat-based specialties will be difficult and very important. So long as the urban gentry continues to celebrate in public with game meat, the aspiring classes and generations will be enticed to follow suit whenever they can afford it. Furthermore, the implication is that the rich celebrate in private by consuming endangered species. Whether this is true or not, the perception that the elite eat illegal meat undercuts arguments against the illegal trade. Perhaps proactive public campaigns in which restaurants and respected leaders declare 'we serve grasscutters, not gorillas' would be one way to make a difference.

To address this issue, marketing and advertising experts should be brought to the table with applied social scientists and representatives from African urban, rural, and wilderness societies. With the necessary data about individual and community preferences, taboos, and aspirations, a keen market professional can figure out how to turn people away from one product (bushmeat) and towards another. Negative advertising risks audience backlash but can turn focus towards positive alternatives. Many people concerned about the bushmeat crisis assume that chicken and pork preparations, along with game ranch or farm products, would be a way to reduce bushmeat demand. Market promotions of domestic meat that include recipes and on-the-spot samples can shift housekeeper choices in urban supermarkets. Similar but idiosyncratic culture-specific programs may work in rural areas.

Bushmeat alternative development

Alternative meat and vegetable protein products that look and taste like bushmeat seem to be a promising way to reduce bushmeat market share. With sufficient financial and developmental support, domesticated game products could be subsidized and promoted as mid- and low-priced African food-lines across the continent. Nearly every treatise on bushmeat commerce points to alternative protein development as a solution (Rose, 1999a; Wilkie and Carpenter, 1999). Producers of domesticated game animals claim success in limited experiments (see, for example, Jori, Mensah and Adjamohoun, 1995). Why then do we not see major players in global food markets being solicited to underwrite and organize such ventures? In part it may be that traditional conservation and wildlife professionals are not interested in going into the game ranch business. It is not their 'calling'. It could also be that conservationists are afraid of the 'upside risk' inherent in this kind of venture. Rightly so.

If profitability becomes the driving force in game ranching, rather than capture of market share from bushmeat, the success of such ventures could whet public appetite for 'the real thing' and stimulate corporate food marketers to enter the bushmeat business. It is clear that commercial 'harvesting' of wildlife devastates species and ecosystems, provides 'boom and bust' profiteering, and is ultimately not sustainable. What we do not need is more organized and efficient bushmeat commerce in Africa. The growing market in wildlife products for meat, fur, and medicine in Asia is already so well organized that many conservationists believe there is no hope for preferred food species there (A. Eudey, personal communication). The urban market logic, which holds that wildlife species can be conserved best when they are given value in the food marketplace, appears to be dead wrong. Promoting markets for ranched bush pigs and grasscutters might add customers for wild game to the consumer force.

The provision of subsidized domestic protein – vegetable products, chicken, goat, pork, beef – seems a safer approach. Such a development may not increase taste for bushmeat. But in many rural societies in central Africa, farm animals are seen as economic security. Live chickens may be held for barter or given as gifts, eaten only when the bushmeat supply is gone. A solution might be to breed, butcher, and sell chicken meat in the same channels as bushmeat, at lower prices and prepared to taste like wild game. This would likely supplant bushmeat in the lower sector of the market. An attempt to test this hypothesis is beginning in a logging concession in eastern Cameroon where chicken farming is being set up and subsidized (P. Auzel, personal communication, 2000). A concern remains: will the chicken farmers use the cash they earn to buy more expensive bushmeat for their own table? As discussed earlier, the importation of new business into forest economies has complex implications.

If putting real cash into rural and wilderness economies disrupts these societies, how will they manage to maintain conservation programs? How can new converts to cash economy avoid being bought off by the highest bidder? Some would declare that Africans to survive must enter the modern, legalistic, individualistic, market-driven, utilitarian world. At best, they argue, small islands of native peoples and parks can remain as reminders of what once was. Others suggest that, without strengthening the rich human social and cultural heritage of Africa, African people and African wildlife will vanish together. This author holds the latter view, and believes we must set our sights high and strive to preserve as much of the social and natural landscape of Africa as we can (Rose, 2000c). The question to answer now is: what must we do to succeed in this biosynergy-focused conservation movement?

Conclusion: the conservation movement must expand to succeed

Renewal is needed

The conservation movement needs a vast renewal: new premises and players, new organizing principles, new strategies and tactics, new values and disciplines, new goals and objectives, new levels of effectiveness (Rose, 1999). The new conservation movement's mission will be the promotion of biosynergy: the harmony of humanity and nature. To pursue that mission, methods for achieving synergy among diverse conservation biologists, social change professionals, and countless other experts and stakeholders must be invented and installed. The barriers and prejudices that keep us apart and in conflict must be overcome first. Interdisciplinary battles must be stopped. Common language and ground must be created. Conservation organizations, small and large must stop competing and join forces (Rose, 1996a).

We have no choice. Basic science approaches will not work in conservation, and most stakeholders and donors know it. Long term action-research is needed. The free enterprise model with conservation and animal welfare NGOs competing for limited market share has failed. Coalitions are needed. Conservationists need the courage, will, and ability to collaborate with strange bedfellows in corporate headquarters and in field locations where exploitation, migration, and conflagration are destroying people, wildlife and environment. Facilitation is needed. The first step is to accept that conservation cannot succeed without undergoing a profound renewal.

Strategies must escalate

To succeed in the face of an exploitation revolution that is causing rampant resource consumption, regional conflagration and local anarchy, the conservation movement must escalate its strategic imperatives (Rose, 1998c). There are five strategic goals to be attained: (i) social and moral leaders must promote humanity's profound obligation to conserve wildlife and wilderness and to restore nature; (ii) political and economic authority must place conservation on a par with human rights and welfare; (iii) conservationists must go beyond protecting biodiversity to assuring the biosynergy of human social systems and natural ecology; (iv) public demand for intrinsic and spiritual values of nature must supersede utilitarian exploitation and underwrite massive long-term programs in conservation development; (v) all wildlife habitats must be considered sacrosanct, and human intrusion must be managed in a moral, businesslike, and competent way for the global good. We must affirm and pursue these goals at all levels of the conservation movement, from boardroom to bush.

Capacity must expand

Professional, public, and corporate involvement must be expanded for conservation at any level to succeed (Rose, 2000b). We must identify enterprises, disciplines, and public stakeholders that are missing from the conservation movement. Gaps in competence and understanding must be filled by recruiting the best and brightest talent in a score of new domains to work for this cause. These new conservationists must be organized and molded into collaborative interdependent teams. The list of professional types to be added to this effort is enormous. African specialists and international counterparts will require careful selection from fields such as community development, cross-cultural relations, ethics and applied theology, entrepreneurial agribusiness, small business finance, food marketing, environmental conflict management, peacemaking, law enforcement, environmental justice systems, rural and urban ecology, media advertising, organization development, applied social psychology and social anthropology. Only by this vast expansion of capacity can we expect to create conservation programs that will be grounded in enough domains to be effective in countering the revolutionary exploitative activities that are altering the social, cultural, and ecological terrain of central Africa.

Teamwork must prevail

Fast and lasting success will come to innovative conservationists who work directly with the people involved in wildlife commerce: poachers and traders, suppliers and producers, exploiters and consumers, leaders and rulers. These proactive partnerships will invent socially and ecologically synergistic programs to satisfy the human needs that now drive the commercial extraction and consumption of fauna and flora in Africa. Cadres of devoted ecological and social change practitioners, personally inspired and financially endowed, will join center stage with the lone field biologists and anthropologists who have served as long-suffering crusaders for wildlife. The media will look beyond romantic images of scientists rescuing threatened animals and will celebrate the entrepreneurs, educators, and innovators who help people from forest encampments to corporate board-rooms to improve their quality of life by returning to a reverential and synergistic relationship with the African environment.

The task of living in wild places to track apes and monkeys will take on huge added responsibility as synergistic conservation proliferates. Teams of professionals and community leaders will collaborate to convert poachers to protectors, monitor forest product and service sustainability, and implement eco-social improvement projects. The study of nonhuman biology and behavior will be one of many forest services, sustained in the long term by practical interventions to transform human morality and effect biosynergy. Some lost idols and abandoned adventures will be mourned. But as time passes the sense of loss will be supplanted by satisfaction that will come from saving and enriching the lives of more African people and primates than we can ever know.

Success must be global

This success will be maintained by a general public in Africa and around the world that has claimed its kinship with nonhuman primates through personal interaction and empathetic understanding (Rose, 1994a, 1995, 1996e), and supports the social movement to save wildlife and nature as our moral obligation and spiritual need (Rose, 1994b). It will be known by all that a perpetually rich and thriving African rainforest with its apes and other ancestors alive and well is worth far more now and in the future than bundles of wood and bushmeat. Beyond the oxygen and medicine that the forests produce, and the lush beauty and mystery they provide, they give us profound insight into our identity. It is, after all, out of Africa that we hominids came. It is in Africa that we discover who we are and thus face our potential for being more than selfish humans ruling and consuming a vanishing natural world.

The success of this great new conservation movement will do more than save wildlife and wilderness. It will safeguard the world ecology, restore biosynergy, and re-inspire the natural spirit of humanity itself. As founders of the movement we must work together with a wealth of colleagues and fellow travelers, always in reverence, to celebrate the fulfillment of our natural origins and human destiny in the vast and wonderful creation that unfolds and evolves on this remarkable planet.

References

Alvard, M. (1993). A test of the 'ecologically noble savage hypothesis': interspecific prey choice by neotropical hunters. *Human Ecology* **21**, 355–87.

Ammann, K. (1993–4). Orphans of the forest, parts I & II. *SWARA*, Nov–Dec: 16–19; Jan–Feb: 13–14.

Ammann, K. (1994). The bushmeat babies. *BBC Wildlife*, Oct., pp. 16–24.

Ammann, K. (1996a). Primates in peril. *Outdoor Photographer*, February.

Ammann, K. (1996b). Timber and bushmeat industries are linked throughout west/central Africa. Talk at *Seminaire sur l'impact de l'exploitation forestiere sur la faune sauvage*, Bertoua, Cameroon, April.

Ammann, K. (1996c). Halting the bushmeat trade: Saving the great apes. Talk at World Congress for Animals, Washington, D.C., June.

Ammann, K. (1997). *Gorillas*. Insight Topics. Hong Kong: Apa Publications (HK) Ltd.

Ammann, K. (1998a). The conservation status of the bonobo in the one million hectare Si forzal/Danzer logging concession in Central D.R.Congo. http://biosynergy.org/bushmeat/

Ammann. K. (1998b). Conservation in central Africa: A more business-like approach. *African Primates* (Journal of IUCN/SSC African Primate Specialists Group), Winter **3**, 3–5.

Ammann, K., Pearce, J. (1995). *Slaughter of the Apes: How the tropical timber industry is devouring Africa's great apes*. London: World Society for the Protection of Animals.

Ammann, K., Pearce, J. and Williams, J. (2000). *Bushmeat: Africa's Conservation Crisis*. London: World Society for the Protection of Animals.

Ape Alliance (1998). *The African Bushmeat Trade: A Recipe for Extinction*. In Bowen-Jones (ed.), London: Ape Alliance.

Auzel, P. and Wilkie, D.S. (2000). Wildlife use in northern Congo: Hunting in a commercial logging concession. In *Hunting for Sustainability in Tropical Forests*, ed. J.G. Robinson and L. Bennett, pp. 413–54. New York: Columbia University Press.

BSI (Biosynergy Institute/Bushmeat Project) (1996–2000). Website at http://www.bushmeat.net/

BCTF (Bushmeat Crisis Task Force) (1999–2000). Website at http://www.bushmeat.org/

Cartmill, M. (1993). *A View to a Death in the Morning: Hunting and Nature through History*, Cambridge, Massachusetts: Harvard University Press.

Eltringham, S.K. (1984). *Wildlife Resources and Economic Development*. New York: John Wiley and Sons.

Eves, H.E. and Ruggiero, R.G. (2000). Socioeconomics and sustainability of hunting in the forests of northern Congo (Brazzaville). In *Hunting for Sustainability in Tropical Forests*, ed. J.G. Robinson and L. Bennett, pp. 427–54. New York: Columbia University Press.

Fossey, D. (1983). *Gorillas in the Mist*. New York: Houghton Mifflin.

Gao, *et al.* (1999). Origin of HIV-1 in the chimpanzee *Pan troglodytes troglodytes*. *Nature* **397**, 436–41.

Hahn, B.H. (1999a). Origin of HIV-1 in *Pan troglodytes troglodytes*. Talk at Keystone Symposium on HIV Vaccine Development: Opportunities and Challenges/AIDS Pathogenesis, Denver, January 1999.

Hahn, B. (1999b). The origin of HIV-1: A puzzle solved? Keynote speech at 6th Conference on Retroviruses and Opportunistic Infections, Chicago, January 1999.

Hennessey, A.B. (1995). *A Study of the Meat Trade in Ouesso, Republic of Congo*. Brazzaville: GTZ.

Horta, K. (1992). Logging in the Congo: Massive fraud threatens the forests. *World Rainforest Report No. 24*.

Incha Productions / ZSE-TV. (1996). *Twilight of the Apes*. (Video: 25 min.) Johannesburg: ZSE-TV.

IUCN (1996). *1996 Red List of Threatened Animals*. Gland: IUCN.

Jori, F., Mensah, G.A. and Adjamohoun, E. (1995). Grasscutter production: an example of rational exploitation of wildlife. *Biodiversity and Conservation* 4, 257–65.

Juste, J., Fa, J.E., Del Val, J.P. and Castroviejo, J. (1995). Market dynamics of bushmeat species in Equatorial Guinea. *Journal of Applied Ecology* 32, 454–67.

Kano, T. and Asato, R. (1994). Hunting pressure on chimpanzees and gorillas in the Motaba River area, northeastern Congo, *African Study Monographs* 15(3), 143–62.

Kellert, S.R. (1996). *The Value of Life: Biological Diversity and Human Society*. Washington, D.C.: Island Press.

Lahm, S.A. (1993). Utilization of forest resources and local variation of wildlife populations in northeastern Gabon. In *Tropical Forests, People, and Food: Biocultural Interactions and Applications to Development*, ed. C.M. Hladik *et al.*, pp. 213–16. Paris: Parthenon.

Lawrence, E.A. (1993). The sacred bee, the filthy pig, and the bat out of hell: animal symbolism as cognitive biophilia. In *The Biophilia Hypothesis*, ed. S.R. Kellert and E.O. Wilson, pp. 301–44. Washington, D.C.: Island Press.

Maslow, A. (1993). *The Farther Reaches of Human Nature*. New York: Penguin-Arkana.

McNeil, D.G. and Ammann, K. (1999). The great ape massacre. *New York Times Magazine*, May 9.

McRae, M. and Ammann, K. (1997). Road kill in Cameroon. *Natural History Magazine* 106, 1, 36–47, 74–5.

Mordi, R. (1991). *Attitudes toward Wildlife in Botswana*. New York: Garland.

Oates, J.F. (1996a). Habitat Alteration, hunting, and the conservation of folivorous primates in African forests. *Australian Journal of Ecology* 21, 1–9.

Oates, J.F. (1996b). *African Primates: Status Survey & Action Plan* (Revised). Gland: IUCN.

Owens, D. and Owens, M. (1992). *The Eye of the Elephant: Epic Adventure in the African Wilderness*. Boston: Houghton Mifflin.

Robinson, J.G. and Bennett, E.L. (eds) (2000). *Hunting for Sustainability in Tropical Forests*. New York: Columbia University Press.

Rose, A.L. (1994a). Description and analysis of profound interspecies events.

Proceedings of XVth Congress of International Primatological Society, Bali, Indonesia.

Rose, A.L. (1994b). New paradigms for personhood in the age of atonement. *Proceedings of XVth Congress of International Primatological Society,* Bali, Indonesia.

Rose, A.L. (1995). Talking to the animals: The role of personal experience in primate research, caretaking, and conservation. *Proceedings of the 5th Annual ChimpanZoo Conference,* Jane Goodall Institute / University of Arizona.

Rose, A.L. (1996a). Orangutans, science and collective reality. In *The Neglected Ape,* ed. Nadler, Galdikas, Sheeran and Rosen, pp. 29–40. New York: Plenum Press.

Rose, A.L. (1996b). Commercial exploitation of great ape bushmeat. In *Rapport du seminaire sur l'Impact de l'Exploitation Forestiere sur la Faune Sauvage,* ed. R. Ngoufo, J. Pearce, B. Yadji, D. Guele and L. Lima, pp. 18–20. Bertoua: Cameroon MINEF & WSPA.

Rose, A.L. (1996c). The African forest bushmeat crisis: Report to ASP. *African Primates* **2**, 32–4.

Rose, A.L. (1996d). The bushmeat crisis is conservation's first priority. Talk at IUCN Primate Conservation Roundtable Discussion on an Action Agenda, XVIth Congress of International Primatological Society/American Society of Primatologists, Madison.

Rose, A.L. (1996e). Epiphanies with animals and nature transform the human Weltbildapparatur. Symposium on human–animal interaction. *Proceeedings of the International Society of Comparative Psychology,* Montreal.

Rose, A.L. (1996f). The African great ape bushmeat crisis. *Pan Africa News* **3**(2), 1–6.

Rose, A.L. (1998a). Finding paradise in a hunting camp: Turning poachers to protectors. *Journal of the Southwestern Anthropological Association* **38**(3), 4–11.

Rose, A.L. (1998b). On tortoises, monkeys, and men. In *Kinship with the Animals,* ed. M. Tobias and K. Solisti, pp. 21–46. Hillsboro, Oregon: Beyond Words.

Rose, A.L. (1998c). Growing commerce in African bushmeat destroys great apes and threatens humanity. *African Primates* **3**, 6–10.

Rose, A.L. (1999). Bushmeat commerce can be controlled: Organizing to confront a complex crisis. Talk at African Bushmeat Crisis Workshop/Symposium, American Society of Primatologists, New Orleans, August.

Rose, A.L. (2000a). Treatise on social values and social change applications: Capacity expansion for bushmeat commerce management. *Proceedings of Bushmeat Workshop.* Accra, Ghana: Center for Applied Biodiversity Science/ CI.

Rose, A.L. (2000b). Why is conservation failing? The question of capacity. Paper presented at Symposium on Ethics, Culture and Social Responsibility, Conference on Apes: Challenges for the 21st Century, Brookfield Zoo, Chicago, May.

Rose, A.L. (2000c). Gorillas on the menu. *ZooView Magazine: Special Gorilla Edition* **34**, 2. Los Angeles: Los Angeles Zoo.

Rose, A.L. (2000d). Establishing the gorilla protection society: Keys to longterm success. Report to managers and donors on site visits to a gorilla research and tourism project in eastern Cameroon, *Biosynergy Reports* (in press).

Rose, A.L. and Ammann, K. (1996). The African great ape bushmeat crisis. Talk and workshop at XVIth Congress of International Primatological Society/ American Society of Primatologists, Madison, August.

Rose, A.L., Ammann, K. and Melloh, J. (1999). Potential impact of hunting practices on cross-species transmission of viruses. Workshop on Cross Species Transmission of Immunodeficiency Viruses, CDC/OAR/NIAID/NHLBI, Atlanta.

Rose, A.L., Bowman, K. and Patterson, F. (2001). Establishing empathy for apes: Conservation values education. Report to managers and donors on start-up of conservation education project in Yaounde, Cameroon. *Biosynergy Reports* (in press).

Splaney, L. (1998). Hunting is greater threat to primates than destruction of habitats. *New Scientist*, March, pp. 18–19.

Stebbins, M., Hawley, J. and Rose, A.L. (1982). Long term action research: a case study. In *Organization Development in Health Care*, ed. Margules and Adams, pp. 105–36. Reading: Addison-Wesley.

Steel, E.A. (1994). *Study of the Value and Volume of Bushmeat Commerce in Gabon.* Libreville: WWF and Gabon Ministry of Forests and Environment.

Stromayer, K. and Ekobo, A. (1991). *Biological Surveys of Southwest Cameroon.* Wildlife Conservation International.

Vabi, M.B. and Allo, A.A. (1998). The influence of commercial hunting on community myth and ritual practices among some forest tribal groups in southern Cameroon. Paper presented at Regional Workshop on the Sustainable Exploitation of Wildlife in Southeast of Cameroon, Bertoua, August.

Weiss, R.A. and Wrangham, R.W. (1999). From Pan to pandemic. *Nature* **397**, 385–6.

Wilkie, D.S. and Carpenter, J.F. (1999). Bushmeat hunting in the Congo Basin: an assessment of impacts and options for mitigation. *Biodiversity and Conservation* (in press).

Wilkie, D.S., Sidle, J.G. and Boundzanga, G.C. (1992). Mechanized logging, market hunting, and a bank loan in Congo. *Conservation Biology* **6**(4), 570–80.

11 A cultural primatological study of Macaca fascicularis on Ngeaur Island, Republic of Palau

BRUCE P. WHEATLEY, REBECCA STEPHENSON,
HIRO KURASHINA AND KELLY G. MARSH-KAUTZ

Introduction

The most eastward location of the long-tailed macaque, *Macaca fascicularis*, is on the Island of Ngeaur. The monkey population was estimated to be about 400 in 1994 (Wheatley *et al.*, 1999), a population reduction of about half of the 825–900 estimated by Farslow (1987) in 1981. Ngeaur is one of sixteen States that make up the Republic of Palau. It is a small, 830 ha, raised coral atoll, one of the Republic's 350 islands (Karolle, 1993). This atoll is the only home for monkeys of any kind in the entire Pacific region of Oceania. The 150 people of Ngeaur (*Rechad er Ngeaur*), however, feel more cursed than blessed by the presence of monkeys. The people of Ngeaur hunt and trap monkeys, although they are not eaten. A previous paper examined the hunting and trapping techniques as well as the behavioral responses of monkeys to hunting (Wheatley *et al.*, 1999). This chapter describes the rationale of the people for their disdain of these monkeys.

This species is a problem to local farmers throughout Southeast Asia (Wheatley, 1999). Despite problems that monkeys can cause, however, some possibilities for their conservation exist (Wheatley, 1999). Balinese conservation efforts at a Monkey Forest in Ubud are a success because the local community of Padangtegal eventually supported them. The first step in our investigation of these monkeys in Palau therefore, was to interview the people of Ngeaur and elicit information on how they felt about monkeys. Our other questions included how these monkeys arrived on Ngeaur. Preliminary comparisons can then be made between the people of Padangtegal and Ngeaur on how they view this species of monkey.

Primate commensalism is generally viewed as a symptom of disturbance and as an unnatural situation unworthy of research. The study of cultural primatology or ethnoprimatology examines the interactions between humans and nonhuman primates. Cultural anthropologists, an archaeologist, and a primatologist thus combined efforts in this research, which occurred between July 7 and July 17, 1994.

The 1994 study

On our way to Ngeaur, the Minister of Cultural Affairs, Mr Riosang Salvador, explained to us in Koror, the Capital of Palau, that there are two kinds of people on Ngeaur: 'There are the humans and then there are the monkeys.' Furthermore, he said, the latter were the masters of humans. We thus felt instant justification for our cultural primatological approach. He further elaborated that legislation for the eradication of monkeys to the point of extinction was enacted in 1975 and a reward for their tails was offered. The Constitution, however, banned firearms in 1979 and the eradication campaign officially ended.

Subsistence gardening supports most of the people on the Island, although some rice and canned foods are eaten. There are only about seventeen paying jobs on the Island. Almost everyone we interviewed, from government officials in Koror to the local people of Ngeaur, hated monkeys. Here is just a small selection of their quotes. 'To me, it is a pest and I would like to see them eradicated.' 'They are the worst animals.' 'Only children like monkeys. People here don't find them cute. They want to kill them.' 'We hunt monkeys because they are bad.' 'Everything monkeys do is bad.' 'Their bad habit is they don't take what they need, they take everything.' However, not everyone hates monkeys. One individual said that the lengthy presence of the animals made them a part of the Island's culture and history and she couldn't imagine the Island without them. Some young children treat the pets as babies, for example by giving them baths. There is one teacher who advises her students to respect all life, including that of monkeys.

On July 4, 1994, three days prior to our arrival, a fifty-four year-old man was found dead in the harbor. There are no police stationed on the Island. On July 13, the police arrived with their M15 assault weapons and shotguns. Although the man's death appeared to precipitate the arrival of the police, the Ngeaur woman's association had a long-standing request for the police to come and control the monkeys. They patrolled the Island and shot three monkeys. A boy on a bicycle dragged a shot juvenile male monkey by his tail down the road. When the boy approached some of our students, he said 'You want to study monkeys, here, study this one,' and he threw the animal towards them. We thus learned that some individuals, the police, could carry weapons and shoot monkeys. The police made some inquiries about the death at the harbor and left four days later.

Why such hatred towards monkeys? They are considered to be the most serious of all the pests listed by the people of Ngeaur. Out of a total of fourteen different pests that they listed in a survey conducted by Marsh-Kautz and Singeo (1999), monkeys were the most destructive and the most irritating. They were the hardest to keep out of the garden. The people of Ngeaur wanted to know of new ways to keep them out of the gardens, because the '*mongkii*' always returned 'to pillage and destroy' no matter what. Some of the other pests they listed were shrews, rats, mice, introduced fruit flies, hermit crabs, caterpillars, mosquitoes, red ants, chickens, a taro virus, tapioca blight, and a disease that affects banana plants.

There were numerous complaints about monkeys. They are destructive of the indigenous flora and fauna. They eat *Megapodius laperouse* (*bekai*) eggs. Perhaps too they are responsible for damaging the nests of pigeons, *Ptilinopus pelewensis* (purple-capped or Palau fruit doves) and the Micronesian starling (*Aplonis opaca*) that nests in betel nut trees. We also heard that monitor lizards and a black shrew were predators on many of these birds.

Despite the ubiquity of land crabs, monkeys do not apparently live up to their common name (crab-eaters) on Ngeaur Island. The large land crabs on the Island are not eaten. One of the residents said that even starving monkeys wouldn't eat them. The monkeys also steal food from pigs, and eat chicks and eggs.

The Islanders mentioned a wide variety of crops that were eaten by monkeys. The most important of these is taro. Women are responsible for taro production, the primary food on Ngeaur. There are several different types of taro that the women garden. Great prestige is attached to women who have the best taro gardens, especially those that are artfully gardened and those that produce well. The best way to state the importance of taro is to quote them. '*Mesei a delal a telid.*' ('The taro garden is our life') and 'One cannot breathe without taro.' The monkeys don't eat the itchy kind of taro according to one informant. They do eat baby *kukau* (taro tubers), said another, but they were not destructive to the plant. The monkeys prefer to eat the new growth on the taro plants while they are young and sweet, said another. One woman stated that partly eaten taro effectively damaged the remaining taro. Although the damage to taro was slight during our visit, probably because the fruits of other trees were available, the perceived threat is great. Giant taro, *Crytosperma chamissonis*, known as *brak*, and true taro, *Colocasia esculenta*, known locally as *dait*, grows in the swamps. A woman brought in a taro stem from the gardens and pointed out monkey tooth marks on it. The marks were not obvious tooth marks from incisors, for example, that the first author could detect. During our short stay on the Island, the animals only ate a small leaf of one out of a thousand taro plants in the swamps beneath their sleeping tree. All of the other plants were completely untouched. Rats, shrews, and starlings eat taro and other root crops, and chickens, according to one informant, may do more damage than monkeys by scratching the plants.

One common story we heard was that monkeys liked to irritate the women by pulling recently planted taro and cassava up and sticking them back into the ground upside down, thereby killing them. The women, however, outwitted the monkeys by planting their crops upside down so that the monkeys replanted them in the correct manner. The women therefore fooled these monkeys and re-established their control over the gardens. Another aspect of this story, however, is that the monkeys are also happy because they can claim credit for planting and growing the crops (Mo-Bo Morei, Palau Community College, 1998, personal communication to Helen McMahon). Monkeys are therefore seen as clever, because they 'imitate everything people do,' and 'smart,' because they are hard to keep out of the gardens. Although monkeys were referred to as 'humans' that are taking over the island, the pets are not given names. 'They don't really look like humans,' said one individual, but they 'exhibit human-like behavior,' said another. 'They are on private property,' said one individual. The people of Ngeaur were moving into the middle of the Island and even coming to Koror to get away from the monkeys, we were told. There are no legends or myths of monkeys as there are for other animals, such as the dugong and spider.

Another story about the monkeys of Ngeaur involves a man who shaved with a straight-edge razor while a monkey watched. He ran the dull side of the razor over his throat and then walked away, leaving the materials behind. A monkey then imitated the man, but used the sharp edge of the razor with disastrous results. This story is from an old Aesopic fable of Western origin (Husband, 1980).

Out of the more than fifty items in the diet of the monkeys on Ngeaur (Poirier and Smith, 1974; Farslow, 1987), two of their favorite foods were in fruit during our stay on the Island. The first is the wax or Palauan apple or *rebotel* (*Eugenia javanica*). They have a thin, red, flexible skin, 4–8 mm thick. The greatest width of this fruit was measured to be 31–36 mm. The seeds are 19–56 mm long and 18–39 mm wide. Children like to eat this fruit salted.

The second favored food (Farslow, 1987) is the Polynesian or tropical almond, *Terminalia catappa*, known locally as *miich*. It is a one-seeded drupe up to 8 cm long. The thin (approximately 1–2 mm) flexible skin is eaten but the seed or nut, approximately 18 mm long and 9 mm wide, is spit out and thus dispersed wherever the monkeys happen to be. The macaques are not seed predators of this species. Only one nut out of hundreds of eaten fruits was damaged. The cortex of *miich* is tough and fibrous. People commonly eat the dry nuts and they are, or were, extremely important in a cultural sense to the people of Belau.

The macaques also eat betel nut (*Areca catecho*), locally known as *buuch*. The betel nut from Ngeaur is supposedly famous for its taste, but the monkeys are said to drink the juice of the unripe nut. Thus, betel nut cannot be sold or used in feasts. Also eaten are coconuts, cassava, oranges, lemons, bananas, and papaya. Other plants such as pandanas, are damaged and the 'hearts' are eaten.

Both the young and the old shoot monkeys on Ngeaur. Younger people use BB guns and slingshots, while the older people use old ·22 rifles. They sometimes bring dogs into the gardens to chase the monkeys away. Some dogs have killed monkeys and some monkeys will bite dogs.

Trapping monkeys appeared to be widespread in 1994. The baited traps resemble fish traps: monkeys enter into a funnel-shaped entrance, but cannot escape. The young, trapped monkeys are sometimes sold as pets off-island for US$100. Permission is needed and a tax is assessed for importing monkeys and bats, but we heard they are smuggled. To help prevent the spread of monkeys to other Islands, only males and not females are allowed off the Island. Baby monkeys are fed with sugar water on a cotton applicator. The caged or tethered captive animals are fed garden fruits such as bananas, oranges, and papaya. Children, more so than adults, like pet monkeys. One woman said 'their poop stinks'. An official in Koror told us to bring him a pet monkey. People cry when their pet monkeys die. The people of Ngeaur may be making the crop-raiding problem worse because the animals sometimes escape – one of about five animals that we saw did escape during our stay – and they have learned to feed off these garden foods. They therefore hang around and raid the gardens.

The people of Ngeaur confirmed that Germans introduced these monkeys. The most common explanation for their introduction that we heard was that they were pets. Ucherbelau Masao Gulibert of the top-ranking matriclan, sometimes called Chief Endo, explained in Japanese that the Germans brought five young monkeys from Indonesia. No one wanted to care for them when the Germans left and the monkeys escaped. The most interesting explanation of how monkeys got to Ngeaur was that they might have been brought in for medical testing. An older woman told us that her mother says that the Germans brought in two monkeys as pets to the hospital for testing medicines. One person said that it was possible that monkeys were brought in on several different occasions for various reasons.

The trade in pet monkeys is also the most common explanation for the occurrence of monkeys on other Islands in Palau. We heard that there was a group of about five monkeys near the airport in Koror and in a village north of there, Ngisaol. Mr Salvador, the Minister for Cultural Affairs, stated that he has seen monkeys in three States in the Republic of Palau.

The Palauan word for monkey is interesting because there does not seem to be any obvious derivation for it. The Palauan word is *sikerii* or *sikou*, but people say *sukerii*. People over 40 years old use either of these terms, whereas the younger people may use the English word *mongkii* more frequently (Y. Singeo, personal communication). There is perhaps a remote similarity to the Indonesian word *kera*, and, perhaps, an even more remote possibility of a connection to Sugriwa, a Monkey King mentioned in the Ramayana. The letter 'w' is not in the Palauan alphabet and the 'wa' would be dropped. Malaysians have been reported in Palau. In 1782 for example, a Malaysian man was abandoned in Palau by a Chinese junk (Rechebei and McPhetres, 1997).

1999 Update

Marsh-Kautz and Singeo (1999) completed a more recent study on Ngeaur. Most of the people they interviewed supported the hunting of monkeys, which has drastically escalated between 1997 and 1999. Additional hunting permits were issued to three *rubaks* (respected male elders and/or traditional chiefs) in order to control the monkey population. Two of the *rubaks* said they subsequently shot 83 monkeys. The hunters are not allowed to use new weapons, only their old pre-1979 ones. The rationale for the increase in hunting is that monkeys are more numerous in the garden areas than ever before and they are eating the young coconuts and betel nuts and destroying pandan trees. Marsh-Kautz and Singeo (1999) also report that the attitudes of some of the young adults are becoming a little more accepting of monkeys on their Island.

Monkey repellents of hot peppers and bilimbi (*oterbekii*) fruits were applied to trees and other plants for a few months, but were apparently unsuccessful. It washed away and didn't seem to work, they said. 'Besides, monkeys eat *oterbekii*' so it won't be effective, said one man.

The trapping of monkeys is said to have stopped because the infants are not sold any more. Recently passed legislation imposed a fine of US$500 for having an illegally exported monkey on another Island. Previously, permits were required to export male monkeys. The police are supposed to be contacted if pet monkeys subsequently escaped from their owners on the other Islands. Most of the people interviewed, however, supported the sale of monkeys to foreign researchers. Some method of birth control for the monkeys was also desirable, according to the recent survey.

High tides and flooding have rendered the swamp taro inedible for three years in a row, they say. Insects have also damaged the taro plants and viruses are infecting bananas and cassava. A new destructive fruit fly has also caused some damage.

Discussion

This species of monkey is not indigenous to Ngeaur. Foreigners introduced it during colonial rule. It is commonly stated that all of the monkeys are descended from a pair that escaped during the German occupation (O'Connor, 1992; Stanley, 1992; Poirier and Smith, 1974; Poirier, 1975; Etpison, 1991). The monkeys have inhabited the Island for almost one hundred years and have survived typhoons and the bombing during World War II, but they are not culturally integrated into the local cosmology as many of the other fauna are.

An obvious topic in cultural primatology and primate commensalism is dispersal. The earliest record we could find for the introduction of a pet monkey in Oceania occurred in 1602 (Rogers, 1995). In this case, a Manila galleon from Cavite, Philippines, stopped on the Island of Rota where the Spanish presented a monkey, probably a macaque, among other things to a Chamorro headman. Although this introduction was not successful, introductions onto other Islands were, such as those onto Ngeaur, Mauritius and perhaps Flores in the past. This process of introduction continues today. Pet monkeys are brought to Koror and other Islands, where they eventually escape. Macaques occasionally show up in Guam. On October 2, 1971, for example, a macaque was shot and captured in Guam after it was reported to have attacked a boy and girl (Pacific Daily News). Vietnamese refugees attempted to bring macaques into Guam on June 1, 1977 (Leeke, 1977).

The Germans occupied Ngeaur after they purchased Spain's Micronesian possessions and the United States took the Philippines and Guam in 1898. The Germans discovered phosphates on Ngeaur in 1903, and mining started sometime in 1909 (Irving, 1950). An interesting story about the German discovery of phosphates is told in Pedro (1999). The grandmother of two sisters now in their eighties told them that the Germans were amazed at the white paint on the traditional *Bai* (community house). The *rubak* showed the Germans how they mixed water with the white dirt. The white dirt was of course the guano deposits that date back to the mid-Pleistocene (Otsuki, 1916).

Ngeaur was apparently a regular stop for Imperial mail steamships even before mining operations (Rechebei and McPhetres, 1997). Jan Kubary arrived in Belau in 1871, for example, to make collections for the Godeffroy Museum in Hamburg, Germany, and a steamer stopped by in 1901. Government mail steamers were making regular visits in Palau by 1906 and a government station–post office was opened in Ngeaur in 1909. The steamer *Germania* of the Jaluit Gesellschaft operations – a Godeffroy affiliate – stopped at Ngeaur and Koror six times a year during its three annual round-trips from Hong Kong to Sydney, Australia. Three Lloyd steamers of the Austral-Japan Line probably stopped at Ngeaur twice a month beginning in 1908. Japanese ships of the Hiki Line also made mail runs to Ngeaur. Some of the Jaluit vessels stopped in Indonesia, such as Minado, Halmahera, and New Guinea (D. Ballendorf, personal communication).

Poirier and Smith (1974) state that an old man told them that 'two monkeys were brought by German phosphate mining engineers'. Our work on Ngeaur confirms that the Germans were responsible for the introduction, but our interviews cast doubt as to whether there were just two monkeys. Noria Henry, the mother of one of our informants, is quoted in Pedro (1999) as saying that the 'Germans brought three monkeys to Angaur. The monkeys were kept at the hospital, and people of Angaur went to see these strange animals which they had never seen before.' As already mentioned, Chief Endo gives the number as five monkeys. Although the number of introduced monkeys seems to vary a bit, it appears to be more than two, which would place it more in line with the genetic tests done by Kawamoto *et al.* (1988) and Matsubayashi *et al.* (1989). These researchers further suggest either a mainland Asian origin or a Greater Sunda Islands origin for the monkeys of Ngeaur. There is one other suggestion for the introduction of these monkeys. Robert Bishop at the Palau Community Action Agency says that it is possible that Taiwanese miners brought them in as a potential food item (personal communication to Helen McMahon), but we heard no such reports from the people of Ngeaur. Besides macaques, the Germans also introduced tapioca and rubber trees.

A number of people state that the macaques were brought to Ngeaur as the 'proverbial guinea pigs for testing mine air quality'. Thyssen (1988), for example, says 'Micronesia's only monkeys (were) brought by Germans to test the air in the phosphate mines'. This explanation is dubious. The early mining operations consisted of simply shoveling the phosphate off the top of the ground (Otsuki, 1916). Yawata Aso described the mining operations in 1916 as clearing the jungle, removing 4–5 inches of decayed vegetation covering the ore, breaking it up, and shoveling it into mine cars. The Deutsche Sudsee Phosphat-Aktiengesellshaft was organized in Bremer-haven in 1908 to exploit one of the highest-grade phosphate deposits in the world (Oliver, 1989). This company 'purchased' – see Pedro (1999) – the surface rights to the island for RM13 000. Construction for mining began in February 1909 and eventually kilns, railroad, bridge, residences, and a hospital were built. Labor was imported. There were 55 Chinese laborers, 98 Yapese, and 23 Europeans. The Japanese navy occupied the Island on October 9, 1914 and continued to mine the deposits, with some interruptions, up until 1955 (Hezel, 1995). Subaerial mining was not begun until 1936. Poirier and Smith (1974) mention that the Japanese killed monkeys for sport, food, and medicinal purposes and that some Coast Guard personnel also shot monkeys on occasion.

The reference to monkeys at the hospital is intriguing. We were unable to locate information on using these animals for medical research. There is a report by the Palau Community Action Agency (1977) that mentions Dr. Buse, a government doctor on Yap, who heard of 494 cases of *missillepik* between 1911 and 1912, six of whom died, on Ngeaur. Hezel (1995) states that the cramped living conditions of the long wooden barracks for the miners of Ngeaur 'facilitated the spread of epidemics, as when typhus broke out in 1912'. Several years earlier, a medical report of the German Protectorate reported hundreds of deaths from a 'great plague', 'the bacilli of catarrh, influenza, etc.'.

Introduced species often have deleterious effects on the local flora and fauna. It is possible that monkeys have contributed to the decline of some of the birds on Ngeaur, especially some of the pigeons, doves, starlings and megapodes. This species of macaque was also introduced onto the Island of Mauritius in the sixteenth century, where they may be nest predators on two of the three rarest bird species in the world, the kestrel and the pink pigeon (Sussman and Tattersall, 1981). They also damage the local sugar-cane.

The fact that monkeys damage taro is probably the most serious of all the problems caused by monkeys. Most of the hunting of monkeys appears to be around the taro gardens in the swamps on the south side of the Island. The animals are extremely wary there, in contrast to other locations (Wheatley et al., 1999). Only men hunt monkeys. The people of Palau practice matrilineal descent so it is understandable that the women are more upset over the damage that monkeys do than the men. The status of older women seems to be at most risk because damage to taro would affect their ability to meet household food requirements and exchange obligations at, for example, marriages, funerals, and house buildings (Force and Force, 1981). The younger women may be more integrated into the cash economy. The condition of the taro garden is said to be a measure of a woman's standing in the community (Rechebei and McPhetres, 1997). Taro growing is considered an art form and taro itself was in some cases treated as traditional money in the past. Men are not supposed to touch the taro patch. If women are upset, however, and 'women are the way to wealth', by being the 'channel through which money can be collected' (Rechebei and McPhetres, 1997) then the entire community is affected.

It may also be possible that the damage to crops caused by the monkeys is perceived to be greater than it really is. Poirier and Farslow (1984) reported that 90% of some of the taro plots sustained some damage by monkeys, but that reports of extensive crop damage were unsubstantiated. Poirier and Smith (1974) also state that their interviews 'established that the natives had no desire to rid the island of the monkeys'. One important issue is that the previously lower-maintenance taro garden is now higher-maintenance. In the experience of the first author, a garden that is not constantly looked after day after day, and thus, indirectly 'guarded', would be at risk of possible damage. Balinese farmers, for example, maintain a constant vigil at their gardens near the Monkey Forest at Ubud, Indonesia. These farmers, if necessary, are also quick to respond to monkeys in their gardens by throwing rocks at them or chasing them with sticks or sickles.

There are a few 'benefits' from the presence of monkeys on Ngeaur. Many tourists such as Japanese, Spanish, Germans, Americans, Australians, and the people of Palau, come to see monkeys. Tourism provides some jobs and income, such as cooking meals, renting the guesthouse and snorkeling gear, etc. The people of Ngeaur formed a *Sakurasai* organization to look after all visitors. The beauty of the Island and its monkeys were two of the most popular topics that visitors wrote about in the guesthouse journal. The families of deceased former villagers, especially Japanese, also visit the Island on a regular basis. The seed dispersal of the tropical almond by monkeys might also be considered to be a benefit, however slight. The tropical almond is of great traditional, cultural significance. For example, the nuts are sugared and shaped like a mermaid to be eaten at the ceremony when a chief becomes a *reklai* (one of the two highest chiefs in all of Palau). There is also a goddess called *Tibitibikmiich* who flies from tree to tree collecting the almonds. The word *miich* is sometimes used to refer to a genuine blood relative as well as an original inhabitant of the village. A phrase, *Techel-a-miich* (the essence of the tropical almond), describes the quality of a family – and only through matrilineal descent – with deep roots in the community. It is analogous, perhaps, to the phrase describing a family as solid as oak. New immigrants are sometimes referred to as driftwood. *Miich* has anti-bacterial properties. The juice of pounded leaves is applied to cuts (Salcedo, 1970). Another 'benefit' was the selling of infant monkeys, a 'cash crop' of sorts, but this is said to have stopped.

Although monkeys are perceived to be a threat to exchange systems, and thus could influence the distribution of resources and status as well as affect the social order, there are other influences operating as well. The Germans appear to be the first colonialists to make an impact on the Republic of Palau. The mining eventually ruined the agriculture in that area, as well as in other areas, (by causing salt water to intrude into the swampy taro-growing areas), displaced people, introduced a wage economy, and brought in foreign workers and ideas (Pedro, 1999). The Germans or perhaps the imported laborers also introduced cassava, new types of taro and coconuts. The Japanese, and later Americans, continued this process of exposure to a wage economy and other aspects of industrialization, etc., which may have more of an impact on the economy and social organization of Ngeaur than monkeys. The mining royalties to the eighteen land-holding families of Ngeaur apparently ran out in November of 1994.

The monkeys of Ngeaur are considered to be a serious pest. The Palau Legislature passed a law in 1975 calling for the complete eradication of the animal as a pest. At that time, Robert Owen, the conservation chief, suggested that the monkeys be poisoned, trapped for medical research labs, or eaten by released wild tigers (Whaley, 1992). Most of the trapping occurred in the north where there are no gardens. A Sustainable Food Production Project on Ngeaur, part of the Cooperative Research and Extension and the Pacific Commission, identified the destructiveness of monkeys to crops as one of the top three problems. Some of the solutions suggested were: kill them all; ask the Germans to take them back; sell them for medical research; and ask conservation groups to help them reduce the population. Speakers at the seventh annual women's conference held in March 2000 also discussed the 'monkey problem.' Some of the women present were upset that outsiders did not understand the numerous problems caused by monkeys. Both houses of the legislature also recently discussed and appropriated US$25 000 for fiscal year 2000 to study how to eliminate or mitigate the monkey and shrew problems and to assess the economic damage that they caused. The funds come from investment earnings in the Compact and the Governor of Ngeaur allots them. They also issued a few monkey-hunting permits to the *rubaks*.

There are more monkeys than people on the Island of Ngeaur. Non-human primates, with their human-like appearance, seem to be treated as an invading foreign army: they are thus killed and not used as a food source. A larger issue then may be the attitude of the people of Palau towards foreigners and a new way of life. *Rechad er Belau* have a great love for their environment, but 'Palau today is still not in control of the reins of development' that pose a serious threat (Ueki, 2000). Monkeys could, therefore, be a constant symbol for foreigners and the destruction that they have caused to native flora, fauna, and the destruction of the islanders' traditional way of life begun by the Germans. The conservation officer, Demei Otobed, for example, says that foreigners brought in the monkeys and that foreigners are now protecting them. 'What about the people of Ngeaur?' said Otobed. 'They get nothing.' (Whaley, 1992). One interesting idea from these observations is that the attitude of the people of Ngeaur towards monkeys might change when their attitude towards foreigners – including anthropologists – changes. It would thus seem important to establish a long-term rapport with the people of Ngeaur and to help alleviate the many problems that monkeys cause.

Government authorities advise people to obey the Convention on International Trade and Endangered Species of Wildlife, but the problems that monkeys cause are still there. It is obvious that the attitudes of local people are crucial with regard to monkey conservation. The same species of monkey is perceived and treated differently in different locations. Monkey conservation efforts are much more successful in Ubud, Bali, for example. In contrast to those of Ngeaur, the monkeys of Ubud do not outnumber the people and they are relatively confined to a temple area. Bali also has a long, rich, cultural tradition of commensalism with these monkeys (Wheatley, 1999). The monkeys on the Island of Ngeaur, however, are a recently introduced species that continue to be hunted and killed.

Acknowledgements

We are very grateful to the *Rechad er Belau*, People of the Republic of Palau, and especially the *Rechad er Ngeaur*, People of Ngeaur. The generous help and sharing of knowledge by Ucherbelau Masau Endo Gulibert, Chadmeseb Setsuo Tellei, Ucherkemul Agusto Naruo Michael,

Dirremasech, Renguul Orrenges Thomas, Hesus Belibei, Lorenzo Pedro, David Idip, Ben Gulibert, Riosang Salvador, Mario S. Gulibert, Theodosia Blailes, Governor V. Ben Roberto, Faustina Rehuhr, Grace and Leon Gulibert, and Yvonne Singeo is gratefully appreciated. Our study team included nine students from the University of Guam, M. Akapito, R. Barrett, J. Drake, W. Johnson, J. Matter, Y. Singeo, D. Tibbetts, and C. Ogo.

References

Etpison, M.T. (1991). *Palau. Portrait of Paradise.* Koror, Palau: NECO Marine Corp.

Farslow, D. (1987). The behavior and ecology of the long-tailed macaque, *Macaca fascicularis* on Angaur Island, Palau, Micronesia. Ph.D. thesis, Ohio State University.

Force, M. and Force, R. (1981). The persistence of traditional exchange patterns in the Palau Islands. In *Persistence and Exchange*, ed. R. Force and B. Bishop, pp. 77–89. Honolulu, HI: Pacific Science Association.

Hezel, F.S.J. (1995). *Strangers in Their Own Land: A Century of Colonial Rule in the Caroline and Marshall Islands.* Honolulu: University of Hawaii Press.

Husband, T. (1980). *The Wild Man.* New York: Metopolitan Museum of Art.

Irving, E.M. (1950). Phosphate deposits of Angaur Island, Palau Islands. Prepared by the Military Geology Branch, US Geological Survey, for Intelligence Division.

Karolle, B.G. (1993). *Atlas of Micronesia*, 2nd edn. Honolulu: Bess Press.

Kawamoto, Y., Nozawa, K., Matsubayashi, K. and Gotoh, S. (1988). A population-genetic study of crab-eating macaques (*Macaca fascicularis*) on the Island of Angaur, Palau, Micronesia. *Folia Primatologica* **51**, 169–81.

Leeke, J. (1977). One-of-a-kind refugee finds new home on Guam. *Pacific Daily News*, June 1, p. 20.

Marsh-Kautz, K., and Singeo, Y. (1999). Discerning Rechader Ngeaur community perceptions of their island Mongkii (monkey). Unpublished draft report to Ngeaur State, Republic of Palau.

Matsubayashi, K., Gotoh, S., Kawamoto, Y., Nozawa, K. and Suzuki, J. (1989). Biological characteristics of crab-eating monkeys on Angaur Island. *Primate Research* **5**, 46–57.

O'Connor, C. (1992). Southern Comfort: Palau's Angaur Island. *Guam and Micronesia Glimpses* **32**(1), 23–5.

Oliver, D.L. (1989). *The Pacific Islands*, 3rd edn. Honolulu: University of Hawaii Press.

Otsuki, Y. (1916). Phosphate deposits of Angaur Island. *Jour. Geogr.* **28**(328), Tokyo.

Pacific Daily News (1971). Monkeys moving in. *Pacific Daily News*, Oct. 2, p. 3.

Palau Community Action Agency (1977). *A History of Palau*, Vol. 2, *Traders and Whalers, Spanish Administration and German Administration.* Koror, Republic of Palau: The Ministry of Education.

Pedro, L. (1999). The effects of foreign culture and school on Angaur, Palau, 1899–1966. A special project submitted in partial fulfillment of the requirements for the MA Degree, University of Guam.

Poirier, F. (1975). *The Human Influence on Subspeciation and Behavioral Differentiation Among Three Nonhuman Primate Populations.* Denver: AAA.

Poirier, F. and Farslow, D. (1984). Status of the crab-eating macaque on Angaur Island, Palau, Micronesia. *Primate Specialist Group Newsletter* 4, 42–3.

Poirier, F. and Smith, E. (1974). The crab-eating macaques (*Macaca fascicularis*) of Angaur Island, Palau, Micronesia. *Folia Primatologica* 22, 258–306.

Rechebei, E.D. and McPhetres, S.F. (1997). *History of Palau: Heritage of an Emerging Nation.* Koror, Republic of Palau: Ministry of Education.

Rogers, R.F. (1995). *Destiny's Landfall: A History of Guam.* Honolulu: University of Hawaii Press.

Salcedo, C. (1970). The search for medicinal plants in Micronesia. *Micronesian Reporter* 18(3), 10–17.

Stanley, D. (1992). *Micronesia Handbook.* Chico, CA: Moon Publications.

Sussman, R.W. and Tattersall, I. (1981). Behavior and ecology of *Macaca fascicularis* in Mauritius: a preliminary study. *Primates* 22, 192–205.

Thyssen, M. (1988). *A Guidebook to the Palau Islands.*

Ueki, M. (2000). Eco-consciousness and development in Palau. *The Contemporary Pacific* 12, 481–7.

Whaley, F. (1992). Palau wants monkeys off its back. *Pacific Daily News*, Dec. 6, pp. 1, 4.

Wheatley, B. (1999). *The Sacred Monkeys of Bali.* Prospect Heights: Illinois: Waveland Press.

Wheatley, B., Stephenson, R. and Kurashina, H. (1999). The effects of hunting on the Long-tailed Macaques of Ngeaur Island, Palau. In *The Nonhuman Primates*, ed. P. Dolhinow and A. Fuentes, pp. 159–63. Mountain View, California: Mayfield Publishing.

12   Monkeys in the backyard: encroaching wildlife and rural communities in Japan

DAVID S. SPRAGUE

Introduction

Wildlife conservation is unavoidably a major issue in Japan, a highly industrialized nation with high human population density. However, the conservation issues faced by Japan today cannot be explained as simple cases of recent habitat destruction by expanding human activity, or wildlife populations decimated by excessive hunting. The Japanese people have repeatedly reformulated their relationship with the fauna and flora of their islands. Wildlife conservation in Japan must be analyzed in the background of radical changes in the historical ecology of the Japanese archipelago.

Humans and monkeys have co-existed for much of the history of both species in the Japanese archipelago. The Japanese monkey, *Macaca fuscata*, is descended from an ancestral species of the macaque genus that crossed to the Japanese islands during the middle Pleistocene Epoch (Dobson and Kawamura, 1998) and spread to the three main islands of Honshu, Shikoku and Kyushu, and the offshore islands of Yakushima and Tanegashima. The archeological record of Japan contains remains of monkeys in the shell mounds of the Neolithic Jomon Period (Mito and Watanabe, 1999). Throughout this time, natural resource utilization by humans has shaped monkey habitats and pressured monkey populations. Forest utilization altered the distribution and composition of the forests that constituted monkey habitats. Monkeys were often hunted. Monkeys survived, nevertheless.

During the half-century following the Second World War, the Japanese people once again radically altered the way they utilize the natural resources of their archipelago. For Japanese monkeys, the turn of the new century has brought the best of times and the worst of times. The best of times may have arrived for many Japanese monkey populations because the disappearance of traditional forms of natural resource utilization has released monkeys from many threats to their survival. The worst of times

254

may have arrived under the recent land use regime of the Japanese archipelago that has created many new threats to the ecological security of Japanese monkeys.

Primatologists confront a bewildering array of issues surrounding human–animal interactions at the turn of the twentieth century in Japan. Whereas many primatologists believe that conservation policies need to be strengthened, many local governments and farmers argue that monkey populations are expanding to the point that monkeys are now serious agricultural pests. Government has responded to crop-raiding monkeys by permitting the culling of an ever-increasing number of wild monkeys. To argue against further culling, and propose a conservation strategy for Japanese monkeys, primatologists confront the dual task of explaining why monkey populations can still be endangered, while also explaining the increasing damage to crops caused by monkeys raiding fields and orchards all over Japan. This dual task requires primatologists to delve deeply into how Japanese life and culture has repeatedly reformulated how it uses or abuses Japanese wildlife.

Traditional forest resource utilization

Far from being a post-Second World War problem, habitat destruction and hunting must have threatened monkey populations in the past. The Japanese people have utilized forest resources very intensively for millennia. Pollen analysis reveals the widespread increase in pine species, indicating human activity, beginning about 1500 years ago in central Japan (Yonebayashi, 1998). The agricultural societies that followed found many more uses for forest resources.

Timber resources supplied Japan's raw materials for architecture and fuel (see Totman, 1989). Massive construction projects to build palaces, temples, and shrines exploited the large timber resources every time the emperors moved their capital. Timber resources were further depleted as Japan built large cities and castles, and then rebuilt them following the periodic wars and fires. The pottery industries consumed large amounts of fuel wood. Kitchens consumed wood for cooking. Rooms were heated by the charcoal in hibachi.

The Japanese utilized a vast array of non-timber forest resources as well. Agriculture depended on the woodlands and grasslands surrounding villages for a variety of natural resources. Until recently, farmers and foresters recognized large areas of a class of woodland called the 'agricultural use woodland' or *noyorin* (Okutomi, 1998). The trees in the *noyorin* were often species favored by farmers or hardy species that could tolerate intensive and repeated utilization, especially the red pine (*Pinus densiflora*) and coppice broad-leaf species (such as *Quercus* spp.). Grasslands were often maintained by human activity. Natural grasslands are mostly coastal or alpine. Rural grasslands were maintained by the repeated cutting and firing that kept the vegetation from succeeding to woodland (Tsuchida, 1998). All such forms of resource utilization created a band of human activity around human communities that existed until as late as the 1950s (Kamada and Nakagoshi, 1997).

Many woodlands and grasslands were common-lands subject to elaborate rules of exploitation negotiated by the villages that used them, since traditional agriculture required large amounts of green fertilizer and fodder (Dore, 1978; McKean, 1982). Green fertilizer refers to the leaves and grasses plowed into fields to provide organic material to the soils, or mixed with manure to make compost. Farmers often maintained much larger areas of woodland and grassland than the areas of fields and paddies (Sprague, Goto and Moriyama, 2000). In addition to fodder for farm animals, the grasslands supplied the thatch necessary for roofing farm houses. Rural communities used kindling and leaves for fuel, but fuelwood and charcoal could also be sold to towns. For villages in mountains, the woodlands supplied important fall-back foods, such as acorn, chestnut, and horse chestnut.

Farmers did not limit cultivation to fields close to villages. Less known outside Japan, but very important for Japanese historical ecology, is the fact that swidden agriculture was very widespread in Japan. Mizoguchi (1988) proposed that the area of slash-and-burn fields also grew as population grew during the early modern period when Japan was ruled by the Tokugawa Shogunate (1603–1868). Sasaki (1972), in his classic study on swidden agriculture in Japan, summarized the government statistics available on swidden agriculture. He found that in 1920, 152 000 households practiced swidden agriculture in 77 414 cho (1 cho is equal to about 1 hectare). In 1950, the last year the Forest Agency reported statistics on swidden agriculture, there were still 110 500 households practicing swidden agriculture in 9533 cho.

Despite Buddhist prohibitions against hunting or eating mammals, the Japanese utilized the skins, feathers, and meat of most Japanese fauna for food, clothing, decoration, and medicine. Farmers were sometimes part-time hunters, and a professional class of hunters called the *matagi* hunted large mammals by means of firearms. In the mountainous rural parts of Japan, monkeys were hunted for their meat. According to Mito (1992), hunting may have exterminated monkeys from much of their range in northeastern Honshu Island. Monkeys were especially heavily hunted in this part of Honshu, where winters were more severe, crop failures more frequent, and no taboo existed against hunting them (Hirose, 1979; Mito and Watanabe, 1999). In southwestern Honshu and Kyushu, monkey hunting was left largely to professional hunters because of a taboo against killing monkeys. The hunting pressure on monkeys may have become heavier after the fall of the Shogun in 1868, and the establishment of a new government under the Emperor Meiji that set Japan on a course towards Westernization and modernization. The new government lifted prohibitions against farmers or commoners owning firearms, an inexpensive Japanese-made rifle became readily available, and the commercial hunting of monkeys expanded (Mito and Watanabe, 1999).

Monkeys were prized for their medicinal properties. In traditional medicine, animal parts possessed curative powers according to the similarity of each species to humans and the functions of their respective organs (Gose, 1967). Monkey gall-bladder and small intestine worked against various stomach and intestinal ailments, monkey fetuses eased the ailments of women and especially those following childbirth (Gose, 1967; Hirose, 1979; Mito and Watanabe, 1999). Monkey gall bladder also cured various ailments of horses. Perhaps the most famous monkey medicine was the *kuro-yaki*, black-roasted monkey. The *kuro-yaki* was made from a monkey head that was roasted black in an urn, then ground into a powder (Hirose, 1972). *Kuro-yaki* was said to be effective against ailments of the head, brain and mind (Gose, 1967; Hirose, 1979; Mito and Watanabe, 1999).

Most forms of traditional resource utilization practiced by the Japanese for centuries have now virtually ceased, especially since the end of the Second World War. Japanese agronomists refer to the 'fuel revolution' of the 1950s, when petroleum fuels replaced charcoal, firewood, and dried leaves as the commonly used fuels. Statistics on charcoal and firewood production reflect the dramatic decline in forest utilization (Fig. 12.1). Charcoal was a common fuel through the 1950s. In the post-war period, charcoal production peaked in 1957 and thereafter declined precipitously. Fuelwood production had already declined starting in 1950.

The fuel revolution was accompanied by a fertilizer revolution, as chemical fertilizers replaced compost. Traditional agricultural practices, and the extremely intensive forest utilization on which it depended, have all but disappeared today. Rural communities are often surrounded by *noyorin* that have been 'abandoned', in the parlance of agronomy. Swidden fields died out in the 1960s, largely because of the opportunities for nonagricultural work (Mizoguchi, 1988).

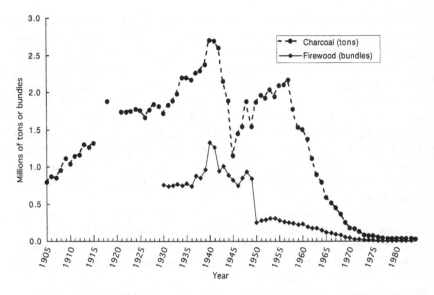

Figure 12.1. Charcoal and firewood production in Japan. Source: Historical Statistics of Japan (JSA, 1988).

The Japanese people also have ceased to look upon monkeys as a natural resource. The hunting regulations have protected monkeys from commercial and sport hunting since 1947. The use of monkey-based medicines ceased at about the same time (Hirose, 1979; Mito and Watanabe, 1999). Most Japanese today would be horrified by the thought of eating a monkey, although a market exists for wild boar meat and bear gallbladder. The urban public supports stronger conservation policies, and monkeys are popular icons in hip culture.

New threats to monkeys' ecological security

The disappearance of traditional forest resource utilization released monkeys from past threats. However, two major events form the background to a new and severe period of history for the monkeys of Japan: the expansion of conifer plantations and the increasing agricultural depredation by monkeys.

After the Second World War, Japan's forestry policy turned its attention to providing lumber for a rapidly growing economy. In addition to reforesting the mountains logged during the war, a policy of 'expanded afforestation' (*kakudai zorin*) replaced existing forests with commercially

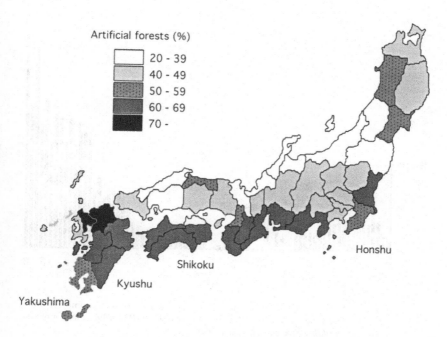

Figure 12.2. Percentages of artificial forests in the monkey habitat prefectures of Japan. Source: MAFF (1991).

important plantations of conifer (Manome and Maruyama, 1978). The conifer plantations were often large, single-species stands, most important-ly Japanese cedar (*Cryptomeria japonica*), hinoki (*Chamaecyparis obtusa*) and larch (*Larix kaempferi*), which occupied large areas within monkey habitats.

Large conifer plantations diminish the quality of monkey habitats. The Japanese monkeys depend for food primarily on the fruits and new leaves of broad-leaf trees (Agetsuma and Nakagawa, 1998; Hill, 1997). As of 1990, about half of Japan's forest area within the range of Japanese monkeys had been occupied by forests classified as 'artificial' forests, which are almost entirely plantation conifers (MAFF, 1992; Fig. 12.2). The proportion of artificial forests exceeded 60% for the whole islands of Shikoku and Kyushu, and exceeded 70% in the two prefectures of Saga and Fukuoka in northern Kyushu.

Monkeys are the third most damaging mammalian agricultural crop-raider in Japan today after deer and wild boar. As pests, an ever increasing number of wild monkeys are culled (Agetsuma, 1999; Watanabe, 2000). Under the Law for Wildlife Protection and Hunting, current regulations

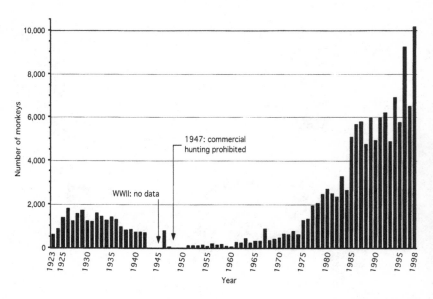

Figure 12.3. Japanese monkeys hunted or removed as nuisance animals from 1923 to 1998. Sources: 1923–1972, Watanabe (1978); 1972–1998, Environment Agency, *Wildlife Statistics*, for each year.

place monkeys on the list of nonhunted animals, thus prohibiting the commercial or sport hunting of Japanese monkeys. However, the same law that protects wildlife also provides for the removal of nuisance animals. The crop-raiding monkeys are removed with the nuisance animal permits issued by prefectural or local government in response to demands from farmers and agricultural authorities to control monkey populations that damage crops.

Japanese primatologists are alarmed by the rapid rise in the number of monkeys removed as nuisance animals (Fig. 12.3), especially since the early 1970s. The number of monkeys removed exceeded 2000 per year for the first time in 1978. Crop damage continued to worsen through the 1980s into the 1990s, and the number of monkeys culled finally exceeded 10 000 per year for the first time in 1998, the most recent year for which data are available (Environment Agency, 2000). The average number of monkeys removed for the ten-years from 1989 to 1998 is 6666 monkeys per year.

Can monkey populations survive such high rates of 'harvest' in habitats degraded (at least from the monkey point of view) by conifer plantations? This is an open question that has been debated by Japanese primatologists for many years. I join many colleagues in considering the combination of conifer plantations and excessive culling to be the major current threat to the ecological security of Japanese monkey populations (Agetsuma, 1999; Manome and Maruyama, 1978; Maruhashi, Sprague and Takezawa, 1992). What should primatologists do about these threats? Both forests and forestry policy are slow to change. Many primatologists have chosen to focus on the immediate crisis of crop damage by monkeys. Primatologists have plunged themselves into research on human–animal interactions in Japan to try and explain why crop damage by monkeys has become such a serious problem in recent years (Agetsuma, 1999; Knight, 1999; Mito and Watanabe, 1999; Wada, 1998; Watanabe, 2000).

Crop damage: an old problem in a new guise

When asked when the monkeys had started to raid their crops, farmers often said something like, 'about 10 years ago', which put the beginning of the problem in the mid-1970s at the time I started having conversations with farmers about crop damage. At first, I had taken for granted that habitat destruction must have become particularly severe during the 1970s, causing monkeys to spill into farms. On further thought, the farmers' claims seemed quite mysterious. Japanese farmers must have fought wild-life crop depredation for centuries.

Mito and Watanabe (1999) reviewed how farmers during the Tokugawa period protected their fields from wildlife as described in contemporary agricultural handbooks written and kept by local authorities and land-lords, or published for sale to farmers. Farmers set up scarecrows, fencing, and noisemakers, scattered wolf feces, or burned the hair of animals. Farmers often stood guard at their fields, building seasonal houses to remain close to more distant fields and swidden fields. Farmers used the help of dogs or hired hunters to shoot animals. Farmers were not permitted to own guns. The local authorities sent soldiers with guns, or loaned guns to landlords and local leaders who then used the guns themselves or employed hunters to guard fields.

Moriyama (1999) argued that traditional farmers understood and manipulated the ecology of predators to protect their community from undesirable wildlife. He stated that villagers even habituated Japanese wolves (*Canis lupus*) by feeding them to discourage deer, boar, and perhaps monkeys as well, from approaching farms and villages. The habituation of predators sometimes took the form of religious offerings of food to predators, including wolves, as messengers of the mountain gods. The origin of fox figurines guarding the Shinto shrines may have been real foxes drawn to villages by offerings to hunt mice. Gose (1967) related a story of a person habituating a wolf to guard a shiitake mushroom patch from monkeys, and pointed out that the wolf was the guardian animal of several shrines in the Yoshino Mountains of Kii Peninsula, where wolves, now extinct, were last seen in Japan in 1905.

In the present form of agricultural human–wildlife conflict, the farmer has become a hapless victim of wildlife depredation. The crops eaten by Japanese monkeys run the gamut from orchard fruits to vegetables and even grain crops including rice. The total area damaged by monkeys reported to the Japanese Ministry of Agriculture, Forestry, and Fisheries (MAFF) was about 6000 ha in 1994, and occurred in all prefectures where monkeys live (MAFF, 1996). The farm enterprises harmed by monkeys range from large commercial orchards to the elderly lady's solitary vegetable garden. The conflict between farmers and monkeys is characterized as a war (Knight, 1999), a war that farmers feel that they are losing.

In conversation, I have heard farmers express surprise, anger and helplessness about monkeys feeding with apparent impunity on their crops. Farmers claim that they attempted to defend their fields by various means, fences, dogs, or guarding, but no method worked well. Further discussion with farmers often reveals technical problems with how each method was implemented. For example, electric fences can be shorted out by an entangling vine. However, proximate technical issues do not answer the broader question that applies across Japan. Why are modern farms vulnerable to monkey attacks and why do modern farmers have such difficulty defending their fields?

Where have all the farmers gone?

Primatologists have proposed many hypotheses to explain why fields and orchards are so easily raided by monkeys today (Agetsuma, 1999; Knight, 1999; Manome and Maruyama, 1978; Mito and Watanabe, 1999; Wada, 1998; Watanabe, 2000). Here, I will focus on two complementary factors that form the land use and demographic background to rural Japan that cannot be ignored by any stakeholder in Japanese wildlife management today. First, the modern fields have lost the buffer zone of human activity that may have protected the permanent fields from damage by wildlife under traditional agriculture. Second, the agricultural population is declining, rapidly aging, and may no longer be able to mount the scale of defense against wildlife once mounted by farmers in the past. These two hypotheses are relevant for explaining what I would like to call the 'last 10 meters' problem, i.e. the problem of why monkeys are able to cross the last 10 m from a forest into fields and orchards.

I came to believe that the last 10 m problem was a critical issue after participating in surveys of agricultural areas suffering from monkey crop damage. Several years ago, colleagues pulled me out of the deeper forests of my original study site to join in one such survey. I was assigned 1 km² survey blocks where student helpers and I were to search for monkeys. In one block I was assigned to, the topographic map showed that much of the block was occupied by fields and houses. I set out with the students, expecting to spend a dull time seeing no monkeys and explaining repeatedly to passing people what we were supposed to be doing. I was wrong on both counts.

We found many monkeys, and hardly any people at all. I rushed to and fro through the survey block as the students radioed in that they had seen monkeys, and I found that the rural communities of modern Japan had become the farmland equivalent of ghost towns, at least on weekdays. In my naive vision of rural life, I had imagined farmers walking about their fields, children playing outside their homes, tractors driving up and down the now well-paved roads among the fields. In fact, I spent many hours without seeing any farmer. Most fields were deserted, though well tended. The few passing vehicles drove straight through without stopping. There were no people around the houses, and I found it difficult to tell whether anybody was at home. The few people I met in the survey block were carrying out their own tasks in their specific locations, including a farmer throwing stones to chase away monkeys eating his water melons. These people went out of sight the moment I walked around a corner. When they finished their task and went away, that location became completely deserted.

I was also astonished by the clarity of the division between the human world and the forest. As I stepped off the asphalt into the trees, I felt I was stepping into a world completely out of sight and out of mind of the human community, as evidenced in some places by the trash accumulated just a few steps into the forest. To be sure, signs of human activity dotted the forest. Some farmers had set up pipes drawing water to their fields from streams in the forest. Plots of conifer plantations or recent logging appeared along logging roads. These activities, however, probably would not take people into the forest very often, and I did not meet anybody in the forest. The monkeys traveled deep into the survey block through the forest belts along rivers. They skirted along the forest edge, invisible to humans except when they stepped out into the open, popping up on to the plateau to raid orchards.

264   *D.S. Sprague*

The agricultural buffer zone

According to the buffer zone hypothesis, wildlife crop damage has increased in Japan because the buffer zone of human activity no longer exists between fields and wildlife habitats. The hypothesis proposes that the foothills surrounding villages were once used, occupied, and defended by humans and their domesticated animals under traditional agriculture. The buffer zone may have been poor habitat for monkeys, often grasslands or pine woodlands with little for monkeys to eat. Feral dogs and hunters frightened away monkeys. Wild monkeys lived in the deep mountains.

In modern times, the foothills have been 'abandoned' by humans who no longer need woodlands or grasslands to support agriculture. Broad-leaf woodlands near villages recovered, even while expanding conifer plantations reduced the quality of forests as a whole for wildlife. Most dogs are now on leashes. Even when dogs are tied up near a field, the monkeys know exactly how closely they can approach the dogs without being bitten.

Unfortunately, no single study has succeeded in examining all aspects of the buffer zone hypothesis. The most critical part of the hypothesis, the changes in land use around villages, is perhaps the most difficult to study. This requires detailed, case-by-case analyses of how vegetation and human activity have changed in the past several decades for each of the many forms of rural land use once practiced in various parts of Japan.

Fortunately for primatologists, rural land use is an extremely active area of research among geographers in Japan (Kamada and Nakagoshi, 1996, 1997; Mizoguchi, 1996; Okahashi, 1996). Although most such studies do not directly address wildlife, they do provide models for how animal ecologists may be able to study changing wildlife habitats. Kamada and Nakagoshi (1997) present a model of landscape change in the Chugoku Mountains of Hiroshima Prefecture in western Honshu Island, based on their study of land registers, aerial photographs, and interviews with residents. They state that farmers maintained grasslands around paddy fields by mowing and burning. The grass was used for organic fertilizer in the paddy fields. The grassland area was double that of paddy fields. Large pastures maintained at sites far from settlements were shared by community members. After the 1960s, grassland area decreased drastically. Abandoned grasslands were replaced by conifer plantations or naturally regenerating, secondary oak forests. Their figure (Fig. 12.4) depicts forests replacing grasslands to surround paddy fields. If applicable to other parts of Japan, this model may explain why forests, and thus monkey habitats, have come to immediately surround fields.

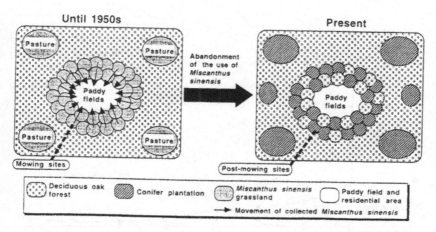

Figure 12.4. Model of changes in land use around Hiwa Township in the Chugoku Mountains of western Honshu Island, Japan, by Kamada and Nakagoshi (1997). Reproduced with the permission of Elsevier Science.

The declining agricultural population

The agricultural demographic transition is the second hypothesis explaining the vulnerability of modern farms to wildlife depredation (Wada, 1998). As in many nations that have undergone rapid economic development, the rural population has declined in Japan. The agricultural population has been in a slow, steady decline although farmland area has increased (Agetsuma, 1999). Consequently, the remaining farmers must carry out all agricultural activities, including defending fields from wildlife.

Statistics compiled by MAFF show not only that the Japanese agricultural population has declined but that the agricultural population is less focused on agriculture, and aging (MAFF, 1992). The total farm population of all persons that engaged in agriculture (*nogyo jyuji-sha jinko*) declined from 1960 to 1990 (Fig. 12.5). Furthermore, the population declined steadily for the occupational farm population (*nogyo shugyo jinko*) for whom agriculture was the only or main form of work. Within the occupational farm population, the proportion of men held steady at about 40%, but the total number of these men declined as a whole with the decline in the total farm population. Farmers were spending less time in the fields, as indicated by the declining number of persons spending more than 150 days working on farming within the occupational farm population. The only rising variable in this data was the number of persons aged 60 years and older. As the farm population declined in the same period, the 60 years

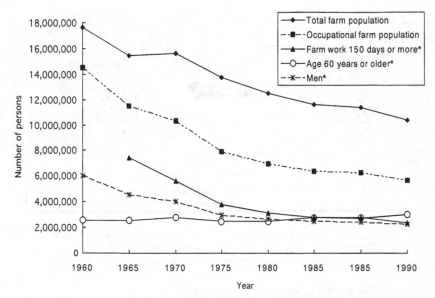

Figure 12.5. Changes in the farm population of Japan from 1960 to 1990.
Asterisks, within the occupational farm population. Source: MAFF (1992).

and older age group rose from 18% to 54% of the occupational farm population from 1960 to 1990.

Of the households engaged in agriculture, fewer are classified as full-time farming households in MAFF statistics. Figure 12.6 shows changes in the number of full-time and part-time farm households from 1960 to 1990 using 1960 as the baseline. The total number of farm households had been in steady decline from 1960 to 1990. More importantly, the number of full-time farm households declined precipitously by about 70% between 1960 and 1975, then continued a slow, steady decline. Part-time farm households increased initially, presumably because full-time farmers moved into the part-time category, but later joined full-time farmer households in decline. Part-time farmers are classified into Type 1 households, which gained the majority of household income from agriculture, and Type 2 households that gained the majority of household income from non-agricultural sources. The Type 1 part-time farm households declined below their 1960 level by 1970. The decline in Type 2 part-time farm households after 1980 implies that many households were giving up farming entirely. In 1960, the three household types had each accounted for approximately one third of farm households.

The agricultural demographic transition was just as severe in the hilly

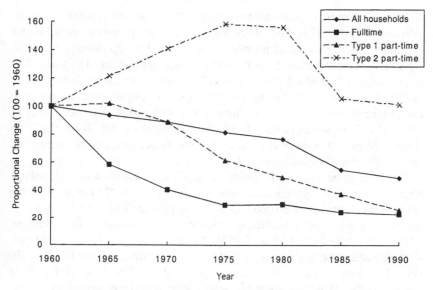

Figure 12.6. Changes in the proportions of full-time and part-time farm households in Japan as a proportion of the 1960 figure. Source: MAFF (1992).

and mountainous communities of Japan located near monkey habitats. MAFF uses a combination of population and geographic criteria to define the hilly and mountainous regions (*chusankan chiiki*) that are more distant from urban markets, located at higher altitudes with steeper topography, and have fewer fields and lower populations. The 1995 Agricultural Census (MAFF, 1998) provides age composition and the number of days engaged in agriculture for the hilly and mountainous regions of the prefectures (*Tofuken*) comprising the monkey habitat region of Japan. These show that 43% of the total farm population was aged 60 years and older, and only 19% had engaged in agriculture for 150 or more days. The proportion of full-time farm households was higher in the hilly and mountainous regions compared with the 15% of the flatland region, but still only 17%. Even in the hilly and mountainous regions, Type 1 part-time farm households accounted for 13%, and Type 2 part-time households 71% of farm households.

Ito, Okumura and Kawashima (1993) carried out one of the few studies linking the modern agricultural demographic transition to crop damage by monkeys in a specific village. They interviewed villagers in Natasho Village in Fukui Prefecture to determine when crop damage by monkeys had started at settlements in the village. They first point out that crop damage did not start until 15 years after the rapid expansion of conifer reforestation in the early 1970s, and proposed that 'human pressure' would better explain the actual onset of monkey crop damage. Crop damage by monkeys had started in 1982 and spread to all settlements by 1989. The population of Natasho Village had declined from its peak of 4855 persons in 1955 to 3041 in 1990. The farming and forestry population had declined even more drastically. In census data on the occupation of persons over 15 years old, farm workers had declined by more than 90% from the 1005 persons in 1970 to a mere 99 persons in 1990. Forestry workers declined by 70% in the same period from 259 to 78 persons.

Crop damage occurred earlier in small settlements where monkeys already lived in the surrounding forests. Ito *et al.* (1993) identified three small settlements where crop damage had started early in 1982. These three settlements had experienced the sharpest drops in population in the village, about 70% between 1960 and 1980 to only 15, 16, and 35 residents. Consequently the population densities of the three settlements in 1980 (based on non-forest area) had also declined to 32%, 22%, and 35% of those in 1960. However, even the settlements with large populations had experienced crop damage by monkeys by 1989. Ito *et al.* (1993) concluded that monkeys started to raid crops in settlements where 'human pressure' had diminished the most rapidly, but monkeys extended their forays to settlements with stronger 'human pressure' as they became accustomed to humans.

Ito *et al.* (1993) also provided suggestive details supporting the buffer zone hypothesis. They proposed that earlier logging and reforestation in the deeper forest plots may have encouraged monkeys to range closer to settlements because forestry records showed that forest plots closer to settlements contained much higher proportions of broadleaf trees than those located further away. Villagers had ceased to collect fodder for cattle with the decline in cattle husbandry after the 1960s. With the aging of the private forest owners, only 166 forest owners engaged in forestry work for more than 30 days in a year, and many said during interviews that they went to the forest much less often.

Discussion

Primatologists working in Japan are attempting to formulate a set of hypotheses to explain the dual question of why monkey populations can still be endangered even though monkeys are increasingly raiding crops (Agetsuma, 1999; Mito and Watanabe, 1999; Wada, 1998; Watanabe,

2000). According to these hypotheses, the duality stems from the transformations since the Second World War in how the Japanese people utilize forests and farmland (Mito and Watanabe, 1999; Wada, 1998). Modern forestry has replaced half of the monkey habitat region of Japan with conifer plantations although many broad-leaf forests may be recovering from the intensive utilization of the past. The commercial hunting of monkeys was prohibited in 1947 but nuisance animal removal has increased dramatically. Japan boasts the second largest national economy in the world, but its rural population is in decline. These contradictory events may be leading to degraded habitats for monkeys overall, even as some monkey populations find a habitat among the depopulated fields and orchards in the hilly and mountainous regions of Japan.

Although some monkey populations may have increased or migrated undisturbed after the ban on commercial hunting of monkeys, the contemporary history of agriculture must be factored in to explain the actual onset of crop damage. Academic researchers cannot ignore the ecological ramifications of the decline in rural communities. Ignoring the plight of farmers will lead to less public support for conservation, especially in the rural communities that must actually carry out the new conservation policies (Knight, 1999).

The historical evidence suggests two broad reasons why monkeys may have found it difficult to cross from their habitats into the human realm in the past. First, under traditional agriculture, monkeys may have had to cross a buffer zone of human activity perhaps hundreds or even thousands of meters wide to reach the fields near villages. The buffer zone is now gone, and villages in the hilly and mountainous regions today are often surrounded by a mosaic forest consisting of conifer plantations and maturing broad-leaf forests. Second, the entire rural population would have been devoted full-time to the primary industries of farming, forestry, and hunting.

Ironically, monkey populations are being threatened by the modernization and decline of agriculture and forestry. Monkey populations may be redistributing themselves over a landscape containing broad-leaf forests recovering after the fuel and fertilizer revolutions led farm communities to cease maintaining forests and grasslands for local agricultural use, or for the extraction of the products of broadleaf trees, especially charcoal. Today, forestry utilizes hills and mountains for intensive commercial forestry of conifers. Forestry for products of broad-leaf trees has diminished drastically. Most importantly, forestry for local agricultural use has virtually ceased, creating a mosaic forest landscape near fields and villages where some patches are preferred monkey habitat.

270   D.S. Sprague

The twenty-first century farm community must devise new means to defend fields (Watanabe, 2000). The farm community must find a way to mark the human presence in and around villages when the members of most households hold non-farm jobs, a majority of farmers are elderly, and the population density as a whole is a fraction of that of the past.

Acknowledgements

I thank all my colleagues who provided many insights and references on wildlife conservation and historical ecology in Japan.

References

Agetsuma, N. (1999). Primatology and the protection and management of wildlife. In *For People Who Wish to Study Primatology* [Reichorui-gaku wo Manabu Hito no Tame Ni], ed. T. Nishida and S. Uehara, pp. 300–26. Kyoto: Sekaishiso-sha.

Agetsuma, N. and Nakagawa, H. (1998). Effects of habitat differences on feeding behaviors of Japanese monkeys: comparison between Yakushima and Kinkazan. *Primates* **39**, 275–89.

Dobson, M. and Kawamura, Y. (1998). Origin of the Japanese mammal fauna: allocation of extant species to historically-based categories. *Quaternary Research* [Daiyonki Kenkyu] **37**, 385–95.

Dore, R. (1978). *Shinohata: A Portrait of a Japanese Village*. London: Allen Lane.

Environment Agency (2000). *1998 Wildlife Statistics* [Chojyu Kankei Tokei]. Tokyo: Nature Protection Bureau, Environment Agency.

Gose, K. (1967). Livelihoods and nature in Oku Yoshino. In *Natural History: Ecological Studies* [Shizen: Seitaigakuteki Kenkyu], ed. M. Morishita and T. Kira, pp. 249–83. Tokyo: Chuokoron-sha.

Hill, D.A. (1997). Seasonal variation in the feeding behavior and diet of Japanese macaques (*Macaca fuscata yakui*) in lowland forest of Yakushima. *American Journal of Primatology* **43**, 305–22.

Hirose, S. (1979). *Monkey* [Saru]. Tokyo: Hosei University Press.

Ito, E., Okumura, N., and Kawashima, M. (1993). Agro-economic changes in a mountain area causing crop damage by *Macaca fuscata*. *Bulletin of the Gifu University Faculty of Agriculture* [Gifu Daigaku Nogakubu Kenkyu Hokoku] **58**, 17–26.

JSA (1988). *Historical Statistics of Japan,* Vol. 2. Tokyo: Japan Statistical Association.

Kamada, M. and Nakagoshi, N. (1996). Landscape structure and the disturbance regime at three rural regions in Hiroshima Prefecture, Japan. *Landscape Ecology* **11**, 15–25.

Kamada, M. and Nakagoshi, N. (1997). Influence of cultural factors on landscapes of mountainous farm villages in western Japan. *Landscape and Urban Planning* **37**, 85–90.

Knight, J. (1999). Monkeys on the move: the natural symbolism of people-macaque conflict in Japan. *Journal of Asian Studies* **58**, 622, 647.

MAFF (1991). *1990 World Agriculture-Forestry Census Results Summary* [Sekai Noringyo Sensasu Kekka Gaiyo], Vol. 5, *Forestry Region Study*. Tokyo: Statistics and Information Department, Ministry of Agriculture, Forestry and Fisheries.

MAFF (1992). *Agricultural Census Cumulative Statistics* [Nogyo Sensasu Ruinen Tokei-sho]. Tokyo: Statistics and Information Department, Ministry of Agriculture, Forestry and Fisheries.

MAFF (1996). *Plant Protection Annual* [Shokubutsu Boeki Nenpo]. Tokyo: Plant Protection Division, Agricultural Production Bureau, Ministry of Agriculture, Forestry and Fisheries.

MAFF (1998). *1995 Agricultural Census* [Nogyo Sensasu], Vol. 9, Part 2, *Statistics by Agricultural Region*. Tokyo: Statistics and Information Department, Ministry of Agriculture, Forestry and Fisheries.

Manome, H. and Maruyama, N. (1978). Agriculture, forestry and the protection of Japanese monkeys. *Japanese Monkey* [Nihonzaru] **4**, 112–21.

Maruhashi, T., Sprague, D. and Takezawa, T. (1992). What future for Japanese monkeys? *Asian Primates* **2**, 1–4.

McKean, M. (1982). The Japanese experience with scarcity: management of traditional common lands. *Environmental Review* **6**, 63–88.

Mito, Y. (1992). Why is Japanese monkey distribution so limited in the northern Tohoku Region? *Biological Science* [Seibutsu Kagaku] **44**, 141–58.

Mito, Y. and Watanabe, K. (1999). *Social History of People and Monkeys* [Hito to Saru no Shakai-shi]. Tokyo: Tokai University Press.

Mizoguchi, T. (1988). Slash-and-burn field cultivation in pre-modern Japan: with special reference to Shirakawa-go. *Geographical Review of Japan* **62**, 14–34.

Mizoguchi, T. (1996). Studies in the historical geography of Japan, 1988–1995. *Geographical Review of Japan* **69**, 14–34.

Moriyama, H. (1999). Ethnological aspects of agroecosystems. *Environmental Information Science* [Kankyo Joho Kagaku] **28**, 28–31.

Okahashi, H. (1996). Development of mountain village studies in postwar Japan: depopulation, peripheralization and village renaissance. *Geographical Review of Japan* **69**, 60–9.

Okutomi, K. (1998). Conservation of secondary forests. In *Handbook of Nature Conservation* [Shizen Hogo Handobukku], ed. M. Numata, pp. 392–417. Tokyo: Asakura Shoten.

Sasaki, K. (1972). *Swidden Agriculture in Japan* [Nihon no Yakihata]. Tokyo: Kokon Shoin.

Sprague, D., Goto, T. and Moriyama, H. (2000). GIS analysis using the rapid survey map of traditional agricultural land use in the early Meiji Era. *Journal of the Japanese Institute of Landscape Architecture* [Randosukeipu Kenkyu] **63**, 771–4.

Totman, C. (1989). *Green Archipelago: Forestry in Preindustrial Japan.* Berkeley: University of California Press.

Tsuchida, K. (1998). Conservation of semi-natural grassland. In *Handbook of Nature Conservation* [Shizen Hogo Handobukku], ed. M. Numata, pp. 432–76. Tokyo: Asakura Shoten.

Wada, K. (1998). *Getting Along With Monkeys: Feeding, Habituation and Monkey Damage* [Saru to Tsukiau: Ezuke to Engai]. Nagano: Shinano Mainichi Newspaper Publications.

Watanabe, K. (2000). *Crop Damage by Japanese Macaques and Wildlife Management* [Nihonzaru ni yoru Nosakumotsu Higai to Hogokanri]. Tokyo: Tokai University Press.

Watanabe, R. (1978). On the capture of Japanese monkeys as recorded by the hunting statistics and wildlife statistics. *Japanese Monkey* [Nihonzaru] **4**, 91–5.

Yonebayashi, N. (1998). Pollen analysis and nature conservation. In *Handbook of Nature Conservation* [Shizen Hogo Handobukku], ed. M. Numata, pp. 309–16. Tokyo: Asakura Shoten.

Part 4
Government actions, local economies and nonhuman primates

National governments often become involved in controlling access to wildlife sanctuaries and reserves. In the first chapter in this section (Chapter 13), Ardith Eudey discusses conservation projects in Thailand and other nearby countries where local people are evicted or hired to act as guides. Eudey was a graduate student when she began her research in the Huai Kha Khaeng Wildlife Sanctuary in the Uthaithani province in Thailand. She found herself at a center of the discord between the ethnic hill folk and the sanctuary's officers from Bangkok. True to her identity as an anthropologist, she acted as a buffer between the government of Thailand and their desire to move the Hmong from the sanctuary to another village site. As vice-chair for Asia of the IUCN Species Survival Commission's Primate Specialist Group, she also expresses her concern with the situation in both Myanmar and Vietnam. She argues that conservationists should be sensitive to the needs of both the indigenous people and the local wildlife. She also presents the case that the best park rangers are often the local people because they are familiar with the forest. She argues, moreover, that it is more humane to employ indigenous people in their native areas than to move them to a new area that may be less suited to their needs. She maintains that in order to have successful conservation programs there must be political stability that is lacking in many Asian countries. We concur with Eudey on these points.

The interactions between nonhuman and human primates are sometimes of an economic nature. The use of nonhuman primates may supplement household or village incomes. Sponsel, Ruttanadakul and Natadecha-Sponsel (Chapter 14) illustrate this with their discussion of the use of macaques to harvest coconuts and other tree crops in Thailand. In this chapter the authors describe the process and context of training macaques to pick coconuts. The family income from the use of macaques to harvest coconuts is substantial and the macaques are well treated by their handlers, although they are chained to trees when not working. Local people consider that it is safer for macaques than for people to climb the tall trees. Macaques are occasionally injured or even die from falling out of the trees or from snakebites. Sponsel *et al.*, hypothesize that the survival of macaques in the face of the destruction of their natural habitat may depend, in part, on their cultural and economic value as tree crop harvesters.

Sponsel *et al.*, make reference to the concept of the macaque as a 'weed species' based on the 1989 article by Richard, Goldstein and Dewar. One of us (LDW) finds the use of the term 'weed species' to characterize macaques problematic, the other (AF) does not. This points out the complexities, and limitations, of language and labels when attempting to discuss human and nonhuman cultural and economic interconnections. Because of the relative newness of the field of ethnoprimatology, terminology from cultural anthropology, zoology, resource management, wildlife management, and conservation paradigms are incorporated by the various practitioners. The term 'weed' has a scientific definition that is not derogatory and professional primatologists understand the scientific use of the term. However, this term can be pejorative when used in everyday speech. If misunderstood, the term weed can, incorrectly, carry the implication that macaques are at fault for the problems between macaques and people. Actually, these problems frequently begin when people modify the habitats in which monkeys live and/or when monkeys begin to exploit humans' gardens as primary food sources (see Part 3 of this book for specific examples of habitat conflict between human and nonhuman primates). While the terms mutualism or commensalism are alternate possibilities, they are also embedded with their own paradigmatic values and contexts. By examining the linkage between human economies and nonhuman primates Sponsel *et al.*, clearly illustrate the importance of the ethnoprimatological approach in understanding these cross-species interconnections.

Governmental agencies sometimes determine that some animals are undesirable and set about to remove them from their habitat. The final chapter by Linda D. Wolfe (Chapter 15) describes two populations of rhesus monkeys and the interaction between humans and nonhuman primates. The first population to be discussed is located in Jaipur, India. As the forest around the walled city of historic Jaipur has been destroyed for new housing projects, the monkeys have become more dependent on humans for a substantial part of their food. The link between the rhesus macaques and people of Jaipur is the Ramayana, the Hindu legend, and its monkey hero, Hanuman. Wolfe argues that in Jaipur people associate Hanuman with the rhesus monkey and not the gray langur. She suggests that people living in different places and at different times may associate

Hanuman with a different cercopithecoid species. The rhesus macaques of Silver Springs, Florida, are the other population discussed. The Florida Game and Fresh Water Fish Commission (FGFWFC) tried unsuccessfully to remove the monkeys from the Silver Springs area. Wolfe argues that the FGFWFC was unsuccessful because they failed to take into account the cultural beliefs and desires of the local people and tourists to see the monkeys well treated. In general, Wolfe argues that, for conservation programs to be successful, the local people, primatologists, cultural anthropologists, and conservationists need to be involved.

References

Richard, A.F., Goldstein, S.J. and Dewar, R.E. (1989). Weed Macaques: The Evolutionary Implications of Macaque Feeding Ecology. *International Journal of Primatology* **10**(6), 569–94.

13 *The primatologist as minority advocate*

ARDITH A. EUDEY

Introduction

As a graduate student preparing to enter the field for the first time, I was admonished by a senior faculty member to remember that I was an anthropologist as well as a primatologist. His words struck a very responsive chord because I never forgot them during my initial fieldwork on primates, or subsequently. As a consequence, the conceptual framework of this book appeared to offer me an opportunity to incorporate my own experience into my contribution, almost as if I were utilizing myself as an anthropological informant. There may even come a stage in one's career when the weight of professional experience begins to require more personal expression: Oates' (1999) examination of the loss of West African rainforest ecosystems is a recent example.

My career as a 'monkey watcher' may have begun somewhat differently from that of the other contributors to this volume. As a graduate student, I proposed to examine the process of speciation (in mammals) by looking at behaviors such as ecological preference that were thought to keep closely related sympatric species – in this case macaque monkeys (*Macaca* spp.) in west-central Thailand – reproductively isolated from one another (Eudey, 1980, 1981). At that time, although not subsequently, actual elucidation of an aspect of the monkeys' behavior was not a motivating factor. At this time, in my capacity as vice-chair for Asia of the IUCN Species Survival Commission's Primate Specialist Group, the ecological requirements of nonhuman primates throughout Asia have become my concern as well as the effects of human exploitation and habitat encroachment on other primates (Eudey, 1987, 1999).

Thailand

From the moment that I entered the field in Thailand in 1973, I found myself confronted by the interplay between minority and majority peoples.

277

I interpreted and responded to the complex situations in which I found myself a player largely in terms of, I believe, my anthropological training. Huai Kha Khaeng Wildlife Sanctuary, where I have continued to conduct field research in what was then the relatively remote Uthaithani province, had been gazetted only one year earlier; and the sanctuary's officers from Bangkok, none of whom actually had been trained in wildlife biology or management, undervalued the knowledge and skills of the local people, primarily ethnic Thais and Karen hill folk, who had been recruited to be workmen. Karen formerly had inhabited many areas included within the sanctuary (Eudey, 1986) and to retain their self-respect would resort to such tactics as falling back while leading a forest reconnaissance so that the sanctuary chief or another officer suddenly became leader and had to stop out of ignorance of the environment. The success or progress of my research was largely dependent on the assistance of the hill people, and the more I worked with them the more I opened myself up to official discrimination, which frequently manifested itself as indifference. Once during the hot, dry season when the temperature hovered around 34°C, I found myself having to walk more than 20 km both to the highway and back, mostly through dry dipterocarp forest, in order to get supplies, although motorcycles were available that could have provided me with transportation. I had anticipated such a response in a society in which discrimination against women was, at that time at least, highly institutionalized.

By the end of the 1970s and early 1980s, younger, better trained officers were serving in the sanctuary and exhibiting much less social distance from the local people. But these junior officers likewise were subject to discrimination and found themselves the 'hands and feet' of the sanctuary while the 'heads' remained in Bangkok. Although social stratification may explain in part the inherent disdain exhibited in the situations described above, judgments based upon stereotypes rather than first-hand interaction or observation may underlie some of the discrimination shown against both rural Thais and cultural minorities. A former director of the Royal Forest Department Herbarium once commented to me, for example, on the ingenuity of the Hmong hill folk who had worked with him in collecting botanical specimens.

Although I might have been able to offer only moral encouragement in the instances mentioned above, I made a conscious effort to intervene on behalf of Hmong when they were confronted by involuntary relocation from Huai Kha Khaeng Wildlife Sanctuary (Eudey, 1988, 1989, 1991). The decision to evict the Hmong had been reached by the Royal Forest Department and the Third Army and was subsequently endorsed by a World Bank Global Environment Facility (GEF) analyst. Independent support for the relocation came from an American wildlife biologist who had only recently initiated a field study in the area. He represented an international conservation organization from which I had received financial support. He publicly expressed the sentiment that in any conflict between people and wildlife the animals must take priority and that there was no room for compromise. His commentary on conservation in Thailand, including the role of the Royal Forest Department, is contained in Rabinowitz (1991).

The highlanders in question, some 176 inhabitants of Yoo Yee village, were part of about 3000 Hmong living in ten villages in Uthaithani and Tak provinces in western Thailand, less than 300 km northwest of Bangkok. These Hmong were virtually unstudied, and their remoteness, strategic proximity to Myanmar (Burma) and cultivation of opium as an easily transportable cash crop had made them especially vulnerable to political action (Eudey, 1986, 1988, 1989). Commitment to the growing of opium poppy (*Papaver somniferum*) may have varied from one village to another: unpublished data collected by me in three villages in 1986 suggest that the amount of opium produced in a specific village, given the appropriate climatic conditions for poppy cultivation, may have been influenced by the extent of contact with Karen and ethnic Thai traffickers, the latter group not inconceivably including some local government authorities.

In 1982, the same year that the Royal Forest Department began to express increasing concern over the presence of cultural minorities in protected areas and the apparent threat of their shifting cultivation to the watershed and forest habitat, I became the first non-Thai to contact the people of Yoo Yee, probably the most traditional settlement in the entire region (Fig. 13.1). As such, I was able to refute claims – and did so at the invitation of the English-language press in Bangkok – that these self-sufficient Hmong farmers were 'illegal squatters' who had encroached on the forest to poach animals and raise opium. With the exception of two households, all heads of household (17 of 19, or 89 per cent) had established residence in the village 4–5 years before the sanctuary was gazetted in 1972. During the initial confrontation with the Yoo Yee villagers, military and forestry personnel expressed surprise at the diversity of the Hmong's agriculture and animal husbandry. Incidental hunting of primates and other wildlife for food, including great hornbills (*Buceros bicornis*), had not resulted in their extirpation. Technically all land in Thailand belongs to the state, however; and areas such as protected forests and wildlife sanctuaries or even hill land legally cannot be owned or occupied by anyone, including tribal minorities (Ratanakorn, 1978). An account of the confrontation on 14 April 1986 may be found in Eudey (1989).

Figure 13.1. First encounter of Ardith Eudey, accompanied by Forest
Department workers, with the Hmong of YooYeeVillage in 1982.

Although I was unable to stop the eviction of the Yoo Yee villagers, I
was able to discourage plans to hold them in a detention or indoctrination
center before their relocation; to alert personnel of the United Nations
Fund for Drug Abuse Control (UNFDAC), a potential source of funding
for the resettlement, that the relocation was involuntary and therefore in
violation of the UN charter; and to prepare the Hmong for resettlement in
a lowland area with inadequate water and exhausted soil to the north in
Tak province, which they voluntarily had left about 20 years earlier at the
request of the Thai government because of insurgency in the area (Eudey,
1988, 1989).

In 1990 during my first visit to Village 11, the Yoo Yee resettlement site
accessible from kilometer 53 on the Mae Sot–Umphang road in Tak
province, I was able to confirm that a potable source of water had not been
made available for three years, that the electricity system was unfinished
and that the soil remained arid and infertile, as had been reported earlier
by the cultural geographer J. McKinnon *(in litt.* 1989). I noted the failure
of the volunteer Thai teacher to appear at the schoolroom even though all
the children had been attempting for almost an hour to sweep away its
dusting of red earth with twig brooms and how the hard-working women,
especially, had aged, almost to be unrecognizable. Everyone old enough to
remember me asked, without exception, if I had been back to Yoo Yee, the

place foremost in their thoughts. During my second visit in 1994, there were significant improvements in the infrastructure and the men were beginning to think seriously about how they might be able to return to the highlands. Six years later, I have not abandoned my commitment to convince the Royal Forest Department to establish a pilot project using these Hmong to combat the ongoing poaching in Huai Kha Khaeng Wildlife Sanctuary.

Vietnam

Differential access to funding and technical knowledge also may reduce researchers and resource managers to minority status within their own country. Vietnam is a case in point. A consortium originally composed primarily of zoos has invested considerable sums of money in a specialized facility – a rescue center for endangered primates (EPRC) – in the first national park in the country established during the American War in 1965 (Baker, 1999). Volume 4 of the *EPRC-Newsletter*, published in 1999, reviews the history of the rescue center. Professional Vietnamese have been excluded to a large part from participation in the center, thereby losing an opportunity for training, and from contributing to the formulation of policy for the rescue program. At times there were even suspicions that foreign personnel were ignoring Vietnamese law in the acquisition (supposedly the confiscation) of endangered primates by offering compensation, a practice that is not unknown to the Vietnamese themselves (Anon, 1993; Baker, 2000). This is not to say that the rescue center has not been effective; it is impossible to enforce protective legislation without appropriate facilities to receive confiscated primates or other wildlife and the center has been able to maintain and breed rare colobine monkeys (Baker, 1999). The Vietnamese are rumored to admire the style of the foreign personnel although questioning the substance of the program in what would appear to be a 'love–hate' relationship. But the opportunity to empower Vietnamese through collaborative efforts, a goal that I have set throughout Asia for the Primate Specialist Group (Eudey, 1996/1997), would appear to have been lost.

Elsewhere in Vietnam, a different international project, but one largely funded by the same sources as the above, has been more successful as a collaborative effort, on both ministerial and institutional levels, to protect a critically endangered endemic species, the Tonkin snub-nose monkey (*Rhinopithecus avunculus*). A small number of local people have been recruited to ranger patrols (*Tuan Rung*) to enforce hunting bans decreed by district and provincial People's Committees and to monitor monkey populations, and educational campaigns have begun to inform local people of the regulations that protect primates and their habitats.

A level of insensitivity comparable to that of the first example continues to manifest itself, sometimes very subtly, among zoo professionals in the North. Recently a zoo worker, in conjunction with an international conference on primates held in the United States in 2000, remarked that participation by Southern primatologists would give them an opportunity to learn what Northern zoos are doing for conservation; one is led to believe that the opportunity for the zoos to learn about the conservation needs and objectives of the habitat countries in question was overlooked. Another zoo professional sought to change the wording and intent of a resolution on apes proposed by a senior Asian primatologist, and approved by acclamation at the conference; he appears to have been motivated to make the resolution more expedient from his perspective. Even a specific attempt (Koontz, 1995) to establish ethics for zoo biologists (with reference to the acquisition of wild animals) fails to recognize the potential to strengthen and enrich conservation action in habitat countries through collaborative efforts (Eudey, 1995).

Myanmar

African Primates (IUCN, 1996), the revised edition of the *Action Plan for African Primate Conservation* (Oates, 1986), recognizes the extent to which civil war or similar political instability has prevented the completion of conservation objectives set forth in the original action plan. Politically sensitive countries and regions in Asia, literally 'political hot spots', may have diminished since publication of the *Action Plan for Asian Primate Conservation* (Eudey, 1987). Duckworth, Salter and Khounboline (1999) summarise the results of recent fieldwork, for example, sponsored by the US-based Wildlife Conservation Society (WCS) in the formerly politically closed Lao Peoples' Democratic Republic. But countries such as Afghanistan, where the westernmost population of the rhesus macaque (*Macaca mulatta*) has probably been extirpated, and Burma (Myanmar), where as many as 12 nonhuman primate species were recorded earlier (Eudey, 1987), remain virtually inaccessible to scientific study.

In their enthusiasm to initiate or support environmental programs or conservation action in politically closed countries, it would be possible for foreign aid agencies and conservation organizations alike to lose sight of the fact that their help could be construed as endorsement of a repressive regime, such as the State Law and Order Restoration Council (SLORC), the ruling military junta in the country formerly known as Burma and now known as Myanmar. Perhaps such enthusiasm might be motivated in part by the competitive desire to be the first organization to establish a 'sphere of influence', but even ignorance of sociopolitical conditions within a country could well be a contributing factor. In the case of Myanmar there has been considerable caution, however. This may be due in no small part to the restraint of Aung Sun Suu Ky, daughter of Burma's independence hero Aung San as well as the 1991 Nobel Peace Prize laureate and Chairman of Burma's National League for Democracy Party. Although the League won approximately 82 per cent of the vote in the 1990 elections, SLORC never has allowed the opposition to take their seats in Parliament and Suu Ky has spent more than a decade under surveillance or some form of imprisonment, including six years of house arrest from 1989 to 1995 (Wright, 1998; Anon., 1991, 1995).

In April 1997, US President Bill Clinton banned new investments in Myanmar under mounting pressure from Congress and human rights advocates; this was accomplished in spite of lobbying efforts by US firms, rallied by California-based Unocal Corp., against the imposition of further economic sanctions. The United States already had cut off bilateral aid to Myanmar and banned the issue of visas for the ruling SLORC (see Anon, 1997). On 19 June 2000, the US Supreme Court ruled 9–0 that cities and states may not boycott companies that do business with the regime in Myanmar, stating that only the president and Congress have the authority to set foreign policy for the nation (Savage, 2000). The Supreme Court had agreed to review a Federal Appeals Court ruling declaring unconstitutional a 1996 Massachusetts state law banning agencies from doing business with companies involved in Myanmar. Massachusetts decided to pursue the appeal based on strong backing from more than 20 other states and localities that had imposed similar laws because of Myanmar's human rights violations. The Massachusetts selective purchasing law was challenged by the Washington-based National Foreign Trade Council, a coalition of about 580 multinational firms (Savage, 2000), while the European Union (EU) and Japan also protested the Massachusetts law and joined in the legal action (Anon, 2000). Activists have pledged that they will initiate new legal efforts to stop foreign investment from flowing to SLORC (Savage, 2000).

Although the potential danger of conducting field research in Myanmar ultimately may be responsible for the lack of current information on the distribution and conservation status of primates within its borders (R.A. Mittermeier, personal communication, 2000), it is noteworthy that the

1999 'Environmental Grassroots Hero' and recipient of the US$125 000 Goldman Environmental Prize for Asia, Ka Hsaw Wa, is a member of the Burmese Karen minority (Anon, 1999). Among his other activities, he co-founded Earth Rights International (ERI) in 1995, with the express purpose of exposing and raising awareness of the inextricable links between human rights and environmentalism in Myanmar and elsewhere. He reportedly has documented thousand of cases of forced labor, execution, rape and confiscation of property carried out by the military in support of the US$1.2 billion Yadana pipeline project to supply natural gas to Thailand. This is the largest foreign investment in Myanmar and has been built by a consortium of petroleum companies including transnational corporations Unocal Corp. [USA] and Total [France] (Anon., 1998, 2000). The pipeline traverses the Tenasserim rainforest, which is inhabited by diverse ethnic peoples and contains the habitat of such rare and endangered species as tiger (*Panthera tigris*) and Asian elephant (*Elephas maximus*).

That the status of women and children in society, specifically in Southern countries but also elsewhere, may be used as an indicator or predictor of a country's commitment to conservation is an idea that I have been exploring in informal contexts for several years. An historical connection between animal protection and child protection is well established in Northern countries. The galvanizing factor in the movement against cruelty to children in the United States, the case of Mary Ellen Wilson in 1874, was brought to court by the counsel of the president of the New York Society for the Prevention of Cruelty to Animals, in an atmosphere in which a publicly perceived link between the defenselessness of animals and that of children already existed (Costin, 1991; Shelman and Lazoritz, 1999). Following this logic, the widespread human rights abuses in Myanmar would call into question any commitment to conservation on the part of SLORC.

Although in a country such as Myanmar there may be considerable evidence of habitat degradation, illegal logging and wildlife exploitation, activities in which neighboring countries have also been implicated, it appears to present one of those difficult situations in which an absence of outside assistance may be necessary to avoid undermining efforts of democratically elected officials to take control of the government.

Conclusions

A 1992 workshop sponsored by the American Zoo and Aquarium Association (AZA) brought together zoo professionals, animal activists,

conservation biologists and philosophers in an attempt, not necessarily
satisfactory, to explore ethical issues associated with captive breeding
programs in broader conservation and social contexts (Norton *et al.*,
1995). In the United States, growing concern with ethics, or the 'reasoning
behind questions of morality', as a consequence of rapid advances in
technology, science and medicine, is resulting in an increasing number of
trained ethicists. Some universities are establishing specialized ethical centers and others are requiring ethics as part of degree programs in such
diverse subjects as engineering, healthcare, business, journalism, law and,
significantly, environmental studies (Rourke, 2000). It would be tempting
to postulate that training as an anthropologist endows primatologists (and
conservationists) with a sense of morality. In reality, individual life history
is probably as important, if not more so. Indeed, I have heard a colleague
apologize for his degree in anthropology as an easy route to become a
biologist (although the majority of my colleagues might feel the opposite is
more accurate). However, a set of personal traits may preselect an individual to study anthropology.

On reflection, I think it would be fair to say that in no way can my
cross-cultural sensitivity or advocacy be considered comparable to that of
the local people with whom I, or any other primatologist for that matter,
may work in the field. We come and go, seemingly at will, seeking answers
to what frequently must appear to be esoteric if not foolish questions and
are usually received with at least toleration. How could I help but look
with warmth at the Thai master carpenter turned forest ranger who, with
great pride, explained to a woman at the local market that I was so fearless
of the wild animals that I was able to go into the forest alone and unarmed;
or the sure-footed Hmong woman who, looking at the telltale signs of the
sticky and slippery soil of Yoo Yee on my trousers, asked me with
compassion if the people in my country traveled only by airplane rather
than walked. Perhaps they are the true minority advocates.

Acknowledgments

Research on the Hmong and primates during 1986 was made possible by a
grant from the National Geographic Society. From 1977 to 1983, field
work on *Macaca* spp. and conservation in Huai Kha Khaeng and Thung
Yai Naresuan Wildlife Sanctuaries was supported by the Wildlife Conservation Society, New York Zoological Society. I am very grateful to the
Wildlife Conservation Division of the Royal Forest Department for
having sponsored my research in the sanctuaries since 1973, and I am

permanently indebted to the workmen and other personnel of the Khao
Nang Rum Wildlife Research Station who assisted me in the field and to
the people of Yoo Yee village for their hospitality and trust.

References

Anon. (1991). Developments of interest. *Asian Primates* 1(3), 8.
Anon. (1993). Meeting on endangered primates in Vietnam. *Asian Primates* 2(3,4),
    1–2.
Anon. (1995). Developments of interest. *Asian Primates* 4(4), 19.
Anon. (1997). Developments of interest. *Asian Primates* 6(1,2), 33–4.
Anon. (1998). Developments of interest. *Asian Primates* 6(3,4), 29.
Anon. (1999). The 1999 Goldman prize winners. *Ouroboros* 10(1), 1.
Anon. (2000). Awards: Goldman Environmental Prize. *Asian Primates* 7(1,2),
    21–2.
Baker, L.R. (1999). The plight of Vietnam's primates. *IPPL News* 26(3), 15–21.
Baker, L.R. (2000). Endangered Primate Rescue Center update. Cat Ba monkey
    tragedy: 11% population drop in one day. *IPPL News* 27(1), 25–6.
Costin, L.B. (1991). Unraveling the Mary Ellen legend: Origins of the "cruelty"
    movement. *Social Science Review* 65, 203–23.
Duckworth, J.W., Salter, R.E. and Khounboline, K. (compilers) (1999). *Wildlife in
    LAO PDR: 1999 Status Report*. Vientiane: IUCN-The World Conservation
    Union; Wildlife Conservation Society; Centre for Protected Areas and Water-
    shed Management, Lao PDR.
Eudey, A.A. (1980). Pleistocene glacial phenomena and the evolution of Asian
    macaques. In *The Macaques: Studies in Ecology, Behavior and Evolution*, ed.
    D.G. Lindburg, pp. 52–83. New York: Van Nostrand Reinhold.
Eudey, A.A. (1981). Morphological and ecological characters in sympatric popula-
    tions of *Macaca* in the Dawna Range. In *Primate Evolutionary Biology*, ed.
    R.S. Corruccini and A.B. Chiarelli, pp. 44–50. Berlin: Springer-Verlag.
Eudey, A.A. (1986). Hill tribe peoples and primate conservation in Thailand: A
    preliminary assessment of the problem of reconciling shifting cultivation with
    conservation objectives. In *Primate Ecology and Conservation*, ed. J.G. Else
    and P.G. Lee, pp. 237–48. Cambridge University Press.
Eudey, A.A. (compiler) (1987). *Action Plan for Asian Primate Conservation: 1987–
    91*. Gland, Switzerland: IUCN.
Eudey, A.A. (1988). Hmong relocated in northern Thailand. *Cultural Survival
    Quarterly* 12(1), 79–82.
Eudey, A.A. (1989). 14 April 1986: Eviction orders to the Hmong of Huai Yew Yee
    village, Huai Kha Khaeng Wildlife Sanctuary, Thailand. In *Hill Tribes Today*,
    ed. J. McKinnon and B. Vienne, pp. 249–58. Bangkok: White Lotus and
    Orstom.
Eudey, A.A. (1991). Conservation's moral dilemma: A case study from Thailand.
    In *Primatology Today*, ed. A. Ehara, *et al.*, pp. 59–62. Amsterdam: Elsevier

Science Publishers (Biomedical Division).
Eudey, A. (1995). Counterpoint: To procure or not to procure. In *Ethics on the Ark: Zoos, Animal Welfare, and Wildlife Conservation*, ed. B.G. Norton *et al.*, pp. 146–152. Washington, D.C.: Smithsonian Institution Press.
Eudey, A.A. (1996/1997). Asian primate conservation – The species and the IUCN/SSC Primate Specialist Group network. *Primate Conservation* **17**, 101–10.
Eudey, A.A. (1999). Asian primate conservation: my perspective. In *The Nonhuman Primates*, ed. P. Dolhinow and A. Fuentes, pp. 151–8. Mountain View, CA: Mayfield Publishing Co.
IUCN (1996). *African Primates: Status Survey and Conservation Action Plan.* (Revised edn.) Gland, Switzerland: IUCN.
Koontz, F. (1995). Wild animal acquisition ethics for zoo biologists. In *Ethics on the Ark: Zoos, Animal Welfare, and Wildlife Conservation*, ed. B.G. Norton *et al.*, pp. 127–45. Washington, DC: Smithsonian Institution Press.
Norton, B.G., Hutchins, M., Stevens, E.F. and Maple T.L. (1995). *Ethics on the Ark: Zoos, Animal Welfare, and Wildlife Conservation.* Washington, DC: Smithsonian Institution Press.
Oates, J.F. (compiler) (1986). *Action Plan for African Primate Conservation: 1986–90.* Gland, Switzerland: IUCN.
Oates, J.F. (1999). *Myth and Reality in the Rain Forest: How Conservation Strategies are Failing in West Africa.* Berkeley: University of California Press.
Rabinowitz, A. (1991). *Chasing the Dragon's Tail: The Struggle to Save Thailand's Wild Cats.* New York: Doubleday.
Ratanakorn, S. (1978). Legal aspects of land occupation and development. In *Farmers in the Forest*, ed. P. Kunstadter, E.C. Chapman and S. Sabhasri, pp. 45–53. Honolulu: East–West Center Book, University Press of Hawaii.
Rourke, M. (2000). Morality Inc. *Los Angeles Times*, May 20.
Savage, D.G. (2000). Justices ban many city, state boycotts. *Los Angeles Times*, June 20.
Shelman. E. and Lazoritz, S. (1999). *Out of the Darkness: The Story of Mary Ellen Wilson.* Lake Forest, CA: Dolphin Moon Publishing.
Wright, R. (1998). Myanmar dissident forced to end protest, Albright says. *Los Angeles Times*, July 30.

14 Monkey business? The conservation implications of macaque ethnoprimatology in southern Thailand[1]

LESLIE E. SPONSEL, NUKUL RUTTANADAKUL AND
PORANEE NATADECHA-SPONSEL

Ethnoprimatology

Three main reasons for studying nonhuman primates have long been recognized by anthropologists and others: our closest living relatives in the animal kingdom are primates; our ancestors were primates; and we are also primates (see, for example, Darwin, 1871, 1872; Huxley, 1863; Morris, 1967). Recently, a fourth reason for studying nonhuman primates is increasingly being recognized: humans interact with nonhuman primate species in multifarious, fascinating, and important ways, especially where they are sympatric.

The term *ethnoprimatology* was coined to refer to research on interactions between human and nonhuman primate populations in regions where they are sympatric species. Ethnoprimatology operates at the interface of cultural anthropology (ethnography) and primatology (field study of primate behavior, ecology, and conservation) (Sponsel, 1997a)[2].

The recency and neglect of ethnoprimatology reflects several factors. Among these is the paradox that, whereas humans have long been accepted by most scientists as a product of organic evolution, seldom have biologists and others accepted any human populations as a natural and integral part of ecosystems. This even applies to those 'traditional' indigenous societies who live sustainably with their environment. Primate fieldwork has mostly focused on the unintrusive naturalistic observation of the behavior and ecology of free-ranging primates, ideally in areas where human influence or interference is absent or minimal (Carpenter, 1964). Also most ethnographers, and even cultural ecologists, have simply ignored the importance (ecological, economic, cultural, and/or religious) of nonhuman primates in the habitat of the human societies they have studied (Sponsel, 1997a; Wheatley, 1999, pp. 147, 150).

Wherever *Homo sapiens sapiens* and other species of primates are sympatric, however, they likely interact over time developing one or more varieties of symbiosis (commensalism, predation, parasitism, competition, mutualism, etc.). Such interactions have gone on for centuries, millennia, or longer, depending on the region (Sponsel, 1997a). For example, at the archaeological site of Niah Cave in northern Sarawak, *Macaca fascicularis* and *M. nemestrina* subfossils are found from 40 000 to 2000 years ago; apparently they were used as food by hunters (Burton, 1995, pp. 155, 163).

Human societies may have very elaborate cultural beliefs, values, attitudes, customs, and rituals regarding nonhuman primates. This human factor can be a significant influence on primate behavior, ecology, and conservation, even though primatologists had only just started to recognize, appreciate, and study human–nonhuman primate interactions at the end of the twentieth century. (Incidentally, another interesting aspect of ethnoprimatology is how cultural differences between researchers affect their work in primatology, such as primatologists from Japan compared with those from the USA (see, for example, Asquith, 1991)).

Early explorations in what amounts to ethnoprimatology include various intellectual histories of the relationship between humans and other primates as well as cultural representations of nonhuman primates by Janson (1952) for the Middle Ages and Renaissance in Europe, McDermott (1938) for European antiquity, Morris and Morris (1966) for apes, and van Gulik (1967) on the gibbon in China. A recent survey is provided by Carter and Carter (1999), who give examples of cultural representations of primates in religions, folklore, literature, and movies.

Especially pertinent to the present essay, Carter and Carter (1999, p. 271) note that the sacred status of the hanuman langurs (*Semnopithecus entellus*) in India, where they are tolerated and not harmed in human settlements, has contributed to their widespread distribution. Cultural beliefs and practices may have inadvertently served primate conservation elsewhere, as in ancient Egypt with the sacred hamadryas baboon (*Papio hamadryas*), in Japan with the macaques (*Macaca fuscata*) (Ohnuki-Tierney, 1987) and in Bali for the long-tailed macaque (*M. fasicularis*) in sacred forests (Wheatley, 1999). Ohnuki-Tierney (1987) provides a symbolic analysis of the role of macaques in the history and ritual of Japan which demonstrates how a nonhuman primate species may serve as a heuristic microcosm of a culture. Wheatley (1999) describes the sacred monkeys of Bali, their habitat, behavior, and ecology, and how they are culturally viewed and treated by Balinese society. Also Galdikas (1995) should be mentioned since her book on fieldwork with orangutans includes many enticing comments about relations between orangutans and Dayak people. Examples of ethnoprimatology in Thailand include the pioneering work of Eudey (1986, 1999) on macaques as crop pests for the swidden farmers of the so-called hill tribes in the north, and the detailed survey of temple monkeys by Aggimarangsee (1992).

Many species of nonhuman primates are endangered or threatened because of human predation and/or habitat disturbance or destruction; thus there is also a linkage between ethnoprimatology and conservation concerns (see, for example, Johns and Skorupa, 1987). Likewise, ethnoprimatology relates to the expansion of animal rights to include nonhuman primates (Cavalieri and Singer, 1993; Goodall, 1991). As Dolhinow and Fuentes (1999) observe:

> Biodiversity and conservation-related themes have become a critical part of primate studies. It is no longer possible to study a group or population of free-ranging nonhuman primates without coming into contact with human disturbance, manipulation, or destruction of the habitat (p. 146). It has become readily apparent that no form of conservation action is possible without taking into account the human role in local utilization of protected areas (p. 147).

Now, let us turn to a specific example of ethnoprimatology, macaques trained and used to harvest tree crops in southern Thailand, and then its relevance for conservation.

Macaque crop pickers

The use of trained macaques for picking coconuts in Southeast Asia is briefly mentioned as a curiosity in many different kinds of publications[3]. Usually the reference is only to Southeast Asia in general, and no countries or species are specified. However, in some of the literature India, Thailand, Malaysia, and Indonesia (especially Sumatra and Borneo) are identified[4]. There are only a few descriptions of this phenomenon with any substance, but they are mostly anecdotal accounts for a popular audience (Bertrand, 1967; Gudger, 1919, 1923; LaRue, 1919; Pfeffer, 1989; Sirorattanakul, 1997; Sitwell and Freeman, 1988)[5].

In comparison to these previous reports, our research offers a more holistic and detailed survey of the subject, even though it is far from comprehensive and in many ways superficial. During the summers of 1994–1995, while the senior author was on a Fulbright Fellowship in Biology at Prince of Songkla University in Pattani, Thailand, he collaborated with zoologist Nukul Ruttanadakul in some exploratory research on the ethnoprimatology of macaques, although this was incidental to our main research project. This essay draws on information from three sources: our field observations of macaque monkeys trained to pick tree crops; our interviews with the trainers and handlers of the monkeys among Muslim and Buddhist villagers; and relevant background and complementary literature. [Other aspects of this phenomenon are discussed in a previous complementary essay (Sponsel, Nattadecha-Sponsel and Ruttandanakul, 2000)]. Our fieldwork was conducted within a 100 km radius of Pattani, a port city in southern Thailand near the border with Malaysia.

The antiquity and geographic range of this phenomenon is unknown. It probably extends back at least into the early 1800s in Sumatra and Borneo (Gudger, 1923). However, it may have spread from Sumatra into Malaysia and then Thailand along with Islam in the 1300s because it is concentrated mainly in countries and areas populated mostly by Muslims. Although we have not found any historical documentation for its antiquity in Thailand, one informant in his seventies said that at least as far back as the generation of his grandparents macaques were used in this manner. Their use likely increased markedly with the growth of commercial coconut plantations during the twentieth century.

In southern Thailand almost all of the macaques used for harvesting tree crops are initially captured from free-ranging populations, either with a baited wooden box trap or when a hunter shoots a mother monkey and then removes a clinging infant from her dead body. In the latter case the trauma may have long-lasting and profound effects on the infant, such as psychological and physical withdrawal symptoms (see Galdikas, 1995, p. 211). The optimum age to train a monkey is from two to five years old. Usually only about two to three weeks are adequate for training. They learn to respond to at least six verbal commands, choose coconuts in different phases of ripeness, and twist the stem to tear the nut from the tree. The trainer applies or withholds physical punishment (e.g. hitting with a stick) and rewards appropriate behavior with food (e.g. coconut milk). Although the trainer and subsequent handler dominate the monkey with fear of punishment, an affectionate bond develops between them. The domination over the monkey takes advantage of the dominance hierarchy found in macaque society. The monkey is handled by only one person, almost always a man because an adult male monkey can be aggressive and dangerous.

Each monkey is kept separately rather than released into a group in an enclosure, although individuals may be within sight of one another. The monkey is usually attached by a chain or rope to a heavy pole, about 5 m in length, which leans against a tree. This allows the animal to move around

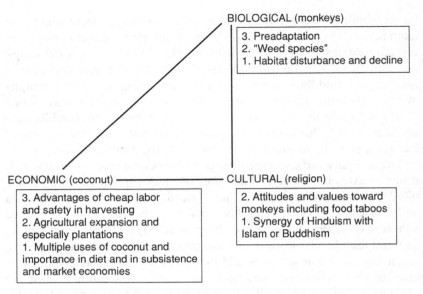

Figure 14.1. Synergy of facilitators.

on the ground, up and down the pole, and into the branches of the tree. Three times each day the monkey is given food such as rice, eggs of chicken or duck, bananas and other local fruits, and coconut juice.

Synergy

This extremely rare if not unique phenomenon of monkeys trained to harvest tree crops may be largely explained in southern Thailand by three sets of factors, which facilitate it by combining in synergy: biological, economic, and cultural (Fig. 14.1).

Biological facilitators

Given their arboreal habit, locomotor agility, manual dexterity, and intelligence, higher primates appear to be the only eutherian mammals that might be considered to be 'preadapted' for harvesting tree crops for humans. The macaques have the most appropriate characteristics for this purpose among the many species of primate in tropical Asia. Macaques occur in a greater variety of environments and climates than any other nonhuman primate genus. Their habitat ranges from tropical to temperate forests, and also includes mangrove forests, grasslands, and dry scrub and cactus (Napier and Napier, 1967). They are geographically widespread, ecologically diverse, flexible in arboreal and terrestrial locomotion, possess an opposable thumb, and omnivorous in diet (Lindburg, 1987; Falk, 2000). Some species are unusually large in body size for monkeys, the pig-tailed macaques being among the heaviest. Macaques have been classified into 19 distinct species, and all are found only in Asia, except for the so-called Barbary ape (*M. sylvanus*) of northwest Africa. Thailand has five species of macaque, and in the Western portion of the country they are all sympatric, as in the Huai Kha Khaeng Wildlife Sanctuary (Eudey, 1986; Stewart-Cox, 1995)[6].

Another attribute of macaques is especially critical, some are weed species. Richard, Goldstein and Dewar (1989) distinguish between weed and non-weed species. They tentatively classify the crab-eating macaque among the weed species, but not the pig-tailed, which frequents primary rainforest more than secondary growth. (Wheatley (1999, p. 150) notes that the label 'weed' can be pejorative, but in the present context it does convey a useful meaning that the term 'commensal' does not.)

Richard *et al.* (1989, p. 573) explain the characteristics of weed species:

> Although the metaphor may not be totally apt, there are some striking parallels between weed plants and weed macaques. Weed plants thrive where people leave their mark on the land; they are plants which spread where people travel and settle down. They depend on people and, in fields, they compete with people. The weed macaques often live in and alongside towns and villages, and they exploit the fields of the farmers and secondary growth nearby. Indeed, they often depend directly or indirectly on human activities for a substantial portion of their diet. In short, like weed plants, weed macaques can be construed as human camp followers that may even occupy some habitats only because human disturbance is present.
>
> The distinction between weed and non-weed macaques is a working hypothesis. We are not implying that non-weeds never raid crops or that weed species cannot survive without access to human resources. There are occasional reports of crop-raiding by all of the macaque species and of weed macaques in primary forest, far from human habitation. However, we suggest that there are marked differences in the frequency and success with which species exploit human resources.

Richard *et al.* (1989, pp. 586–8) apply their model to explain the coincidence between the distribution of farming and weed macaques, and the wider occurrence of the weed macaques compared with non-weed species.

Richard *et al.* (1989, p. 570) argue that this relationship between weed macaques and farming does not make these monkeys any less natural or interesting; rather, it is their adaptive strategy to respond effectively to the disturbance of their natural habitat, and accordingly it is worthy of study. Weed macaques are like pioneer species of plants in the early stage of ecological succession (biotic community development). Weed macaques are ecotone species that specialize in taking advantage of the adaptive opportunities afforded by the edge or transition zone between two different environments, such as forest and field (Richard *et al.*, 1989, p. 575).

Some species of macaque are often mentioned in the literature as tolerant of humans and found near settlements, or as crop raiders or pests (see, for example, Rowe (1996)). Burton (1995) refers to the long-tailed macaque as an 'edge' species able to live in disturbed habitats along the edge. Some species, especially the crab-eating macaque, seem to be as well adapted to a mosaic of disturbed and secondary forests and agroecosystems as to primary forests, maybe even more so (Bishop *et al.*, 1981; Johns and Skorupa, 1987). Moderate disturbance in a forest can increase food availability and accordingly stimulate an increase in the size and density of primate populations (Wilson and Johns, 1982).

In southern Thailand in particular, our research has revealed that both pig-tailed macaques from inland rainforests and crab-eating macaques from coastal mangroves are used to harvest several different tree crops (Table 14.1). The pig-tailed species is more likely to be used for climbing coconut palms, the crab-eating one for harvesting various kinds of fruit trees where they have the advantages of lighter weight and a longer tail with some bracing function. Other local tree crops that are harvested by monkeys include ma-toom (Bengal quince, *Aegle marmelos*), sa-to (parkia, *Leguminosae*) and sadao (neem, *Melia* sp.). Also monkeys shake tree branches to harvest mangoes (*Mangifera indica*) and tamarind (*Tamarindus indica*).

In Thailand, beyond the coconut industry in the south, it does not appear that macaques are very important for other human purposes such as bushmeat, sport hunting, medicine, or trade (Eudey, 1999, p. 156).

Economic facilitators

The coconut palm (*Cocos nucifera*) has been called the 'tree of life' because it is so very important to the subsistence and market economies of many coastal areas of islands and continents throughout the humid lowland tropics. It is a reliable source of food and water whenever other sources

Table 14.1. *Comparison of macaques used for harvesting*

Category	Pig-tailed	Crab-eating
Taxon	*Macaca nemestrina*	*Macaca fascicularis*
Malay	berok, beruk	kera
Thai	ling kaang	ling saaem
Characteristics	olive brown above, white underneath, top of head dark brown, face gray, short tail	grey to reddish brown, lighter underneath, face pink, head hairs form crest, long tail
Body mass	4.7–14.5 kg	2.5–8.3 kg
Distribution	Assam, Burma, Thailand, Malaysia, Sumatra, Borneo	Indochina, Thailand, Burma, Malaysia, Sumatra, Borneo, Timor, Mauritius, Philippines
Habitat	inland evergreen and deciduous forests, lowland to montane, rarely along coastal, offshore islands, and in swamp forests	mainly mangroves, also riverine and inland forests, up to 2000 m, woodland patches, farm areas, ecotones, but most common on coasts and islands
Food	fruit and seeds (73.8%), buds, leaves, flowers, insects, nestling birds, termites, eggs and larvae, river crabs, other small animals	fruit (64%), leaves, crabs, crustaceans, shellfish, other small animals
Group size	15–40	10–48, over 100
Group composition	multiple males and females	multiple males and females
Home range	travel widely, 62–828 ha	more limited, 25–200 ha
Population density	18–19 km^{-2}	48–54 km^{-2}
Weaning	12 months	14 months
Sexual maturity	35 months	50+ months
Birth interval	12–24 months	13 months
Life span	26.3 years	37+ years
Dominance	females have matrilineal dominance hierarchy	male dominance hierarchy less marked than other species
Human relations	crop pest	tolerant of humans, may occur near villages, some urban troops, may raid crops, orchards and rubber seedlings on plantations

Sources: Bercovitch and Huffman (1999), Burton (1995), Falk (2000), Humphrey and Bain (1990), Lekagul and McNeely (1988) and Rowe (1996).

fail. It can provide poor people with dependable and quick cash whenever needed, such as from copra (dried coconut meat). Almost every part of the palm has multiple uses. Beyond nutritious food and drink, it may provide material for the construction of houses, furniture, and fences, for charcoal as fuel, and for dozens of other products for local and external consumption (Persley, 1992).

There are two types of coconut palms, the tall and the dwarf. The tall one grows up to 30 m tall; starts producing nuts after five to seven years; is most productive at 15–20 years; and may produce up to 60 years. Under favorable conditions it produces 60–70 nuts annually, which mature within one year of pollination (Persley, 1992).

Coconut is a regular component of Thai cuisine and provides about half of the fat in the diet. Coconut palms grow throughout most of Thailand, but nearly half are grown in the south and most are concentrated along the eastern coast where conditions are optimum. In southern Thailand coconut plantations are an important part of the local economy of villagers, and monkeys are usually an important part of the production process. For example, we discovered one village surrounded by coconut plantations where almost every one of the 200 households had its own monkey. However, in a second village of the same size only seven households had a monkey. Such variations reflect different sizes of plantations and different degrees of villager involvement in harvesting coconuts.

On small plantations, those with less than a hundred coconut trees, the harvesting is usually done by young boys. However, on larger plantations men use trained monkeys. An efficient macaque can harvest 500–1000 coconuts per day, and a handler can earn about US$200 per month by harvesting coconuts and other tree crops. About half of a family's yearly income may be earned in this way. A newly captured monkey one to two months of age sells for about US$20–60, while a trained four-year-old monkey sells for from US$30 to 400, depending on the species, sex, and individual qualities. It costs only about US$12 per month to maintain a monkey, mainly for food. Thus, a monkey is an extremely cheap source of labor after the initial purchase costs. Naturally macaques can harvest tree crops far more productively, cheaply, and safely than humans. Their thick fur protects them from hazards like scorpions, bees, and ants. However, some monkeys may die from the bite of a poisonous snake, or break an arm or worse in a fall.

The development and expansion of coconut palm plantations in southern Thailand is related to several factors that developed at different times, including the transformation from subsistence village to cash market economy; the process of agricultural diversification and regional specialization encouraged by government policies; road, highway, and railroad construction facilitating transport to markets; search by Muslims for new economic opportunities as their fishing yields decline; and decimation of this palm in the central and northeast regions by pest infestations[7].

The expansion of monocrop plantations of rubber (*Hevea brasiliensis*) as well as coconut palm and oil palm (*Elaeis guineensis*) converted vast areas of various kinds of natural forest into a mosaic of diverse agroecosystems. For example, from 1930 to 1990, the area of rubber plantations expanded by 16.4 times and that of coconut plantations by nine times (Wilson, 1983; National Statistical Office, 1991). During this same period forest cover in the south decreased from about 75% to 15%, reflecting a huge loss of habitat for numerous wildlife species including macaques and other primates (see Boulbet, 1995; Leungaramsri and Rajesh, 1992; Sponsel, 1997b).

Cultural facilitators

Southern Thailand is a fascinating region because of the striking contrasts in cultures (about 30% Thai Buddhists and 70% Thai Muslims of Malay heritage) and environments (coastal to inland forest biomes). Furthermore, these contrasts occur in close proximity because of the long and narrow character of this peninsula. The large 'Malay' population results from the arbitrary demarcation of the border between Thailand and Malaysia by the British in 1909. Although many Muslims speak Thai and consider themselves to be Thai, their culture and language (Pattani dialect) originate from Malay, and most refer to themselves as *Jawi*. Their Islamic religious beliefs and practices include five daily prayers in the direction of Mecca, Friday sermons in the local mosque, observance of Islamic holy days, and so on. However, their religion is actually syncretic with a mixture of elements from the earlier religions of Animism, Hinduism, and even Buddhism.

At the present stage of our research it appears that in southern Thailand the cultural (especially religious) meaning of monkeys facilitates their economic use (or more accurately, exploitation), and, in general, their relatively humane treatment. The common attitude toward animals in the various religions of Thailand seem to be mutually reinforcing, Hinduism and Buddhism for the Thai Buddhists, and Hinduism and Islam for the Thai Muslims.

In India Jainism in turn influenced the development of Hinduism, Buddhism, and even Islam (Chapple, 1993). While there are certainly major differences among these four religions, there are important similarities as well. The quintessential similarity is the recognition of the unity of all beings, which should lead to respect and reverence for all life, at least ideally.

The Hindu epic from India, *Ramayana*, remains popular in shadow plays for both Muslims and Buddhists in Thailand as well as for many people elsewhere in Southeast Asia. Among other points, this epic expresses attitudes about the desirable relations between humans and animals, and perhaps it even facilitates the custom of using macaques to harvest tree crops (see Coomaraswamy, 1965; Ludvik, 1994; Majupuria, 1991; McNeely and Sochaczewski, 1995, p. 61).

In Buddhism there is also respect for monkeys because in one of his incarnations the Buddha was supposed to have been a monkey (Buri, 1989; Byles, 1967; Chapple, 1993). A concern for the monkeys may be reflected in one handler's observation that his monkey doesn't work as hard in entertaining tourists as on the plantation (Sirorattanakul, 1997).

Even though monkeys are generally considered to be inferior to humans, they are recognized for their exceptional positive qualities among animals, such as cleanliness as reflected in social grooming of their hair. Recognition of the obvious physical resemblances between monkeys and humans may have contributed to the Islamic prohibition against eating monkey, since it would be close to cannibalism (Vire, 1986). In addition, our Muslim informants believe that, like humans, the monkey has some kind of spirit or soul. (Likewise in Buddhism there is a food taboo on consuming monkeys.)

In general, in Islam animals, as part of Allah's creation, should be treated with appropriate reverence, nonviolence, compassion, and loving care. Nevertheless, humans may also use animals when necessary to satisfy their basic needs and for their convenience. Animals have rights by virtue of being part of creation. There is even a famous Islamic treatise on animal rights from the tenth century called *The Case of the Animals versus Man* (Al-Safa, 1978). Furthermore, it is the responsibility of humans to consider all of this and to treat animals ethically, because as spiritual and moral beings humans are superior to animals even though animals may be superior physically[8].

Some of the above principles are reflected in the relatively humane treatment and care of the macaques by handlers in southern Thailand. The monkeys are given a human name; kept in the village adjacent to the home of the handler and during a storm brought onto the front porch or into a shelter; regularly given water and food; walked to an irrigation canal or other body of water to swim and bathe; groomed with a comb by their

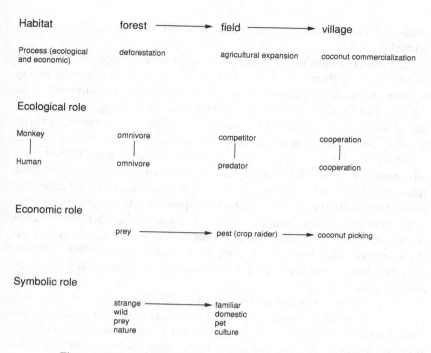

Figure 14.2. Adaptive shifts in the ecology of macaques (after Sponsel *et al.*, 2000).

handler (the monkey may groom the owner as well); and given a ride on the back of the cart, bicycle, or motorcycle of the handler en route between home and plantations. When a monkey becomes too old to work produc-tively it is usually released back into the forest, although some are kept as pets. When a monkey dies it is even buried, but not near human graves and not with any ritual. On the other hand, the monkey is not allowed to roam free, but always tethered on a long rope or chain from around its neck.

Hypotheses

Hypothesis 1: adaptive shift

We hypothesise the following evolutionary sequence in the main type of symbiosis between macaques and humans in southern Thailand, and by implication, possibly elsewhere (Fig. 14.2):

(1) As farming increasingly expands and converts more forest,

macaques increasingly raid crops (Else, 1991; Eudey, 1986; Siex and Strusaker, 1999).

(2) Farmers might try to eliminate or reduce this competitor as a crop pest by hunting them. However, because of the religious taboo on eating monkey the farmer as hunter is less likely to gain energy and nutrients from the kill, and thus the costs of energy expended in hunting may outweigh any benefits or at least not be advantageous. One would expect that the taboo is more often than not honored, even though exceptions can occur.

(3) Under these circumstances, eventually it would seem to be to the advantage of both the macaques and humans to develop a cooperative relationship in which both benefit. Although the monkeys do not enter this relationship voluntarily, this is not a prerequisite for the application of the term cooperation in a biological sense. The human handler provides the macaque with water, food, shelter, protection, care, and affection. In turn the macaques are an economic asset to the handlers since they can pick coconuts faster (and more safely) than humans, thus increasing efficiency and productivity.

The evolutionary sequence of adaptive shifts in the types of symbiosis have a more general implication: they focus more attention on domestication as an *ecological process*. That is, anthropogenic environmental change becomes a catalyst for domestication, rather than the usual emphasis on a particular species as the target for artificial (human) selection replacing natural selection. This implies that, at least in some situations, the earliest stages of domestication may be part of a more or less natural ecological process, and that this may be largely if not entirely inadvertent, instead of involving conscious human agency. This is not to suggest that this evolutionary sequence of types of symbiosis is inevitable and invariant, only that it may fit the particular circumstances of the historical ecology of areas where macaques are used to harvest tree crops. The above does not disregard the fact that this relationship involves opportunistic exploitation of monkeys by humans.

Hypothesis 2. Conservation

In Thailand since the 1960s, numerous government protected areas including national parks and wildlife sanctuaries have been established. These cover about 13% of the country, including most of the remaining forest areas. However, most are not administered adequately (Blockhus, *et al.*, 1992; Eudey, 1999). Outside of these 'protected' areas, conversion of forests to monocrop plantations and other agroecosystems was rampant during the twentieth century. Consequently, the degradation and destruction of primate habitats remains very serious (Sponsel, 1997b, Sponsel *et al.*, 1998).

Since 1976 in Thailand commercial export of all macaque species has been banned (Jintanugool, Eudey and Brockelman, 1982). The capture of primates for town markets appears to be very limited (Robinson, 1994; Round, 1990). Our previous survey of animal trade at two markets in Pattani throughout 1988 did not record any monkeys (Sponsel and Natadecha-Sponsel, 1991). However, some local restaurants specialize in wildlife meat, particularly for Thai Buddhists from the northeast who work on fishing boats, and these include macaques and langurs on their menu. In addition, a few of our informants mentioned that sometimes they kill monkeys while hunting. Perhaps the paucity of primates in the domestic markets reflects their depletion in the forests more than either adherence to food taboos or adequate enforcement of government laws and regulations (see, for example, Dearden, 1995). Presently, for Thailand, conservationists list all species of monkey as threatened and all species of gibbon as endangered (Eudey, 1999; Humphrey and Bain, 1990).

Under such conditions the type of cooperative relationship between humans and macaques described above assumes further importance: it may be of some significance in conserving these two macaque species despite the decrease of their forest habitats. Indeed, the main habitat of the crab-eating macaques, the mangrove forest, is almost gone (Katesombun, 1992). Although no census is available, there may be several thousand monkeys maintained for harvesting coconuts in southern Thailand. The role of macaques trained to harvest tree crops in the conservation of these species assumes greater importance given that the habitat destruction through deforestation suffered by other primate species such as gibbons in southern Thailand leaves them with no adaptive alternative in contrast.

This phenomenon of macaques used to harvest tree crops, however, may be declining for various reasons. The market competitiveness of the coconut is decreasing. Young Thai are less likely to become monkey handlers because through Western education, the media, and other influences they are attracted to more prestigious and higher-paying jobs to try to satisfy the new materialist and consumerist values. Furthermore, in recent years in some locations handlers and their monkeys have shifted from picking coconuts to entertaining tourists. During the peak of the tourist season (January–May) on Samui Island, one monkey owner earns US$24–28 per day, far more than working plantations (Sirorattanakul, 1997). Consequently, if current trends continue, then the macaques may have only a temporary escape from the loss of their natural habitats and they may still be headed for extinction. Nevertheless, as long as coconut is important in the Thai diet and economy it is likely that macaques will still be used as the most efficient means to harvest coconuts. Well over one million tonnes of coconuts are produced annually in Thailand (National Statistics Office, 1991).

Admittedly at least one major weakness in this conservation hypothesis is that most macaques used in tree-crop harvesting in southern Thailand are initially captured in the wild instead of bred in captivity. As a result the captive population is not really reproducing and maintaining itself. Accordingly it is not as relevant for conservation of these two macaque species as would otherwise be the case. Monkeys are trapped in the wild. Informants say that macaques usually don't breed well in captivity, although one 74-year-old man made a career of breeding and training them. He kept them in a group of three males and two females. The usual lack of success in breeding monkeys in captivity may reflect the trauma of their capture as well as the separation from their natural biological and social contexts and other deprivations. Zoos experience similar difficulties with breeding isolated primate individuals when they are introduced to potential mates, and even when births result, maternal care of newborns may be inadequate or even refused. However, as the habitat of wild macaques continues to be degraded and destroyed, thereby reducing the wild population for the recruitment of new individuals for training to harvest tree crops, a greater effort may be made to breed those in captivity as the only alternative source if the coconut plantation industry is to continue to depend on the monkeys for harvesting. Perhaps in this matter science and technology may provide some assistance.

Animal rights

Readers, especially those sensitive to the problem of speciesism in relation to environmental ethics and animal rights, may argue that there is cruelty to animals in this custom because they are trapped in the wild and removed from their natural social and biological contexts; physical punishment is used in training; they are forced to work hard; and sometimes monkeys that are either too aggressive or too old are killed. On the other hand, it appears to us that, in general, compassion prevails in the handler's treatment of his monkey, if for no other reason than the practical matter that he depends on the well-being and cooperation of the monkey to pursue his own economic interests.

As far as we are aware at this stage of our research, for Thai people the use of the macaques as arboricultural labor is not a contested arena. There seems to be essential agreement among villagers from diverse historical, cultural, and religious backgrounds on this matter. We are not aware of any criticism or opposition regarding this practice from any environmental or animal rights organizations, either local, national, or international. There are much larger concerns for threatened and endangered species in terms of habitat degradation and destruction as well as the illegal traffic in wildlife[9]. Although we do not approve of this custom, rather than targeting single species and issues for conservation, we advocate placing priority on the conservation of biodiversity at the ecosystem level, through more effective protected areas, by the government (national parks and wildlife sanctuaries) and by the community (community forests and sacred places) (Sponsel, 2001, Sponsel *et al.*, 1998).

Conclusions

We conclude that ethnoprimatology, as the interface of cultural anthropology and primatology applied to the field study of interactions between human and nonhuman primate populations, is an indispensible component for the effective conservation of primates. This necessity has been illustrated by the case in southern Thailand of macaques trained and used to harvest tree crops. We have offered two hypotheses that need to be refined, explored, and tested through further fieldwork. First, we hypothesize that there has been an adaptive shift from human predator–monkey prey as crop raider to human handler–monkey harvester as a cooperative relationship, as coconut plantations and other agroecosystems have expanded at the expense of the natural habitat of the macaques. Second, we hypothesize that, in the face of decline in the natural habitat of macaques, the survival of this species may in part result from their semi-domestication as trained crop harvesters. In all of this there is a synergetic interaction between three sets of factors that facilitate the development of this phenomenon of using monkeys to harvest tree crops: biological, economic, and cultural. Of course this is only survival; individuals of the species are outside of their natural ecological and social contexts. This unique synergy may also go a long way toward explaining the geographic distribution and rarity of this phenomenon.

Notes

[1] LES is grateful to Agustin Fuentes for inviting an earlier version of this chapter for the session he organized at the American Anthropological Association in

1997. The Center for Southeast Asian Studies at the University of Hawaii provided a travel grant for LES to attend the conference. Financial support for LES's field research in Thailand was provided by the Wenner–Gren Foundation for Anthropological Research, Fulbright–John F. Kennedy Collaborative Research Grant, and the University of Hawaii Research Council. Personnel at the Prince of Songkla University in Pattani greatly facilitated the research, including Preeya Viriyanon and Rawiwan Chaumphruk. Decha Tangseefa (University of Hawaii) searched for Thai government statistics in the library. Lucy Wormser (Executive Director, Pacific Primate Sanctuary, 500 A Haloa Road, Haiku, Maui, HI 96708-9362, http:www.pacificprimate.org) provided penetrating comments that stimulated us to substantially revise some statements that suffered from speciesism.

[2] Wheatley (1999) refers to this subject as cultural primatology. For reviews of cultural anthropological studies of human–animal interactions see Mullin (1999), Noske (1989), and Shanklin (1985).

[3] These are among the sources that briefly mention this phenomenon: Child (1964), Donner (1982), Fraser (1960), Gudger (1919, 1923), La Rue (1919), Lekagul and McNeely (1988), Medway (1978), Roonwal and Mohnot (1977) and Rowe (1996). For example, Rowe (1996, p. 128) states that trained pig-tailed macaques are used to harvest ripe coconuts on plantations, but does not cite any countries. Burton (1995, p. 163) cites the use of the pig-tailed macaque in Malaysia and Sumatra. This custom has also been reported to us in personal communications for southern India by Safia Aggarwal (University of Hawaii) and for Sumatra by Gerard Persoon (University of Leiden) and Sumastuti Sumukti (University of Hawaii).

[4] These articles on coconut-picking monkeys are the only ones we have discovered despite literature searches by LES, the Primate Information Center (University of Washington, Seattle), and the Wisconsin Regional Primate Research Center (Madison).

[5] This phenomenon is one of the extremely rare cases of primates that may have been semi-domesticated. In ancient times the Arabs had direct experience with baboons and macaques, and sedentary people in Yemen and Oman supposedly trained young monkeys to do menial tasks like turning millstones (Vire, 1986, p. 131). The Egyptians are supposed to have used baboons to harvest the fruit of figs and palms as long ago as some 5000 years (Gudger, 1923; Morris and Morris 1966).

[6] Among the some 270 species of primate, 72 are found in Asia (Eudey, 1999). The 13 species of primate in Thailand include: slow loris (*Nycticebus coucang*), rhesus (*Macaca mulatta*), crab-eating macaque (*M. fascicularis*), Assamese macaque (*M. assamensis*), pig-tailed macaque (*M. nemestrina*), stump-tailed macaque (*M. arctoides*), langurs (*Presbytis melalophos, P. phayrei, P. obscura, P. cristata*), and gibbons (*Hylobates lar, H. pileatus, H. agilis*) (Lekagul and McNeely, 1988).

[7] For further information on the regional cultures and economies see Donner (1982), Firth (1946), Fraser (1960), Prachuabmot (1992), Quarto (1992), Ruyabhorn and Phantumvanit (1988) and Stifel (1973).

8 For further discussion on Islamic views of animals see Al-Qaradawi (1960) and Masri (1989).
9 Sources: Brockelman (1987), Graham and Round (1994), Humphrey and Bain (1990), Sponsel and Natadecha-Sponsel (1991), Sponsel *et al.* (1998) and Stewart-Cox (1995).

References

Aggimarangee, N. (1992). Survey for semi-tame colonies of macaques in Thailand. *Natural History Bulletin of the Siam Society* **40**, 103–66.
Al-Qaradawi, Y. (1960). *The Lawful and the Prohibited in Islam*. Indianapolis: American Trust Publications.
Al-Safa, I. (1978). *The Case of Animals versus Man Before the King of the Jinn: A Tenth Century Ecological Fable of the Pure Brethren of Basra* (translated by Lenn Vean Goodman). Boston: Twayne Publishers.
Asquith, P.J. (1991). Primate research groups in Japan: Orientations and east-west differences. In *The Monkeys of Arashiyama*, ed. L.M. Fedigan and P.J. Asquith, pp. 81–98. New York: New York State University Press.
Bercovitch, F.B. and Huffman, M.A. (1999). The macaques. In *The Nonhuman Primates*, ed. P. Dolhinow and A. Fuentes, pp. 77–85. Mountain View: Mayfield Publishing Co.
Bertrand, M. (1967). Training without reward: Traditional training of Pig-Tailed Macaques as coconut harvesters. *Science* **155**, 484–6.
Bishop, N., Hrdy, S.B., Teas, J. and Moore, J. (1981). Measures of human influence in habitats of South Asian monkeys. *International Journal of Primatology* **2**, 153–67.
Blockhus, J.M., Dillenbeck, M., Sayer, J.A. and Wegge, P. (1992). *Conserving Biological Diversity in Managed Tropical Forests*. Gland, Switzerland: International Union for Conservation of Nature Forest Conservation Programme.
Boulbet, J. (1995). *Towards a Sense of the Earth: The Retreat of the Dense Forest in Southern Thailand During the Last Two Decades*. Pattani, Thailand: Prince of Songkla University.
Brockelman, W. (1987). Nature conservation. In *Thailand Natural Resources Profile*, ed. A. Arbhabirama *et al.*, pp. 91–119. Bangkok: Thailand Development Research Institute.
Buri, R. (1989). Wildlife in Thai culture. In *Culture and Environment in Thailand*, pp. 51–9. Bangkok: Siam Society.
Burton, F. (1995). *The Multimedia Guide to the Non-Human Primates*. Scarborough, Ontario: Prentice Hall Canada.
Byles, M.B. (1967). *Footprints of Gautama the Buddha*. Wheaton: The Theosophical Publishing House.
Carpenter, C.R. (1964). *Naturalistic Behavior of Nonhuman Primates*. University Park, PA: Pennsylvania State University Press.
Carter, A. and Carter, C. (1999). Cultural representations of nonhuman primates. In *The Nonhuman Primates*, ed. P. Dolhinow and A. Fuentes, pp. 270–6.

Mountain View, CA: Mayfield Publishing Co.

Cavalieri, P. and Singer, P. (eds) (1993). *The Great Ape Project: Equality Beyond Humanity*. New York: St. Martin's Press.

Chapple, C.K. (1993). *Nonviolence to Animals, Earth, and Self in Asian Traditions*. Albany, NY: State University of New York Press.

Child, R. (1964). *Coconuts*. London: Longmans.

Coomaraswamy, A.K. (1965). *History of Indian and Indonesian Art*. New York: Dover Publications.

Darwin, C.R. (1871). *The Descent of Man*. Princeton, NJ: Princeton University Press.

Darwin, C.R. (1872). *The Expression of the Emotions in Man and Animals*. Chicago, IL: University of Chicago Press.

Dearden, P. (1995). Development, the environment and social differentiation in northern Thailand. In *Counting the Costs: Economic Growth and Environmental Change in Thailand*, ed. J. Rigg, pp. 111–30. Singapore: Institute of Southeast Asian Studies.

Dolhinow, P. and Fuentes, A. (eds) (1999). *The Nonhuman Primates*. Mountain View, CA: Mayfield Publishing Co.

Donner, W. (1982). *The Five Faces of Thailand: An Economic Geography*. St. Lucia, Queensland: University of Queensland Press.

Else, J.G. (1991). Nonhuman primates as pests. In *Primate Responses to Environmental Change*, ed. H.O. Box, pp. 155–66. New York: Chapman and Hall.

Eudey, A.A. (1986). Hill tribes peoples and primate conservation in Thailand: A preliminary assessment of the problem of reconciling shifting cultivation with conservation objectives. In *Primate Ecology and Conservation*, ed. J.G. Else and P.C. Lee, pp. 237–48. New York: Cambridge University Press.

Eudey, A.A. (1999). Asian primate conservation: My perspective. In *The Nonhuman Primates*, ed. P. Dolhinow and A. Fuentes, pp. 151–8. Mountain View, CA: Mayfield Publishing Co.

Falk, D. (2000). Macaques and savanna baboons: sexual politics and human evolution. In *Primate Diversity*, pp. 230–55. New York: W.W. Norton & Co.

Firth, R. (1946). *Malay Fishermen: Their Peasant Economy*. London: Routledge and Kegan Paul.

Fraser, T.M. Jr. (1960). *Rusembilan: A Malay Fishing Village in Southern Thailand*. New York: Cornell University Press.

Galdikas, B.M.F. (1995). *Reflections of Eden: My Years with the Orangutans of Borneo*. Boston, MA: Little and Brown.

Goodall, J. (1991). *Through a Window: My Thirty Years with the Chimpanzees of Gombe*. Boston, MA: Houghton Mifflin.

Graham, M. and Round, P. (1994). *Thailand's Vanishing Flora and Fauna*. Bangkok: Finance One Public Co.

Gudger, E.W. (1919). On monkeys trained to pick coconuts. *Science* **49**, 146–7.

Gudger, E.W. (1923). Monkeys trained as harvesters: instances of a practice extending from remote times to the present. *Natural History* **23**, 272–9.

Gulick, R.H. van (1967). *The Gibbon in China: An Essay in Chinese Animal Lore*. Leiden, The Netherlands: E.J. Brill.

Humphrey, S.R. and Bain, J.R. (1990). *Endangered Animals of Thailand*. Gainesville, FL: Sandhill Crane Press.

Huxley, T.H. (1863). *Man's Place in Nature*. New York: Appleton and Co.

Janson, W.H. (1952). *Apes and Apelore in the Middle Ages and Renaissance*. London: Studies of the Warburg Institute.

Jintanugool, J., Eudey, A.A. and Brockelman, W.Y. (1982). Species conservation priorities for Thailand. In *Species Conservation Priorities in the Tropical Forests of Southeast Asia* (Occasional Papers of the International Union for the Conservation of Nature Species Survival Commission Number 1), ed. R.A. Mittermeier and W.R. Konstant, pp. 41–51. Gland, Switzerland: IUCN.

Johns, A.D. and Skorupa, J.P. (1987). Responses of rain forest primates to habitat disturbance: A review. *International Journal of Primatology* 8, 157–91.

Katesombun, B. (1992). Aquaculture promotion: endangering the mangrove forests. In *The Future of People and Forests in Thailand After the Logging Ban*, ed. P. Leungaramsri and N. Rajesh, pp. 103–22. Bangkok: Project for Ecological Recovery.

La Rue, C.D. (1919). Monkeys as coconut pickers. *Science* 50, 187.

Lekagul, B. and McNeely, J.A. (1988). *Mammals of Thailand*, 2nd edn. Bangkok: Sha Karn Bhaet Co.

Leungaramsri, P. and Rajesh, N. (eds) (1992). *The Future of People and Forests in Thailand After the Logging Ban*, Bangkok: Project for Ecological Recovery.

Lindburg, D.G. (ed.) (1987). *The Macaques: Studies in Ecology, Behavior and Evolution*. New York: Van Nostrand Reinhold.

Ludvik, C. (1994). *Hanuman*. Delhi: Motilal Banarsidass Publishers.

Majupuria, T.C. (1991). *Sacred Animals of Nepal and India*. Bangkok: Craftsmen Press.

Masri, Al-Hafiz B.A. (1989). *Animals in Islam*. Hants, England: The Athene Trust.

McDermott, W.C. (1938). *The Ape in Antiquity*. Baltimore, MD: Johns Hopkins University Press.

McNeely, J.A. and Sochaczewski, P.S. (1995). *Soul of the Tiger: Searching for Nature's Answers in Southeast Asia*. Honolulu, HI: University of Hawaii Press.

Medway, L. (1978). *The Wild Mammals of Malaya (Peninsular Malaysia and Singapore)*. New York: Oxford University Press.

Morris, D. (1967). *The Naked Ape*. New York: Dell Publishing Co.

Morris, R. and Morris, D. (1966). *Men and Apes*. New York: Bantam Books.

Mullin, M.H. (1999). Mirrors and windows: sociocultural studies of human-animal relationships. *Annual Review of Anthropology* 28, 201–24.

Napier, J.R. and Napier, P.H. (1967). *A Handbook of Living Primates*. New York: Academic Press.

National Statistical Office (1991). *Quarterly Bulletin of Statistics*. 39(2), 45, 72–73, 111. Bangkok: National Statistical Office, Statistical Data Bank and Information Dissemination Division.

Noske, B. (1989). *Human and Other Animals: Beyond the Boundaries of Anthropology*. Winchester, MA: Pluto Press.

Ohnuki-Tierney, E. (1987). *The Monkey as Mirror: Symbolic Transformations in*

Japanese History and Ritual. Princeton, NJ: Princeton University Press.

Persley, G.J. (1992). *Replanting The Tree of Life: Towards an International Agenda for Coconut Palm Research*. Oxon, England: CAB International.

Pfeffer, R. (1989). On Malay Peninsula picking coconuts is monkey business. *Smithsonian Magazine* **19**, 111–18.

Prachuabmot, C. (1992). *Socio-Cultural Change In Southern Thailand 1950–1990: A Documentary Study*. Bangkok: Thammasat University/The Thailand Development Research Institute Foundation.

Quarto, A. (1992). Fishers among the mangroves. *Cultural Survival Quarterly* **16**, 12–15.

Rattanapruk, M. (1991). Small scale processing of coconut products in Thailand. In *Small Scale Processing of Coconut Products*, ed. B.B. Pangahas, pp. 169–73. Manila, Philippines: Asian & Pacific Coconut Community.

Richard, A.F., Goldstein, S.J. and Dewar, R.E. (1989). Weed Macaques: the evolutionary implications of macaque feeding ecology. *International Journal of Primatology* **10**, 569–94.

Robinson, M.F. (1994). Observation on the wildlife trade at the daily market in Chiang Khan, Northeast Thailand. *Natural History Bulletin of the Siam Society* **42**, 117–20.

Rodman, P.S. (1999). Whither primatology? The place of primates in contemporary anthropology. *Annual Review of Anthropology* **28**, 311–39.

Roonwal, M.L. and Mohnot, S.M. (1977). *Primates of South Asia: Ecology, Sociobiology, and Behavior*. Cambridge, MA: Harvard University Press.

Round, P.D. (1990). Bangkok bird club survey of the bird and mammal trade in the Bangkok Weekend Market. *Natural History Bulletin of the Siam Society* **38**, 1–43.

Rowe, N. (1996). *The Pictorial Guide to the Living Primates*. East Hampton, New York: Pogonias Press.

Ruyabhorn, P. and Phantumvanit, D. (1988). Coastal and marine resources of Thailand: Emerging issues facing an industrializing country. *Ambio* **17**, 229–32.

Shanklin, E. (1985). Sustenance and symbol: anthropological studies of domesticated animals. *Annual Review of Anthropology* **14**, 375–403.

Siex, K.S. and Strusaker, T.T. (1999). Colobus monkeys and coconuts: A study of perceived human-wildlife conflicts. *Journal of Applied Ecology* **36**, 1009–20.

Sirorattanakul, T. (1997). Monkey see, monkey do. *Bangkok Post,* March 15.

Sitwell, N., and Freeman, M. (1988). Monkey see, monkey pick. *International Wildlife* **18**, 18–23.

Sponsel, L.E. (1997a). The human niche in Amazonia: explorations in ethnoprimatology. In *New World Primates: Ecology, Evolution, and Behavior*, ed. W. Kinzey, pp. 143–65. Hawthorne, New York: Aldine de Gruyter.

Sponsel, L.E. (1997b). The historical ecology of Thailand: Some explorations of thresholds of human environmental impact from prehistory to the present. In *Advances in Historical Ecology*, ed. W.A. Balee, pp. 376–404. New York: Columbia University Press.

Sponsel, L.E. (2001). Human impact on biodiversity: Overview. In *Encyclopedia of*

Biodiversity, ed. S. Levin, Vol. 3, pp. 395–409. San Diego, CA: Academic Press.

Sponsel, L.E. and Natadecha-Sponsel, P. (1991). A comparison of the cultural ecology of adjacent Muslim and Buddhist villages in southern Thailand. *Journal of the National Research Council of Thailand* **23**, 31–42.

Sponsel, L.E., Natadecha-Sponsel, P., Ruttanadakul, N. and Juntadach, S. (1998). Sacred and/or secular approaches to biodiversity conservation in Thailand. *Worldviews: Environment, Culture, Religion* **2**, 155–67.

Sponsel, L.E., Natadecha-Sponsel, P. and Ruttanadakul, N. (2000). Coconut-picking macaques in southern Thailand: ecological, economic, and cultural aspects. In *Wildlife in Asia: Cultural Perspectives*, ed. J. Knight, Curzon Press. (In press.)

Stewart-Cox, B. (1995). *Wild Thailand*. Cambridge, MA: MIT Press.

Stifel, L.D. (1973). The growth of the rubber economy of southern Thailand. *Journal of Southeast Asian Studies* **4**, 107–32.

Vire, F. (1986). Kird. In *The Encyclopedia of Islam*, Vol. 5, ed. C.E. Bosworth *et al.*, pp. 131–4. Leiden, The Netherlands: E.J. Brill.

Wheatley, B.P. (1999). Cultural primatology. In *The Sacred Monkeys of Bali*, pp. 145–51. Prospect Heights, Illinois: Waveland Press.

Wilson, C.M. (1983). *Thailand: A Handbook of Historical Statistics*. Boston: G.K. Hall & Co.

Wilson, W.L. and Johns, A.D. (1982). Diversity and abundance of selected animal species in undisturbed forest, selectively logged forest and plantations in eastern Kalimantan, Indonesia. *Biological Conservation* **24**, 205–18.

15 Rhesus macaques: a comparative study of two sites, Jaipur, India, and Silver Springs, Florida

LINDA D. WOLFE

Introduction

Rhesus macaques (*Macaca mulatta*) are descendants of the cercopithecine monkeys who migrated to Asia from Africa about 3 million years ago and adapted to the range of landscapes and climates of Asia (Fleagle, 1999; Szalay and Delson, 1979). Modern Asian macaques are highly adaptable primates and in the recent past were distributed from eastern Afghanistan through to China and south to the Islands of Indonesia. Asian macaques are also found from Sri Lanka to the islands of Japan (Wolfheim, 1983). Rhesus monkeys were once found from eastern Afghanistan to southern China and from the middle of India to northern Vietnam, but in recent times their distribution has shrunk because of habitat distribution and the removal of monkeys for food and export (Wolfheim, 1983).

Rhesus macaques live in multimale–multifemale troops that have been reported to range from a minimum of 10 monkeys to over 100. There are usually more females than males in a troop because about half of the males choose a solitary lifestyle rather than living in a troop. Females remain in their natal troops and form subgroups composed of females and males based on kinship and friendships. Males leave their natal troops around the age of 4 years and either remain solitary or join a new troop. Mating is promiscuous in that females mate with troop males, males from adjacent troops, and solitary males. There is, however, a distinct fall–winter mating season and a spring birth season. Adult females mate throughout the breeding season independent of their ovulatory status and do not experience a sex swelling.

Rhesus monkeys occupy a wide range of habitats such as rainforests, deciduous dry forests, *Acacia* forests, and mountain areas. The diet of rhesus monkeys consists of young leaves, fruit, roots and tubers, and insects. Rhesus macaques eat many of the same foods as humans except that they do not consume the flesh of vertebrates. They will, for example, raid agricultural fields when available and are also very fond of human

310

snack foods such as peanuts, popcorn, marshmallows, potato and corn chips, ice cream, gum, candy, bread, and so forth. It has long been recognized that many rhesus macaque troops live in complicated commensal and semi-commensal relationships with the people of India (Richard, Goldstein, and Dewar, 1989; Southwick and Siddiqi, 1994). Transported free-ranging and semi-free-ranging rhesus macaque colonies also exist in the United States, where they have adapted to a wide range of New World habitats.

This chapter reports on two free-ranging populations of rhesus monkeys and their interactions with humans. One population lives in and around Jaipur, India, and the other inhabits the Silver Springs area in north central Florida, USA. Because of cultural and religious differences, the treatment of and attitude toward the monkeys differed between the two locations. In order to understand the complex relationship between the Hindus and the monkeys of India we need to consider an ancient Hindu epic, the Ramayana, and related poems. Following the overview of the Ramayana, the lives of rhesus monkeys and their relationships with the people of Jaipur, India, will be discussed. Finally, the lives of the rhesus monkeys of Silver Springs, Florida, will be explored.

The Ramyana

The Hindu epic of the Ramayana and allied poems depict the life of Rama and his relatives, his wife Sita, the evil half-God Ravana, and the monkeys that came to the aid of Rama. The written versions of the Ramayana, or the adventures of Rama, were first recorded between the fifth and first century BC although the story itself is probably much older (Vitsaxis, 1977). Some scholars believe that fragments of the role of the monkeys in the Ramayana are part of an even more ancient Hindu cosmogony (Nivendita and Coomaraswamy, 1985). Rama is believed to be the seventh avatar or incarnation of Vishnu, the preserver of the divine order (Vitsaxis, 1997).

The story of Rama begins when the world is out of balance because of the evil deeds of Ravana. In order to kill Ravana and set the world in balance again, Vishnu is reborn as Rama. He is born into the household of King Dasharatha, who rules a kingdom from its capital of Ayodhya. Because of a misdeed of one of the King's wives, Rama is banished to live in the forest where he stayed for many years accompanied by his wife Sita and brother Lakshman. One day, while Rama is out hunting, Ravana kidnaps Sita and takes her to Lanka. As they wander looking for Sita, Rama and Lakshman come upon a monkey-chief named Sugriva who lived in exile with his people. Rama and Lakshman help Sugriva defeat his evil brother Vali and retake his throne. Sugriva then vows to help Rama and Lakshman find Sita and destroy Ravana. Sugriva sends Hanuman, a monkey whose father is the wind-god, to find Sita.

Hanuman finds Sita on the island of Lanka but Hanuman is captured by Ravana's army, who set his tail on fire. With his tail blazing, Hanuman burns Lanka and returns to Rama. Upon learning of Sita's whereabouts, Rama, Lakshman, Hanuman, Sugriva and his army of monkeys engage Ravana and his demons in a battle. The bridge to the Lanka was built by the army of monkeys under the direction of another monkey-chief named Nala whose father was the 'Architect of the Universe'. Twice Hanuman goes to the Himalayan Mountains and brings them back to Lanka so that herbs can be gathered to save the lives of the wounded monkeys, Lakshman, and Rama. Hanuman, of course, returns the Himalayan Mountains to their rightful place. Eventually, Ravana's army is defeated and Rama kills Ravana. Rama, Lakshman, Sita, Sugriva, Hanuman, and the army of monkeys return to Ayodhya and are welcomed as heroes.

The Ramayana ends when Rama and his followers, including the monkey-chiefs and their families, enter heaven. Hanuman, however, being immortal, remains on earth until such time that people no longer recite the Ramayana (Vitsaxis, 1977; Nivendita and Coomaraswamy, 1985). Hanuman is a special mythological figure for not only can he fly but he can also make himself larger or smaller or turn himself into another animal.

The Ramayana is a popular story in India and is depicted on TV, recited in temples, and appears in children's books. The epic poem is used to illustrate the ethos, morals, and behaviors befitting a Hindu. For example, through the Ramayana, the importance of purity, duty, generosity, self-restraint, gentleness, truth, and honesty are taught to children. Hanuman, more than any of the monkey-chiefs he served, plays a special role in Ramayana and is an important symbol of faithfulness, obedience, and devotion (Nivendita and Coomaraswamy, 1985). Taking actions against monkeys or failing to feed them places Hindus in conflict with their religious beliefs and their desire to use the lessons of the Ramayana, especially the heroic actions of Hanuman, to teach the Hindu ethos to their children.

Because of the antiquity of the story of Hanuman, the absence of written records describing the where or when of the origin of Hanuman, and a lack of a clear understanding of the prehistoric distribution of macaques and langurs, it is difficult to link an extant monkey species with Hanuman. It is not known, moreover, if the people who conceived of Hanuman differentiated between langurs and macaques. For example, in south India the lion-tailed macaque (*Macaca silenus*) is often confused with the Nilgiri langur (*Presbytis johnii*) (M. Singh, personal communication). Once the story of Hanuman spread, different people at different times may have seen a link to a different monkey. All of these factors considered together suggest that we will never know which, if any, living species might have been the basis for the story of Hanuman. There are, however, descriptions of Hanuman having red fur in various translations of the Ramayana and related poems (for examples see Sahai, 1980; Santoso, 1980). It is the rhesus monkey that has a red hue to its fur, unlike other macaques or the gray langur (*Semnopithecus entellus*).

While conducting research in 1987–88 in Jaipur, India, which is located about 258 km South of New Delhi, I asked people questions about Hanuman. Most people said that Hanuman was the 'red-faced' monkey (i.e. the rhesus macaque). A priest told me that the 'black-faced' monkey (as the gray langur is known locally) was the bad monkey in the Ramayana (perhaps referring to Vali). I was also told that black-faced monkeys fought on the side of Ravana. Others suggested that the red-faced monkeys were the chiefs and the black-faced monkeys were part of Rama's army. In fact, there are paintings depicting scenes of the Ramayana in which there are flesh-colored monkeys resembling macaques leading an army of black-faced monkeys. Another priest said that according to the Ramayana the faces of the black-faced monkeys were permanently charred black when Hanuman burned Lanka.

Paintings of Hanuman from Jaipur depict him with the flesh-colored body and face of a rhesus macaque and not with the black face of a gray langur. Moreover, the effigies of Hanuman that are worshiped in and around Jaipur show him with the red face that characterizes rhesus monkeys, especially during the breeding season. All monkeys in India share in the sacredness afforded by the epic of the Ramayana to a greater or lesser degree, depending in part on the local cultural interpretations of the Ramayana. In Jaipur the rhesus monkeys are associated with Hanuman and are treated as such although probably less so now than in the past.

Having explored the reasons modern Hindus tolerate monkeys does not explain how the reverence for monkeys began. Wilson (1984, p. 85) coined the term *biophilia* and suggested that people have an 'urge to affiliate with other forms of life [that] is to some degree innate'. Perhaps the earliest settlers in India tolerated the monkeys because they enjoyed their presence and later added monkeys to their mythology.

On the other hand, understanding why humans might tolerate monkeys does not explain why macaques, including rhesus macaques, are able to survive and, in some cases, seem to prefer living in close association with people. For example, Southwick and Siddiqi (1994) conducted a population survey of rhesus monkeys in 1990–91 in north central India and reported that 86% of the rhesus monkeys were living in association with people. Rhesus monkeys live in villages, towns, cities, temples, and railway stations and along canals. Although in other parts of India rhesus monkeys lived in forested areas away from human habitation, the fact remains that many groups of rhesus monkeys live in a commensal or semi-commensal relationship with people. As suggested by Asquith (1989, p. 152) the food given to Asian monkeys 'could be considered as part of a fluctuating natural resource'.

Although there are many factors that may suggest an answer to the puzzling question, there is no single explanation of why rhesus monkeys often live in association with humans. Several suggestions regarding rhesus macaque commensalism have been offered, and some are listed below. It should be noted that the items in the following list are not necessarily mutually exclusive.

1. Rhesus macaques may be adapted to exploiting plants that live on forest edges and in open patches (Richard et al., 1989; Eudey, 1980). An adaptation to forest edges coupled with their intelligence, sociality, vigorousness, agility, and curiosity may have pre-adapted rhesus monkeys to living in areas disturbed by humans.

2. As the human population expanded and moved into areas inhabited by rhesus monkeys, the monkeys would have become dependent on people. Increased dependency would have taken the form of stealing food from human living areas, taking handouts, and raiding agricultural fields (Southwick and Siddiqi, 1994). In Jaipur, older adults told me that when they were children there were forests outside of the city wall where the monkeys stayed (also see Davar, 1977). Today, however, the forests are gone and have been replaced by suburbs. The rhesus monkeys of Jaipur are now more dependent on handouts, garbage dumps, and the raiding of human homes than they were in the past.

3. In times pasts when tigers and other predators were common, and homes less substantial than today, people may have found that it was to their advantage to have monkeys living nearby and began to tolerate their presence. That is, when monkeys sense danger they make loud warning vocalizations, thereby alerting anyone within hearing distance of the presence of predators. This alarm to the possible presence of danger would have given people more of a chance to protect themselves against predation than if no warning had been forthcoming. On the part of the monkeys, they may have preferred the more open spaces and greater visibility created by people which would have given them the ability to see

tigers more readily and avoid predation. In other words, there were mutual reasons for monkeys and people to maintain an association because of their common desire to avoid predation. Terborgh (2000) has argued that wild animals will associate with humans when they sense that they are safer being around humans than facing the possibility of predation from carnivores.

The rhesus monkeys of Jaipur, India

Between September and December, 1984, I surveyed the rhesus macaques of Jaipur and Galta (Wolfe and Mathur, 1987). I returned to India and conducted a study of the feeding strategies and time budgets of the rhesus macaques of Jaipur and Galta between September 1987, and June 1988 (Wolfe, 1992).

Jaipur has two subsections: first, an old walled city, and, second, is a modern suburb that surrounds the walled city. The walled city has a congested district of living space and businesses, and an open tourist area that includes several temples, the largest of which is devoted to Krishna, the maharaja's palace, and an Observatory, also known as the Jantar Mantar.

Maharaja Sawai Jai Singh II (1688–1743) built Jaipur in 1727 for a population of around 100 000. Today the size of the population is well over two million people. Jaipur is located on the edge of an arid desert, and temperatures range from 45°C in May and June to about 15°C in the winter. The city receives only about 64 cm of rain each year, and the vegetation is dry-deciduous. Water has been a problem since the construction of the city, and in many years the monsoons did not come and thousands of people died of starvation. Today water is carefully parceled out and supplied to Jaipur from deep wells (Roy, 1978). Before the city was built, there was a small lake where wildlife could have found drinking water (Davar, 1977). Whatever natural water sources may have existed 250 years ago have now disappeared. All of the water sources for the monkeys are related to human activities. For example, the macaques drink from water tanks on the tops of roofs, public drinking tanks, public wells, open sewers, and water troughs. Unfortunately, most of the water sources used by the monkeys appeared contaminated to some degree by industrial run-off, human sewage, and/or soap. One juvenile monkey of the open tourist–temple area had the lobster claw deformity on all four limbs; this may be linked to the drinking of contaminated water (Wolfe and Mathur, 1987).

It is generally assumed that the rhesus monkeys of Jaipur were already present when the city was built. The city and the surrounding area does not, however, seem to be a hospitable habitat for a cercopithecine monkey with its high temperature and lack of water. The possibility that Sawai Jai Singh imported the rhesus monkeys in order to have monkeys in the local temples cannot be wholly dismissed.

Rhesus monkeys occupy the open tourist–temple area and the area surrounding the tourist–temple area that is also inhabited by people. Although I was told that there were pockets of rhesus monkeys living in the area outside of the walled city, I could only confirm the existence of two small groups that seemed to be isolated from the main population of monkeys who live in the walled city.

Included in my study were the rhesus monkeys that live at a site known as Galta. Galta is a hilly area located 1 km from Jaipur's Surajpole gate (or 8 km by road). Galta has a long history as a consecrated site with several temples and sacred pools.

The monkeys who live within the tourist–temple complex are arranged into six different troops, each of which have a specific locality where they meet people for their provisions. The people bring or purchase chapatis, roasted chickpeas, carrots, or fruit for the monkeys (Fig. 15.1). The monkeys also feed on the vegetation in the tourist–temple area (see Wolfe (1992) for details). A dominance relationship exists among the troops in the tourist–temple complex. For example, the troop that normally occupies the area in front of the large Krishna temple is able to displace other troops when they leave the Krishna temple to forage on the grass in the Observatory. Each troop also seems to have specific locations into which they disappear when not looking for food from humans. The monkeys are most visible at their feeding locations from sunup, when the early worshipers are going to temple, to around 9 am, when the area becomes congested with workers and tourists. I estimate that around 400 monkeys live in the tourist–temple complex, and in all troops about 53% of the monkeys are immature and juvenile macaques. Based on these figures, I believe that the size of the population of monkeys in this area is stable although some troops look healthier (i.e. appear to have bulkier bodies and smoother hair) than others.

Because of lack of access to the roofs of the homes of the human inhabitants of the city, I do not have reliable counts on the number of monkeys who inhabit the city outside of the tourist–temple complex. Early every morning two men would come on motorcycles with large bags of bananas and feed the city monkeys. A few counts of the monkeys indicate that there are upward of over 100 monkeys in the city. I did not observe the

Figure 15.1. A devotee feeding the monkeys in Jaipur, India. Local goats also eat the food given to monkeys.

city monkeys in the tourist–temple complex or vice versa. There was no apparent trapping of the monkeys while I was in Jaipur, nor did I hear of any plans to trap and remove monkeys.

This lack of trapping was true also of the macaques of Galta, of which there are at least four different troops, each with their own core areas. Because the site is located far from the city, the monkeys of the Galta receive less food from people than the monkeys of Jaipur. Visitors from Jaipur and pilgrims to Galta bring food to the monkeys. There seem to be fewer immature and juvenile monkeys at Galta than in Jaipur, and, in fact, one troop had no offspring in the spring of 1988. There is some vegetation for foraging at Galta but, unfortunately, there was a drought during my study period, and that may account for the lack of food and offspring. Moreover, because of the drought, local villagers brought their goats to Galta, and they destroyed much of the vegetation.

In general, people fed the monkeys of Jaipur and to a lesser extent those of Galta on the following occasions.

- On Tuesdays and Saturdays, the days that are associated with Hanuman carrying the Himalayans to Lanka, the monkeys were well fed by worshipers on their way to temple or by workers as they passed through the area on their way to work. On Mondays very few people fed the monkeys.

- On special occasions such as weddings or local religious celebrations, people brought food to the monkeys.
- In conjunction with national Hindu celebrations such as Diwali (the ceremony of lights) the monkeys were brought food (usually bananas).
- A priest might tell a businessman looking for continued business success to feed monkeys.
- If a Hindu thought that he received special powers from Hanuman, he would feed the monkeys. For example, most days a palmist came to the Observatory with chapatis to feed the monkeys. He told me that Hanuman gave him the power to read palms, and he needed to feed monkeys to keep his power. He was very careful when feeding the monkeys, that each monkey received its fair share of the chapatis.

There are, of course, occasional misdeeds by the monkeys. For example, I observed a monkey tear the sari of a woman who was returning from the Krishna temple because she refused to give him the fruit she was carrying. The monkeys are also very skilled at stealing ice cream cones from people as they enter the Observatory. An older boy at Galta was teasing a monkey, and as a result the monkey bit his finger. The father placed a Band-Aid on the finger and did not seem otherwise concerned. This incident was the only injury I observed during my research period in India. The monkeys who live in the city of Jaipur are problematic because they will get into food pantries and hotel rooms if doors and windows are not securely closed. Local residents also claim that monkeys will steal items of clothing and return the items only in exchange for food. Common sense precautions around monkeys will usually keep people and their belongings safe from harm.

The rhesus monkeys of Silver Springs, Florida

Silver Springs is located in north central Florida near the City of Ocala and is a large limestone artesian springs formation. The springs form the headwater of the Silver River, which, after flowing 11 km, empties into the Oklawaha River near the Ocala National Forest. Native plants and animals including alligators live along the banks of the river.

Silver Springs has been a tourist attraction since about 1860. The first glass-bottom boat rides began at Silver Springs about 1921 and continue

Figure 15.2. A young male at Silver Springs, Florida.

Figure 15.3. A young female and offspring at Silver Springs, Florida.

today. The jungle cruise ride that features rhesus monkeys was added after the advent of the glass-bottom boat ride.

Because there are no records of the beginning of the jungle cruise ride or the release of the rhesus macaques, I searched through old Ocala newspapers from the 1930s and 1940s that are on microfilm at the University of Florida. I had been told that the late Colonel Tooey had started the jungle cruise ride and I included his name in my newspaper search. The first reference I found to Tooey was in the December 3, 1937, issue of the *Ocala Banner*. Likewise, the first mention of the jungle cruise ride in the same newspaper was found in the issue dated May 27, 1938. In the November 11, 1938, issue of the *Ocala Banner* appears the first reference I found to the monkeys of Silver Springs. The article states that Sourpuss, a Silver Springs male monkey, had been shot while drinking water near Anthony, Florida, which is a few miles north of Silver Springs. The account of Sourpuss states that he left behind a family of five. We are also told that Sourpuss was embalmed and put on display and that people came from miles around to see the body of Sourpuss. The local mythology is that the monkeys were released during the filming of a Tarzan movie. However, the only Tarzan film made at Silver Springs was filmed in March 1939, a year after the first mention of the presence of rhesus macaques at Silver Springs. According to Elizabeth Peters (personal communication), who did her PhD dissertation research on the Silver Springs monkeys in 1979, the Tarzan movie made at Silver Springs entitled *Tarzan Finds a Son* contains no rhesus monkeys. The tale of the monkeys being released during the filming of a Tarzan movie is not factual although it made a nice story for the jungle cruise boat drivers to tell their customers.

Local informants told me that Tooey's jungle cruise ride did not actually leave from the Silver Springs Tourist Park. The jungle cruise ride began from a site slightly down river at Paradise Landing, a recreation area for African-Americans. Tooey, I was told, released a small number of rhesus monkeys on an island located near the end of the Silver River in the spring of 1938 in order to enhance the excitement and therefore the earnings of the jungle cruise ride. Unknown to Tooey, rhesus monkeys are very good swimmers and were, reportedly, off the island and on to the banks of the Silver River before Tooey and men could get back in their boat. The monkeys were subsequently kept along the banks of the river by the provisioning of the jungle cruise boat drivers. Although little is known about the monkeys Tooey released, they were probably wild born and had prior contact with people. It is not, therefore, surprising that the monkeys continued to associate with and accept food from people. There are no records of subsequent releases of monkeys. Local informants, however, indicated that they thought that more monkeys were released between 1938 and 1948.

Over the years the size of the population of rhesus monkeys increased, and the length of the jungle cruise ride was shortened. Although there are no records, it would seem that at least some of the monkeys stayed with the jungle cruise ride and moved closer to the headwaters of Silver Springs. Other monkeys, however, did not follow the jungle cruise ride and remained in the lower part of the river near the confluence of the Silver River and Oklawaha River. According to the late William Maples and his colleagues, who first studied the Silver Springs rhesus monkeys, there were three populations of monkeys by 1968 as there were up until the mid-1990s. The names given to the three populations were the Northside troop, the Southside troop, and the Silver River monkeys. The Northside and Southside troops are the monkeys that moved up river and stayed within reach of the jungle cruise ride. The Northside troop occupy the north side of the river and the Southside troop is located on the south side of the river.

The Silver River monkeys inhabit the area down river near the end of Silver River and are independent of the jungle cruise ride. Jungle cruise boat drivers, local people, and some tourists with private boats feed the monkeys of the Northside and Southside troops. In addition, the management of the Silver Springs Tourist Park provides the monkeys with monkey chow daily at feeding stations on either side of the river (Figs. 15.4 and 15.5). Only local people and tourists with boats, however, have passage to the Silver River monkeys (for details, documents, and references see Wolfe and Peters, 1987).

The social system and behaviors of the monkeys of the Silver Springs–Silver River area are similar to those reported for rhesus monkeys in Asia. They live in multimale–multifemale troops based on kinship and rank. Males leave their natal troops between the ages of 4 and 6 years. Some males join another troop and others became solitary. There is a late fall breeding season and a spring birth season. In 1984 before the trapping and removal of the monkeys began there were approximately 250 monkeys in the Northside troop, and 72% were infants, juveniles and subadults. There were 82 monkeys in the Southside troop, and 60% were infants, juveniles and subadults. These figures suggest that the size of the monkey population was increasing. The increase in the size of the Northside and Southside monkeys is at least in part attributable to the daily monkey chow provided to these monkeys by the management of the Silver Springs Tourist Park.

I have a dozen counts of the Silver River monkeys before the trapping began in 1984. There were at least three troops, and each troop had about the same number of infants and juveniles as adults, suggesting that their

Figure 15.4. Monkeys at Silver Springs, Florida, waiting to be fed.

Figure 15.5. Monkeys eating food provided to them.

population size was stable and that they had a lower birth rate than the monkeys of the Northside and Southside troops. In January and February there is a dearth of vegetation along the Silver River. In other words, the Silver River monkeys who lacked access to monkey chow found little to eat in January and February and were reproducing slower than the monkeys of the Northside and Southside troops who were provided monkey chow year around. Some local people who had been observing the monkeys since their childhood would travel up and down the Silver River in January and February looking for the Silver River monkeys. They had dried dog food for the monkeys which the monkeys readily ate.

In 1980, when I first began to observe the rhesus monkeys of Silver Springs, the legal division of the Florida Game and Fresh Water Fish Commission (FGFWFC) was making public pronouncements that the monkeys of Silver Springs had to be removed. In 1983 I had a grant from the University of Florida and began to study the monkeys of Silver Springs and the Silver River on a daily basis. Because they were the most problematic of the three populations of monkeys, I concentrated on the Northside troop. I wanted to know how many monkeys were in the Northside troop and to understand its social structure. I observed the Northside troop and Silver River monkeys from a boat. The Southside monkeys were observed from a boat and on land.

I agreed with the FGFWFC that there were too many monkeys on the north side of the river for the amount of food the management of the Silver Springs Tourist Park was willing to provide the monkeys. The Northside monkeys were crossing Florida Highway 40 to go to a dump to forage and were entering the deer park of the tourist attraction. Both of these movements were cause for concern.

As I was completing my research of the Northside monkeys and ready to make recommendations on how to reduce the size of the Northside Troop with the least social disruption, the FGFWFC threatened the owners of the Silver Springs Tourist Park with jail unless they trapped and removed the monkeys from the park immediately. The trapping began. Shortly after the trapping started, I began working with local animal welfare and conservation groups to prevent the FGFWFC from removing the monkeys. The trapping was stopped after approximately 225 monkeys were removed. However, trapping re-occurred intermittently over the next ten years. The pro-monkey coalition proposed that the monkeys be sterilized and allowed to live out their lives. In the late 1980s, the veterinarians of the University of Florida Veterinary College successfully sterilized 13 females by tubal ligation. It was estimated that by 1991 at least 120 more monkeys would have been born had the tubal ligation not taken place

(Wolfe *et al.*, 1991). The Florida House of Representatives passed legislation to allow further sterilization and re-release of the monkeys, but the
legislation failed in the Senate. The issue of the presence of monkeys in the
Silver Spring area is still not resolved.

I was not opposed to the removal of the monkeys *per se*, but I was
opposed to untrained people trapping the monkeys, holding and caging
the monkeys under inhumane conditions, and selling the monkeys into the
pet trade as occurred with the knowledge of the FGFWFC. Approximately 500 monkeys were removed between 1984 and 1993. The pro-monkey
coalition argued that the monkeys were part of the heritage of Florida,
were not dangerous, and brought tourists and their money into the Silver
Springs area. Other than the FGFWFC, there was no other organized
group opposed to the presence of the monkeys. The Florida newspapers
were generally supportive of sterilizing the monkeys and leaving them to
live out their normal lives. When local people were asked why they favored
the policy of letting the monkeys live out their lives in the Silver River area,
they mentioned the intelligence and curiosity of the monkeys, the way they
live in family groups, and the care they give to their young. The local
people and tourists enjoyed watching the monkeys. It was because of the
popularity of the monkeys that the local people believed that the monkeys
contributed to the financial wellbeing of Silver Springs. There were several
demonstrations on both land and water protesting the removal of the
monkeys.

The FGFWFC based their mandate that monkeys be removed on the
following arguments.

1. The monkeys endanger the lives of the people who visit the Silver
Springs area. For example, a member of the FGFWFC legal division
argued that the monkeys had long canine teeth, were therefore similar to
carnivores, and would kill by going for the jugular vein. Mike Thomas
(1992) writing for the *Florida Magazine* tracked down the boy in one of the
alleged incidents described by FGFWFC as a serious bite on the neck of a
3 year old. The boy's father, according to the reporter, 'laughed when
asked about the report [of the bite on his son]. [The father] says his boy
tried to pet the monkey and it jumped on his back and nipped his shoulder.
"That's the biggest joke I've ever heard of," the [father] said of the
FGFWFC report. "It wasn't very serious. It was like a bad scratch. It
wasn't on the neck but on the back of the shoulder."' Rhesus monkeys
can, of course, deliver severe injuries. However, in this incident the boy was
not seriously injured. In my experience rhesus macaques will run away
from danger unless they are threatened and/or cornered. The solution to

problems such as these is to keep the monkeys out of Silver Springs' Deer Park, where this incident took place.

2. The FGFWFC rightly pointed out that the monkeys occasionally raid people's orange trees and blueberry bushes. However, it seems that this is not a serious problem because local people know how to deal with the occasional incident. For example, one woman told me that she sprays them with a water hose and they leave. Another woman told me that before she saw it was a monkey she 'had her worker shoot him'.

3. The FGFWFC raised the issue that most macaques carry the virus *Herpesvirus simiae* or the herpes B. According to Whitley (1990), herpes B is one of 35 herpes viruses isolated from other primates. Herpes B is widespread among Old World Monkeys, especially the Asian macaques and, although it does not cause disease in most macaques, it is pathogenic for humans. Between 1932, when a physician died from a herpes B infection from a monkey bite, and 1987, 25 people have been infected with herpes B, leading to death in 16 cases. Most of the cases occurred in the 1950s and 1960s at the height of the use of rhesus monkeys to develop and produce the polio vaccine. All of the 25 cases occurred in research centers. The decline in the number of cases is attributed to the development of herpes B-free macaque colonies, revised handling protocols, and the discovery of anti-herpes drugs. The mode of transmission of herpes B from monkey to monkey is unclear. However, transmission from monkey to human occurs when humans contact through an opening in the skin the secretions from a monkey who is shedding the virus at the time of contact. Although most adult macaques are seropositive for herpes B, most are not shedding the virus: that is, they are not capable of infecting another monkey or a human. Open lesions in the month of a macaque occur when the herpes B is active and the individual is shedding the virus. The factors leading to the activation and subsequent shedding of the herpes B virus in macaques are unknown, but stress seems to be the primary cause.

The question is, therefore, was the FGFWFC justified in using herpes B as a reason to eliminate the monkeys from Silver Springs? I will argue that the answer is no; the presence of herpes B in this population of rhesus monkeys is not endangering the lives of the people who visit or work at Silver Springs. Lindburg (1993) argued in an editorial in *Zoo Biology* that macaques should not be eliminated from zoos because of herpes B. He estimates that, since the 1932 reported case of a human herpes B infection, zoo workers have experienced over 11 million exposures to macaques. However, no cases of the transmission of the virus from monkeys to humans in zoos have been reported. In the case of Silver Springs monkeys no one has been seriously bitten, and none of the monkeys sterilized at the University of Florida Veterinary College had oral lesions, which would be indicative of a monkey with an active infection. There is, furthermore, little or no contact between Silver Springs workers and the monkeys. Occasionally monkeys will get on private boats if food is placed within their reach, but, if approached, they run away. There is, therefore, no reason to believe that either the visitors or the workers of the Silver Springs tourist park are in any danger from the herpes B virus that these macaques undoubtedly harbor.

The question has been raised of the possible occurrence of herpes B transmission in India, a country where macaques and humans interact on a daily basis. When I carried out my research in India I talked with people who have been bitten by monkeys or who know people who have been bitten, and no one has ever heard of anyone dying following a monkey bite. I also talked with two Indian primatologists, and they had no knowledge of any persons in India being infected with herpes B from contact with monkeys. Lindburg (1993) reports similar experiences in India and suggests (p. 408) that 'for centuries macaques and humans have rubbed elbows throughout vast regions of Asia. In some instances, the relationship has even been described as commensal. Yet, apparently, humans and macaques pose no significant health threat for one another'. This is not to argue that monkey bites never occur. For example, Singla, Kaur and Lal (1997) reported that in one study of 952 animal bites, 8.8% were the results of monkeys biting people (86% were dog bites and 3.8% were other mammals). The authors mention the possibility of the transmission of herpes B, but none of the monkey bites resulted in herpes B infection. Thus, the FGFWGC was right to point out the possible dangers of monkey bites so that visitors and those who work at Silver Springs could take appropriate actions. The lack of evidence of a monkey-to-human transfer of the virus outside of primate research centers, however, argues that herpes B was not a valid reason for the FGFWFC to call for the immediate elimination of the monkeys from Silver Springs.

4. The FGFWFC stated that the monkeys compete for food with the native fauna of Florida. The monkeys do not, however, significantly overlap in their adaptations with any of the native animals, and cases of harm to any flora or fauna of Florida have not been documented. In fact, the monkeys provide alligators with an occasional meal.

In conclusion, I did not understand why the legal division of the FGFWFC was so eager to remove the monkeys from Silver Springs. The monkeys have lived at Silver Springs for over 40 years without serious incident, and the FGFWFC received criticism from the citizens of Florida for their call for the immediate elimination of the monkeys. If we had continued the sterilization begun in the 1980s, the monkey of Silver Springs would have been eliminated by attrition by the turn of the century. Ideally the role of the anthropologist is to collect and provide scientific information on the primates they study, assist those in authority in protecting both the people and the primates from possible danger, and oversee the humane treatment of the primates. The problem of the monkeys of Silver Springs (real or imagined) could have been solved in conjunction with the management of Silver Springs, the local humane society, and University of Florida researchers had the FGFWFC sought the assistance of these groups. Monkeys are still inhabiting the Silver Spring/Silver River area, and the potential for problems still exists.

Conclusions

Rhesus macaques are an intelligent, curious, agile, and adaptable species. They range from forested areas, where they live independent of humans, to highly populated cities. Those rhesus monkeys who live in association with people are able to survive by accepting handouts, raiding crops, stealing food, and/or eating from the trees, roots, and grasses that grow in their area. Rhesus macaques are omnivores except that they do not consume other vertebrates. Because of their tolerance for a wide variety of foods, they can reside in association with people and thrive on the opportunities offered by that association. Rhesus monkeys need to drink daily; unfortunately, for city monkeys that usually means drinking contaminated water.

The social system and mating pattern of rhesus monkeys is consistent across the various environments they inhabit. That is, they live in multimale–multifemale troops with overlapping home ranges. Females remain in their natal troops all of their lives and mainly associate with female kin and friends. Males leave their natal troops around the age of 4 years and/or remain solitary or join another troop. Mating is promiscuous for both males and females. Males and females mate with troop members and with adults from neighboring troops. Females also mate with solitary males who reside on the peripheries of troops during the breeding season. There is a distinct breeding season in the fall and a spring birth season. Adult females do not have a sex swelling and mate independent of their ovulatory status.

The relationship between humans and rhesus monkeys is complex. In India the epic poem, the Ramayana, and the courageous deeds of Hanuman provide the basis for Hindus to interact with rhesus monkeys.

Moreover, children are taught the values of Hinduism through the story of Rama and Hanuman.

People also seem to be attracted to animals in general (Wilson, 1984) and to primates specifically. Lacking a religious reason to value rhesus macaques, the people who visited or live in Silver Springs still enjoyed seeing, feeding, and protecting the rhesus monkeys that inhabit the area in and around Silver Springs. On the part of the monkeys, some of their association with humans is a necessity brought about by habitat destruction, forced migration, and/or the conversion of their habitat into areas of human habitation. Monkeys may also find that living with humans has some advantages such as predation avoidance.

It is estimated that there are over a half million rhesus monkeys still remaining in India and elsewhere (Southwick and Siddiqi, 1988). Although these monkeys are not in any immediate danger of extinction, there is concern that as the size of the human population increases the monkeys will face increasing difficulties finding adequate shelter, food, and water. I believe that the sterilization of monkeys in highly populated areas can play a role in controlling population growth, but because of the cost and the sentiments of the people of India it is probably not an option. Current conservation efforts are concentrating on the translocation of city monkeys to areas where they will cause fewer problems and on educating people to be careful when feeding monkeys. Malik and Johnson (1994) estimate that there are about 50 000 rhesus monkeys in immediate need of translocation. The future of rhesus monkeys in India depends on the goodwill of the people and on finding places to relocate those rhesus monkeys that are disrupting the lives of people living in densely populated cities or raiding crops that are needed for an expanding human population.

Acknowledgements

I thank all the local people of Silver Springs, Florida, and Jaipur, India, who willingly served as informants without whom this research would not have been possible. I thank by name Russ Bernard, Iqbal Malik, Reena Mathur, Elizabeth Peters, and my husband Gopal Kapoor for all the good conversations and understanding.

References

Asquith, P.J. (1989). Provisioning and study of free-ranging primates: history, effects, and prospects. *Yearbook of Physical Anthropology* 32, 129–58.

Davar, S.A. (1977). Filigree city spun out of nothingness. *Marg: A Magazine of the Arts* **30**(4), 35–58.

Eudey, A.A. (1980). Pleistocene glacial phenomena and the evolution of Asian Macaques. In *The Macaques: Studies in Ecology, Behavior and Evolution*, ed. D.G. Lindburg, pp. 52–83. New York: Van Nostrand Reinhold.

Fleagle, J.G. (1999). *Primate Adaptation and Evolution*, 2nd edn. San Diego: Academic Press.

Lindburg, D.G. (1993). Macaques may have a bleak future in North American zoos. *Zoo Biology* **12**, 407–9.

Malik, I. and Johnson, R.L. (1994). Commensal Rhesus in India: The need and cost of translocation. *Review of Ecology* **49**, 233–43.

Nivendita, Sister and Coomaraswamy, A.K. (1985). *Hindus and Buddhists: Myths and Legends*. London: Bracken Books. 426 pp.

Richard, A.F., Goldstein, S.J. and Dewar, R.E. (1989). Weed Macaques: The evolutionary implications of macaque feeding ecology. *International Journal of Primatology* **10**, 569–94.

Roy, A.K. (1978). *History of the Jaipur City*. New Delhi: Manohar. 260 pp.

Sahai, S. (1980). The Khay Thuaraphi. In *The Ramayana Tradition in Asia*, ed. V. Raghavan, pp. 282–300. Bombay, India: Sahitya Akademi.

Santoso, S. (1980). *Ramayana Kakawin*. New Delhi: International Academy of Indian Culture.

Singla, S.L., Kaur, M. and Lal, S. (1997). Monkey bites: A public health problem in urban setting. *Indian Journal of Public Health* **41**, 3–5.

Southwick, C.H. and Siddiqi, M.F. (1994). Primate commensalism: The Rhesus Monkey in India. *Review of Ecology* **49**, 223–31.

Southwick, C.H. and Siddiqi, M.F. (1988). Partial recovery and a new estimate of Rhesus Monkey populations in India. *American Journal of Primatology* **16**, 187–97.

Szalay, F.S. and Delson, E. (1979). *Evolutionary History of the Primates*. New York: Academic Press. 580 pp.

Terborgh, J. (2000). In the company of humans. *Natural History* **109**, 54–62.

Thomas, M. (1992). Out on a limb. *Florida Magazine, The Orlando Sentinel* **39**(38), 8–13.

Vitsaxis, V.G. (1977). *Hindu Epics, Myths and Legends in Popular Illustrations*. Delhi: Oxford University Press. 98 pp.

Whitley, R. (1990). Cercopithecines Herpes Virus 1 (B Virus). In *Virology*, 2nd edn, ed. B.N. Fields, pp. 2063–75. New York: Raven Press.

Wilson, E.O. (1984). *Biophilia*. Cambridge, MA: Harvard University Press. 157 pp.

Wolfe, L.D. (1992). Feeding habits of the Rhesus Monkeys (*Macaca mulatta*) of Jaipur and Galta, India. *Human Evolution* **7**, 43–54.

Wolfe, L.D. and Mathur, R. (1987). Monkeys of Jaipur, Rajasthan, India: (*Macaca mulatta, Presbytis entellus*). *Journal of the Bombay Natural History Society* **84**, 534–9.

Wolfe, L.D. and Peters, E.H. (1987). History of the freeranging Rhesus Monkeys (*Macaca mulatta*) of Silver Springs. *Florida Scientist* **4**, 234–45.

Wolfe, L.D., Kollias, G.V., Collins, B.R. and Hammond, J.A. (1991). Sterilization and its behavioral effects on free-ranging female Rhesus Monkeys (*Macaca mulatta*). *Journal of Medical Primatology* **20**, 414–18.

Wolfheim, J.H. (1983). *Primates of the World: Distribution, Abundance, and Conservation*. Seattle: University of Washington Press. 832 pp.

Index

Pages in *italics* refer to figures and tables.

340 Index